21. 60

D1426762

METHODS IN
Carbohydrate Chemistry

VOLUME VIII
General Methods

Methods in Carbohydrate Chemistry

METHODS IN

Carbohydrate Chemistry

EDITORS

Roy L. Whistler

Department of Biochemistry
Purdue University
Lafayette, Indiana

James N. BeMiller

Department of Chemistry
and Biochemistry
Southern Illinois University
at Carbondale
Carbondale, Illinois

VOLUME VIII

General Methods

1980

ACADEMIC PRESS

A Subsidiary of Harcourt Brace Jovanovich, Publishers

New York   London   Toronto   Sydney   San Francisco

547·78 M

ACADEMIC PRESS, INC.
111 Fifth Avenue, New York, New York 10003

United Kingdom Edition published by
ACADEMIC PRESS, INC. (LONDON) LTD.
24/28 Oval Road, London NW1 7DX

Library of Congress Cataloging in Publication Data

Whistler, Roy Lester.
    Methods in carbohydrate chemistry.

    Vol. 3–5, editor: R. L. Whistler ; v. 6–   editors:
R. L. Whistler and J. N. BeMiller.
    Includes bibliographies.
    CONTENTS: v. 1. Analysis and preparation of
sugars.––v. 2. Reactions of carbohydrates.––
v. 3. Cellulose. [etc.]
    1. Carbohydrates––Collected works. I. BeMiller,
James N.
III. Title.
QD321.W568   547'.78   61–18923
ISBN 0–12–746208–2 (v. 8)

PRINTED IN THE UNITED STATES OF AMERICA

80 81 82 83   9 8 7 6 5 4 3 2 1

Contents of Volume VIII

Section I. General Methods of Separation and Analysis

Chromatography

Analysis of Polysaccharides by Chemical, Physical, and Enzymic Methods

Other Chemical, Physical, and Enzymic Methods

Section II. Preparation of Mono-, Oligo-, and Polysaccharides and Their Derivatives

Monosaccharides and Their Derivatives

Contributors to Volume VIII

Article numbers are shown in parentheses following the names of the contributors.

EDWARD M. ACTON (28), Department of Bio-organic Chemistry, Stanford Research Institute, Menlo Park, California 94025

SHUNWOO AHN (3), Department of Pharmacology and Nutrition, University of Southern California School of Medicine, Los Angeles, California 90033

S. J. ANGYAL (32), School of Chemistry, The University of New South Wales, Kensington N.S.W., Australia

ABUL KASHEM M. ANISUZZAMAN (5, 31, 45), Department of Biochemistry, Purdue University, Lafayette, Indiana 47907

J. BANOUB* (33, 34, 35), Department of Chemistry, University of Montreal, Montreal, Quebec C.P., 6210, Canada

R. BARKER† (21), Department of Biochemistry, Michigan State University, East Lansing, Michigan 48823

JAMES N. BEMILLER (11), Department of Chemistry and Biochemistry, Southern Illinois University at Carbondale, Carbondale, Illinois 62901

MAURICE J. BERTOLINI (39), Cutter Laboratories, Berkeley, California

SAMUEL P. BESSMAN (3), Department of Pharmacology and Nutrition, University of Southern California School of Medicine, Los Angeles, California 90033

R. J. BEVERIDGE (32), School of Chemistry, The University of New South Wales, Kensington, N.S.W., Australia

ROGER W. BINKLEY (50), Department of Chemistry, Cleveland State University, Cleveland, Ohio 44115

MICHÈLE BLANC-MUESSER (24), Centre de Recherches sur les Macromolécules, Végétales, C.N.R.S., 38041 Grenoble, France

M. BROCKHAUS (46), Chemisches Laboratorium der Universitat Freiburg, Freiburg, West Germany

E. V. CHANDRASEKARAN (11), Department of Biological Chemistry, The Milton S. Hershey Medical Center, The Pennsylvania State University, Hershey, Pennsylvania 17033

TAK-MING CHEUNG* (26, 27), Department of Chemistry, The Ohio State University, Columbus, Ohio 43210

O. S. CHIZHOV (15), N.D. Zelinsky Institute of Organic Chemistry of Academy of Science of U.S.S.R., Moscow

JACQUES DEFAYE (24), Centre de Recherches sur les Macromolécules, Végétales, C.N.R.S., 38041 Grenoble, France

W. M. DOANE (40), Northern Regional Research Center, Science and Education Administration, U.S. Department of Agriculture, Peoria, Illinois 61604

KEVIN L. DREHER (19), Department of Biochemistry and Biophysics, The Pennsylvania State University, University Park, Pennsylvania 16802

DONALD F. DURSO (16), Johnson and Johnson, New Brunswick, New Jersey 08903

MICHAEL E. EVANS (22, 23, 32, 48), The Australian Wine Research Institute, Glen Osmond, South Australia 5064

---

* Present address: Fisheries and Oceans Canada, Microbial Chemistry Section, St. John's, Newfoundland, Canada.

† Present address: Division of Biological Sciences, Cornell University, Ithaca, New York 14853.

* Present address: Gladesville, Sydney, Australia 2111.

P. FANG (4), Forest Products Utilization Laboratory, Mississippi State University, Mississippi State, Mississippi 39762

R. J. FERRIER (36), Department of Chemistry, Victoria University of Wellington, Wellington, New Zealand

L. SCOTT FORSBERG (13, 29), Department of Biochemistry and Biophysics, The Pennsylvania State University, University Park, Pennsylvania 16802

JOHN E. FOX (1, 2), The University of Birmingham Macromolecular Analysis Service, Department of Chemistry, The University of Birmingham, Birmingham B15 2TT, England

BERT FRASER-REID (30), Guelph-Waterloo Center for Graduate Work in Chemistry, Waterloo Campus, University of Waterloo, Ontario, Canada N2L 3G1

HARRIET L. FRUSH (37), Department of Chemistry, The American University, Washington, D.C. 20016

R. H. FURNEAUX (36), Department of Chemistry, Victoria University of Wellington, Wellington, New Zealand

P. J. GAREGG (49), Department of Organic Chemistry, Arrhenius Laboratory, University of Stockholm, S-106 91, Stockholm, Sweden

PAUL J. GEIGER (3) Department of Pharmacology and Nutrition, University of Southern California School of Medicine, Los Angeles, California 90033

ROY GIGG (47), Laboratory of Lipid and General Chemistry, National Institute for Medical Research, Mill Hill, London NW7 1AA, England

CORNELIS P. J. GLAUDEMANS (20, 39), National Institutes of Health, Bethesda, Maryland 20014

LEON GOODMAN (28), Department of Chemistry, University of Rhode Island, Kingston, Rhode Island 02881

S. HANESSIAN (33, 34, 35), Department of Chemistry, University of Montreal, C.P. 6210, Montreal, Quebec, Canada

JAMES HOFFMAN (14), Department of Organic Chemistry, Arrhenius Laboratory, University of Stockholm S-106 91 Stockholm, Sweden

DEREK HORTON (24, 26, 27), Department of Chemistry, The Ohio State University, Columbus, Ohio 43210

YUKO INOUE (42, 43), School of Pharmaceutical Sciences, Kitasato University, 8-1 Shirokane 5 Chome, Manato-Ku, Tokyo, Japan

HORACE S. ISBELL (37), Department of Chemistry, The American University, Washington, D.C. 20016

HAROLD J. JENNINGS (12), Division of Biological Sciences, National Research Council of Canada, Ottawa, Ontario, Canada K1A 0R6

MICHAEL E. JOLLEY (20), Abbott Laboratories, Chicago, Illinois 60600

KICHITARO KAWAGUCHI (38), Mergenthaler Laboratory for Biology, The Johns Hopkins University, Baltimore, Maryland 21218

HIROYASU KAWAI (38), Department of Food Science and Nutrition, Nara Women's University, Nara, Japan

LENNART KENNE (7, 44), Department of Organic Chemistry, Arrhenius Laboratory, University of Stockholm, S-106 91 Stockholm, Sweden

JOHN F. KENNEDY (1, 2), The University of Birmingham Macromolecular Analysis Service, Department of Chemistry, The University of Birmingham, Birmingham B15 2TT, England

N. K. KOCHETKOV (15), N.D. Zelinsky Institute, of Organic Chemistry of Academy of Science of U.S.S.R., Moscow

J. LEHMANN (46), Chemisches Laboratorium der Universität Freiburg, Freiburg, Germany

BENGT LINDBERG (14, 44), Department of

Organic Chemistry, Arrhenius Laboratory, University of Stockholm, S-106 91 Stockholm, Sweden

G. D. McGinnis (4), Forest Products Utilization Laboratory, Mississippi State University, Mississippi State, Mississippi 39762

Patricia A. Manthorpe (née Gent) (47), Laboratory of Lipid and General Chemistry, National Institute for Medical Research, Mill Hill, London, NW7 1AA, England

John P. Marsh, Jr. (28), Department of Bio-organic Chemistry, Stanford Research Institute, Menlo Park, California 94025

Carol W. Mosher (28), Department of Bio-organic Chemistry, Stanford Research Institute, Menlo Park, California 94025

Katsumi Murata (10), Department of Medicine and Physical Therapy, University of Tokyo School of Medicine, Tokyo, Japan

Kinzo Nagasawa (42, 43), School of Pharmaceutical Sciences, Kitasato University, Tokyo, Japan

Frederick W. Parrish* (23), Department of Biochemistry, School of Medicine, University of Miami, Miami, Florida 33124

Hans Paulsen (25), Institute of Organic Chemistry and Biochemistry, University of Hamburg, Hamburg, West Germany

John H. Pazur (13, 19, 29), Department of Biochemistry and Biophysics, Pennsylvania State University, University Park, Pennsylvania 16802

Elizabeth Percival (41), Department of Chemistry, Royal Holloway College, Egham, Surrey, England

Bruno Radatus* (30), Guelph-Waterloo Centre, Waterloo Campus, University of Waterloo, Waterloo, Ontario, Canada N2L 3G1

B. S. Shasha (40), Northern Regional Research Center, U.S. Department of Agriculture, Peoria, Illinois 61604

Volker Sinnwell (25), Institute of Organic Chemistry and Biochemistry, University of Hamburg, Hamburg, West Germany

Ian C. P. Smith (12), Division of Biological Sciences, National Research Council of Canada, Ottawa, Ontario, Canada K1A 0R6

R. J. Sturgeon (8, 9, 17, 18), Department of Brewing and Biological Sciences, Heriot-Watt University, Edinburgh, EH1 1HX Scotland

Sun-Sang Joseph Sung† (6), Department of Biochemistry, Michigan State University, East Lansing, Michigan 48823

Sigfrid Svensson (7), Department of Clinical Chemistry, University Hospital, Lund, Sweden

A. F. Sviridov (15), N.D. Zelinsky Institute of Organic Chemistry of Academy of Science of U.S.S.R., Moscow

C.-G. Swahn (49), Department of Pharmacology, Karolinska Institute, 104 01 Stockholm, Sweden

Charles C. Sweeley (6), Department of Biochemistry, Michigan State University, East Lansing, Michigan 48823

---

* Present address: Eastern Regional Research Center, Federal Research, Science, and Education Administration, U.S. Dept. of Agriculture, Philadelphia, Pennsylvania 19132

---

* Present address: Department of Chemistry, Bishop's University, Lennoxville, Quebec, Canada T1M 1Z7.

† Present address: Department of Cellular Physiology and Immunology, Rockefeller University, New York, N.Y. 10021

STEVE YIK-KAI TAM* (30), Guelph-Waterloo Centre, Waterloo Campus, University of Waterloo, Waterloo, Ontario, Canada N2L 3G1

JOACHIM THIEM (25), Institute of Organic Chemistry and Biochemistry, University of Hamburg, Hamburg, West Germany.

TATSUROKURO TOCHIKURA (38), Department of Food Sciences and Technology, Kyoto University, Kyoto, Japan

D. TRIMNELL(40), Northern Regional Research Center, U.S. Department of Agriculture, Peoria, Illinois 61604

JI-HSIUNG TSAI* (24), National Starch and Chemical Corporation, Bridgewater, New Jersey 08807

T. E. WALKER (21), Los Alamos Scientific Laboratory, University of California, Los Alamos, New Mexico 87544

WOLFGANG WECKERLE (26, 27), Beohringer Mannheim Company, Chemical Research, 8132 Tutzing, West Germany

ROY L. WHISTLER (5, 31, 45), Department of Biochemistry, Purdue University, Lafayette, Indiana 47907

CLIFFORD G. WONG (6), Department of Biochemistry, Michigan State University, East Lansing, Michigan 48823

* Present address: Walker Laboratory. Sloan-Kettering Institute, Rye, New York 10580.

* Present address: National Starch and Chemical Corporation, Bridgewater, New Jersey 08807.

Preface

The widespread use of previous volumes indicates that this series fills a need for reliable methods in carbohydrate chemistry and biochemistry. This volume, as with preceding volumes, was prepared with the goal of presenting procedures that can be used by both specialists and nonspecialists in carbohydrates. In all cases, the descriptions of methods include application to a specific isolation, synthesis, or analysis. However, methods are presented with the intention that they can, and will be, applied to other fundamental and applied research problems and routine and experimental laboratory work.

Volume VIII contains general methods, adding to those previously presented in Volumes I, II, III, V, VI, and VII. Volume VIII highlights methods on automated chromatographic techniques; enzymic and other methods for the analysis and structural characterization of polysaccharides; ^{13}C NMR spectroscopy, which has great potential in carbohydrate chemistry; synthesis of deoxy and branched-chain sugars; synthesis of 1,2-*trans*-glycosides, de-*N*- and de-*O*-sulfation; and de-*N*-acetylation of polysaccharides.

Volumes are being planned that will emphasize methods used in lipopolysaccharide research and enzymic methods.

The editors acknowledge and appreciate suggestions for inclusion of methods from those who have personal experience with them.

Attention is drawn to recognition of purity of purchased carbohydrates as starting compounds. Quite often commercial materials are not pure and, although purity has improved, caution should be exercised in accepting as pure carbohydrates those that are not specifically indicated by the manufacturer as meeting criteria of purity such as those of the U. S. National Research Council as given in their Specifications and Criteria for Biochemical Compounds, 3rd Edition, published by the U. S. National Academy of Sciences, Washington, D.C. (1972).

If investigators accept compounds that are not pure, they may encounter great difficulty in conducting reactions or may suffer significant yield losses.

Roy L. Whistler
James N. BeMiller

Outline of Volume I

ANALYSIS AND PREPARATION OF SUGARS

Outline of Volume II

REACTIONS OF CARBOHYDRATES

Outline of Volume III

CELLULOSE

Outline of Volume IV

STARCH

Section III. Physical Analyses
Whole Starch and Modified Starches; Starch Pastes; Starch Fractions; Starch Hydrolyzates

Section IV. Microscopy

Section V. Starch Degradations

Section VI. Starch Derivatives and Modifications
Reactivity; Esters; Ethers; Oxidation

Outline of Volume V

GENERAL POLYSACCHARIDES

Section I. General Isolation Procedures

Section II. Polysaccharide Preparations

Section III. Chemical Analyses

Section IV. Physical Analyses

Section V. Molecular Weight Determinations

Section VI. Structural Methods

Section VII. Derivatives
Oxidation and Reduction; Esterification and Deacylation; Etherification

Section VIII. Selected Methods Found in Other Collections

Outline of Volume VI

GENERAL CARBOHYDRATE METHODS

Section I. Separation and Analysis

Section II. Preparation of Mono- and Polysaccharides and Their Derivatives

Section III. Oxidation

Section IV. Acylic Sugars

Section V. Etherification

Outline of Volume VII

GENERAL METHODS, GLYCOSAMINOGLYCANS, AND GLYCOPROTEINS

Errata and Additions

Volume I

p. 8, line 3 from bottom. Cross reference to Vol. 1 [37].

p. 3 *i*, lines 2 and 3. This material is no longer manufactured by the Westvaco Chemical Division of the Food Machinery and Chemical Corp. but by the Waverly Chemical Co., Inc., Mamaroneck, N.Y. The Waverly material, however, may have too alkaline a surface and should then be treated to modify this property; see M. L. Wolfrom, R. M. de Lederkremer, and L. E. Anderson, *Anal. Chem.* **35**, 1357 (1963).

p. 57, Table IV, entry 18. For "*gala*" read "*galacto.*"

p. 80, Procedure. Preparation of the calcium salt prior to addition of ferric sulfate and barium acetate is beneficial. A second addition of hydrogen peroxide in the manner described in Vol. I [20], is also useful. Probably, the two most important points are the temperature control (no purple color is obtained if the temperature goes above 45°) and deionization with the ion-exchange resins. The use of a conductivity meter as mentioned in Vol. I [20] is invaluable. If the ion count is too high, a second resin treatment is necessary.

p. 94, subtitle. For "epimerization" read "isomerization."

p. 98, subtitle. For "epimerization" read "isomerization."

p. 171, Introduction, line 2. For "meso-*glycero-gulo*-hepitol" read "D-*glycero*-D-*galacto*-hepitol."

p. 175, subtitle. For "epimerization" read "isomerization."

p. 176, Derivative. This compound, described here as the α-hexacetate of D-*manno*-heptulose, has been reported by E. Zissis, L. C. Stewart, and N. K. Richtmyer [*J. Amer. Chem. Soc.*, **79**, 2593 (1957)] to be the pentaacetate.

p. 176, Derivative, At the end of the first sentence insert "Solution of the pure sugar (pulverized to pass through a 60- or 80-mesh screen) in the acetylating mixture at 0° required 6 to 7 days. No agitation was used. The solution remained colorless during this period and the 48-hr. standing period at 0° which followed. Presence of color in the acetylating mix may indicate the presence of small amounts of plant residues."

p. 178, Procedure, line 3. For "2-Deoxy-*N*-phenyl-D-ribosylamine" read "2-Deoxy-*N*-phenyl-D-*erythro*-pentosylamine."

p. 199, Procedure, line 2. For "Three g," read "Five g."

p. 217, First heading. For "1-Amino-1-deoxy-D-lyxose" read "Lyxosylamine."

p. 257, Label structures "I" and "II."

p. 291, line 4. Read "D-*epi*-inosose-2."

p. 312. Reference 1, Read "R. M. McCready and E. A. McComb, *J. Agr. Food Chem.*, **1**, 1165 (1953)."

p. 312, reference 2. Read "L. R. MacDonnel, E. F. Jansen, and H. Lineweaver, *Arch. Biochem.*, **6**, 389 (1945)."

p. 351, Methyl 4,6-*O*-Benzylidene-α-D-glucopyranoside. Cross reference to Vol. 1 [30].

Volume II

p. 299, line 2. For "thionyl" read "sulfuryl."

p. 333, Procedure, 2nd paragraph, line 1. For "α-D(β-L)" read "α-D(α-L)."

p. 333, Procedure, 2nd paragraph, line 4. For "α-D(β-L)" read "β-D(β-L)."

p. 334, third, paragraph, line 3. For "α-D-poly-O-acylglycosyl" read "poly-O-acyl-α-D-glycosyl."

p. 335, Structure (I). For "R = o-NO$_2$·C$_6$H$_4$—" read "Ar = o-NO$_2$·C$_6$H$_4$—."

p. 346, last paragraph, line 2. For "α-D(β-L)" read "α-D(α-L)."

p. 388, first heading. For "Benzyl 2-O-Methylsulfonyl-β-D-arabinopyranoside (V)" read "Benzyl 3,4-O-Isopropylidene-β-D-arabinopyranoside (III)."

p. 418, last heading. For "6-Deoxy-1,2-O-isopropylidene-α-D-$xylo$-hexofuranoside-5-ulose" read "6-Deoxy-1,2-O-isopropylidene-α-D-$xylo$-hexofuranos-5-ulose."

p. 483, subtitle. For "D-Glucose" read "D-Galactose."

p. 514, Line 7. Read "Kojic Acid   R. Bentley, ref. 2, p. 238."

Volume III

Foreword. The editors regret that an acknowledgment to Dr. T. N. Kleinert in the Foreword erroneously gave his address as "Division of Industrial and Cellulose Chemistry, McGill University." Dr. Kleinert, formerly of the Pulp and Paper Research Institute of Canada, is now retired.

p. 53, third line from bottom. For "oxidation" read "oximation."

p. 135, Preparation of the Chromatographic Column. The preparation can be scaled up by using a column of 15 × 125 cm. and increasing the sample load by a factor of 10. After analysis of the effluent, the tetra- and hexasaccharide fractions are pooled as are the tri- and pentasaccharide fractions. Each of these is then rechromatographed separately. In this way, the original column can be grossly overloaded, but final products of good purity are obtainable in gram lots (K. W. King, personal communication).

p. 135, line 7. For "1.91" read "1.19."

p. 137, lines 4–6. Five to fifteen percent contamination with stearic acid occurs beyond the tetrasaccharide. A single recrystallization is not enough to remove all the stearic acid; however, this can be done by extracting the liquid fraction concentrates with petroleum ether prior to recrystallization until all stearic acid is removed (K. W. King, personal communication).

p. 137, footnote 4. Frequently precipitation occurs immediately on elution for the tetrasaccharide and up. This results in deposition of oligosaccharides on the delivery tip of the column itself which must be cleaned off immediately after elution of each fraction to prevent contamination of subsequent peaks (K. W. King, personal communication).

p. 139, Introduction, line 2. For "Celluloytic" read "Cellulolytic."

p. 141, Cellulase Assay. The reader is warned that this is only one kind of a "cellulase" assay and is not valid for many systems. This assay does not yield zero order kinetics with many cellulases. This system will also detect many enzymes having no action on "native" cellulose.

p. 258, line 13. For "O-methylsulfonycellu-" read "O-methylsulfonylcellu-."

p. 271, Introduction. Add "The methylation of cotton cellulose with diazomethane is described in Vol. II [41]."

p. 369, next to last line. For "Pfleider" read "Pfleiderer."

p. 371, Pressure Vessels, line 2. For "Telflon" read "Teflon."

Volume IV

p. xii, Read "30a.   Inherent Viscosity of   Raymond R. Myers and
                    Alkaline Starch Solutions   Robert J. Smith

p. 320, Teflon. For "Tetrafluoroethylene" read "A polymer of tetrafluoroethylene."

Volume V

p. 47, For "J. K. N. Jones and R. J. Stoodley" read "J. K. N. Jones, R. J. Stoodley, and K. C. B. Wilkie."

p. 174, line 1. For "5" read "50."

p. 289, line 7 from botton. For "[31]" read "[30]."

p. 400, line 6 from bottom. For "N-methyl" read "O-methyl."

Volume VI

p. 197, lines 3 and 19. For "-p-toleunesulfonyl-" read "-p-tolylsulfonyl-."

p. 290, lines 17 and 18. For "potassium acetate" read "fused potassium acetate."

p. 291, line 7. For "0.05 M sulfuric acid" read "0.08 M sulfuric acid."

Section I. General Methods of Separation and Analysis

CHROMATOGRAPHY
Articles 1 through 6

ANALYSIS OF POLYSACCHARIDES BY CHEMICAL, PHYSICAL, AND ENZYMIC METHODS
Articles 7 through 16

OTHER CHEMICAL, PHYSICAL, AND ENZYMIC METHODS
Articles 17 through 21

CHROMATOGRAPHY

[1] Fully Automatic Ion-Exchange Chromatographic Analysis of Neutral Monosaccharides and Oligosaccharides

By John F. Kennedy and John E. Fox

*The University of Birmingham Macromolecular Analysis Service,
Department of Chemistry, University of Birmingham,
Birmingham, England*

Introduction

Analytical separations of monosaccharides and oligosaccharides have been achieved by a number of methods including liquid chromatography on paper, thin layers, charcoal columns, ion-exchange resin columns, and gas-liquid chromatography on a variety of liquid phases on solid supports. Of these methods, two of the most versatile have proved to be ion-exchange chromatography (Vol. VI [9]) and gas-liquid chromatography; their various forms are reviewed annually (*1*). Whereas these two techniques have both different and common advantages, in much of the analytical work of analysis of free monosaccharides and smaller oligosaccharides, liquid phase ion-exchange chromatography is convenient since samples in solution can be loaded directly without prepreparation, derivatization, or removal of aqueous solvent.

A recent review (*2*) of the analytical ion-exchange chromatography of monosaccharides and smaller oligosaccharides shows the good separations which may be achieved conveniently using the borate complexes of the carbohydrates. In terms of automation, this separation and the associated assay is analogous to the oldest of techniques put onto automated equipment, i.e., the ion-exchange chromatography of amino acids and the ninhydrin assay thereof, a system in which a great deal of instrumentation experience has been gained.

The ion-exchange chromatographic system as described herein (*3*) provides excellently for identification and quantitation of the common mono- and disaccharides including the components of the carbohydrate moieties of glycoproteins. The elution pattern is not directly discernable from the structures of the saccharides since it arises from different preferred forms of each reducing unit and from the number of borate ions complexed per molecule.

3

METHODS IN CARBOHYDRATE
CHEMISTRY, VOL. VIII

Trisaccharides and higher oligosaccharides are not easily separable since they are either unabsorbed by the column or show poor separations from other oligosaccharides.

Herein a Jeolco JLC-6AH carbohydrate analyzer in its factory production form is used directly or can be adapted for additional separations, identifications, and measurements of basic carbohydrates (Vol. IX) and higher neutral oligosaccharides (Vol. VIII [2]). Possession of identical equipment is not mandatory to achieve the separations described. Other manufacturers produce analogous equipment, or alternatively a number of the amino acid analyzers already in use in laboratories lend themselves to adaptation. Thus, for the aid of intending workers the mechanics of operation of the analyzer are given in detail.

Procedure

Equipment

A Jeolco JLC-6AH carbohydrate analyzer (Jeol, Japan Electron Optics Laboratory Co. Ltd., Tokyo) is available as a factory converted option on the Jeolco JLC-6AH amino acid analyzer. For use as an ion-exchange chromatographic analyzer for carbohydrates, the factory conversion involves replacement of the ninhydrin reagent piston pump with a piston pump suitable for pumping concentrated sulfuric acid and replacement of the 570 nm and 420 nm absorbance filters in the colorimeters with 510 nm and 425 nm filters. PTFE tubing is standard fitting on the amino acid version and is suitable for the carbohydrate version. A schematic diagram of the analyzer is shown in Figure 1. The analyzer uses a dual-column chromatographic system. The series of borate buffers, labeled A in Figure 1 are selected by two eight-way valves, B_1 and B_2, and are pumped by two double-action, high-pressure pumps, C_1 and C_2. Buffer and sample selection and flow switching are performed by a tape programmer D, and the output from the pumps passes the pressure gauges E_1 and E_2. The samples, previously injected into sample storage loops $F_1–F_{12}$, are pumped onto columns G_1 or G_2, which are maintained at a temperature of 55° by independent heating circulating pumps, H_1 and H_2. The eluant from these columns passes to a two-way selector, I, which enables either of the two columns to be connected to the detection system.

An all-PTFE pressure pump, N, draws the sample stream from I, excess running to waste J, and mixes it with the orcinol–sulfuric acid reagent which is drawn from storage K via a valve, M, which can also admit water L. The reagent-sample mixture is then pumped through a reaction coil, O, at

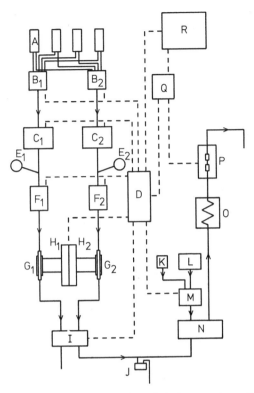

Fig. 1.—Diagram of the modified automatic carbohydrate analyzer. From Kennedy and Fox (3). Reproduced with the kind permission of Elsevier Scientific Publishing Company, Amsterdam.

95°. The color so formed is measured by two flow cells, P, at wavelengths of 425 nm and 510 nm. The data are then printed out onto a point-plot chart recorder, Q. The data may be calculated and quantitated by means of the Nova 1220 computer, R, which is on-line to the analyzer; but this accessory is not essential.

Ion-Exchange Chromatography

The short-type, jacketed columns (13-cm × 0.8-cm id) are packed with LC-R-3 quaternary ammonium ion-exchange resin (Jeol) (or an equivalent resin) (bed height 12 cm), and the columns are regenerated and equilibriated at 30 ml/h (see elution program, Table I). Solutions of samples (1–1000 μg of each component) are centrifuged to remove any insoluble material and loaded into the storage loops (capacity 800 μl) of the analyzer, from which

TABLE I

Recommended Elution Program and Elution Positions for Ion-Exchange Chromatography
of Carbohydrates on a Jeolco JLC-6AH Analyzer

Eluant buffer	Duration of pumping, min[a]	Function of eluant	Elution time of carbohydrate, h[a]	
0.13 M potassium tetraborate buffer, pH 7.50	110	Separation of carbohydrates	2-Deoxy-D-ribose (2-Deoxy-D-erythro-pentose)	0.47
			Sucrose	0.55
			Cellobiose	0.77
			Maltose	1.13
			Lactose	1.38
			L-Rhamnose	1.62
0.25 M potassium tetraborate buffer, pH 9.08	90	Separation of carbohydrates	D-Ribose	2.75
			D-Mannose	3.13
0.30 M potassium tetraborate buffer, pH 9.60	190	Separation of carbohydrates	L-Fucose	4.03
			L-Arabinose	4.12
			D-Fructose	4.33
			D-Galactose	4.46
			D-Xylose	4.75
			D-Glucose	5.38
0.50 M potassium tetraborate, pH 9.60	90	Regeneration of column	—	—
0.13 M potassium tetraborate buffer, pH 7.50	120	Equilibration of column	—	—

[a] These times may vary over a period according to differences in the length, state, and age of the resin bed, but without alteration of the elution order.

they are automatically loaded in turn onto one of the ion-exchange columns and eluted in descending fashion at 55° at the 30 ml/h flow rate using a buffer program (Table I) (Fig. 2). The elution program is fixed to be repetitive using an appropriately punched tape in the elution programmer. Orcinol dissolved in concentrated sulfuric acid (1.5 g/l) is used as assay reagent and is stored in the equipment at 4–6° under which conditions it is stable for at least 5–6 weeks. This reagent (pumping rate 24 ml/h) is mixed with the sample taken from the column eluate (sampling rate 12 ml/h) in the assay manifold. The mixed reagent and column sample are passed through the heating coil maintained at 95° (residence time about 15 min).

The ion-exchange columns are used as a pair running out of phase so that analysis on one column is being conducted while the second column is being regenerated. This gives a turn around time of 6.5 h per analysis.

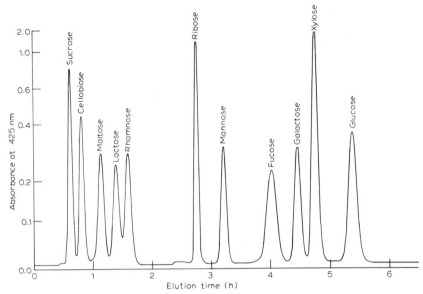

FIG. 2.—Chromatography of neutral carbohydrates on Jeol LC-R-3 ion-exchange resin (borate form) using a modified Jeolco JLC-6AH automatic carbohydrate analyzer. From Kennedy and Fox (3). Reproduced with the kind permission of Elsevier Scientific Publishing Company, Amsterdam.

Sample Preparation

In practice, samples for carbohydrate analysis identification and/or quantitation arise from mixtures of free saccharides from industrial processes, for example, sucrose, D-glucose, and D-fructose production, from starch hydrolysis, from clinical specimens and physiological fluids, from macromolecules such as glycoproteins and polysaccharides, and from synthetic work. For ion-exchange chromatography, if the carbohydrates are already in mono- or disaccharide form, the sample is normally dissolved in 0.13 M boric acid for column loading, a procedure recommended to ensure that the sample is not loaded at neutral or higher pH since this gives rise to peak broadening. Dissolution in this reagent causes no problem if the original sample is in salt-free dry form; if it is in aqueous solution, or very dilute buffer, dilution with the boric acid solution is adequate. There may be an impairment of separation if the sample is dissolved in distilled water (pH 5.5) or other dilute acid. If the sample is received in a higher pH medium, addition of borate plus acidification is normally necessary. Where a hydrolysis is necessary to release the carbohydrate, we have found that the optimum conditions for a glycoprotein are 2.0 M trifluoroacetic acid in a sealed tube at 100° for 6 h, after which the acid is removed by rotary evaporation under reduced pressure. However, it must be appreciated that the most accurate results can only be obtained via a range of hydrolysis conditions because of the different stabilities of monosaccharides and their glycosyl bonds.

Reagent and Sample Purities

All aqueous solutions (buffers) must be pre-filtered, for example through Whatman filter tubes type B_2 (Fisons Scientific Apparatus Ltd., Loughborough, England).

Any deposition of a purple-colored material in the PTFE heating coil which carries the mixed reagent and column eluate is attributable to reagent impurities, and even a double recrystallization of some samples of orcinol may be inadequate. (Orcinol from Fisons Scientific Apparatus Ltd., Loughborough, England, is recommended and may be used without recrystallization.) It is unnecessary to use the purest forms of sulfuric acid. (Standard laboratory reagent grade sulfuric acid from British Drug Houses Ltd., Poole, Dorset, England is recommended.) Standard laboratory solvents are incapable of removing the purple deposit, but by using the reagent grades listed above, a coil will last for at least 1 year without blocking. The fact that sample storage loops are at room temperature warrants flushing with 0.02% w/v sodium azide solution before filling to avoid microbial degradation of samples awaiting analysis. This reagent has no effect on the response of the orcinol–sulfuric acid assay.

Standardization

Unknown components of the sample are identified by their elution position compared with a chart for standards run under the same conditions. Actual elution positions of various carbohydrates are given in Table I, and the separations are shown graphically in Figure 2. The elution positions of a greater range of mono- and oligosaccharides under the same elution program is given in Table II in relative form. It is recommended that one in twelve runs be a standard, that is, one standard run every cycle of the samples.

Components of the sample are quantitated by estimation of the area under the peak on the chart and comparison with that for the appropriate standard using a manual graphic approach or by computing, using the output signal from the colorimeter as input data for the computer.

TABLE II

Relative Elution Positions of Neutral Mono- and Oligosaccharides on
Ion-Exchange Chromatography on a Jeolco JLC-6AH Analyzer

Carbohydrate	Relative elution times[a]	
	Ribose = 1.00	Glucose = 1.00
L-Arabinose	1.50	0.77
Cellobiose	0.28	0.14
2-Deoxy-D-ribose (2-Deoxy-D-erythro-pentose)	0.17	0.09
L-Fucose	1.47	0.75
D-Fructose	1.57	0.80
D-Galactose	1.62	0.83
D-Glucose	1.96	1.00
D-Glucose 1-phosphate	1.03	0.53
Isomaltose	1.75	0.89
Isomaltotriose	1.57	0.80
Lactose	0.50	0.26
Lactulose	1.21	0.62
D-Lyxose	0.96	0.49
D-Mannose	1.14	0.58
Maltose	0.41	0.21
Maltotriose	0.37	0.19
Panose	0.33	0.17
Psicose	1.09	0.56
L-Rhamnose	0.59	0.30
D-Ribose	1.00	0.51
Sucrose	0.20	0.10
Trehalose	0.21	0.11
D-Xylose	1.73	0.88
D-Xylulose (D-threo-Pentulose)	1.00	0.51

[a] The values may vary slightly over a period according to differences in the length, state, and age of the resin bed.

Sensitivity

Using the standard 2-mm path length flow cells, as little as 1 μg of carbohydrate can be detected. Although usually used with the standard 2-mm length flow cells (figures in this article are based on this path length), the sensitivity of response of the analyzer may be increased using 10-mm path length cells. To facilitate a rapid changeover, quick-release couplings should be inserted in the PTFE tubings attached to the flow cells. Increased sensitivity may be achieved using the standard scale expansion facilities (X3 and

X10), although in the case of X10 the increase in background noise/base line absorbance is undesirable.

The adsorption maximum (λ_{max}) for the orcinol chromophore arising from most neutral sugars occurs at 420 nm, and accordingly, the filter to measure absorbance at 425 nm is acceptable. Alternatively, if only the ninhydrin set of filters is available, the 440 nm filter may be used with a 20% loss in sensitivity. Whereas the trace from the 510 nm filter is rarely used, it being intended for low sensitivity/high proportion in sample, it is useful for confirmation of peak identification where fucose or rhamnose or carbohydrate containing these structures are a component of the sample, since the relative absorbances of the chromophore at 425 and 510 nm for these 6-deoxy sugars from those of the other monosaccharides, etc. (Ratios of $A_{425}:A_{475}:A_{510}$ are glucose 1.00:0.45:0.46, fructose 1.00;0.48:0.45, fucose 1.00:0.71:0.34, rhamnose 1.00:0.49:0.27).

Column Elution Programming

The overlap of the elution positions of arabinose, fucose, and fructose in the standard elution program given in Table I is not a disadvantage since it is uncommon for these three compounds or any two of them to occur together in a sample, since L-fucose arises from animal glycoproteins, L-arabinose from plant polysaccharides, and D-fructose from specific enzyme inversions of D-glucose. Complete separation could be achieved by alteration of the eluant program with concomitant lengthening of the turnaround time. The spectrum of mono- and disaccharides identifiable by the technique may, of course, be extended to all such neutral compounds; but in some instances again, the buffer program might have to be modified to achieve effective separations. Shorter runs for the separation of selected carbohydrates may be conducted at will and requires modification of the buffer program by making up a different punched-tape program. For example, D-fructose and D-glucose are regularly estimated in mixtures thereof using a buffer program in which only the pH 9.08 and pH 9.60 buffers (Table I) are used for column elution (regeneration and equilibrium steps unchanged) which gives a turnaround time of less than 4 h.

Equipment Servicing

On the Jeolco JLC-6AH carbohydrate analyzer, the sulfuric acid pumps dispense with the conventional O-ring and the cylinder is packed with PTFE chevrons which can be tightened up at will to give, by the resulting compression, a good fit on the piston and cylinder wall. As wear slowly progresses, the chevrons are simply tightened up further to maintain the fit. Ultimately the chevrons are replaced (lifetime, at least 1 year). Whatever the cause of blocking, blocked PTFE coils (6-m × 1-mm id) are dealt with most simply by replacement (Rowley Plastics, Worcester, England). Since the 6-volt, 18-watt colorimeter bulbs have a very limited life-time, are expensive to replace, and are only available from Japan, we believe it economic to replace them with standard, 12-volt, quartz–iodine bulbs that fit a stabilized power supply (Farnell Instruments Ltd., Wetherby, Yorkshire, England). A matrix programmer is available as an optional extra to replace the punched tape programming unit and is more convenient to operate.

Column Packing Stability

The resin packings are stable for 1 year or more. However, in spite of precautions taken to avoid contamination, including prefiltration of all aqueous solutions through filter tubes, columns gradually accumulate a brown coloration or deposit at the top. This ultimately impairs the separations in the chromatography through peak broadening. Replacement is the only satisfactory remedy, but only the affected parts of the resin bed need be changed.

Comments

The Jeol analyzer is adaptable and can be applied to a number of assays and separations. Although it may not be possible to reproduce precisely the assay condition in terms of the reagent : sample ratio and multiaddition steps recommended in the original literature, many assays can be reduced to a single step [for example, the carbazole assay for hexuronic acids by adding the carbazole (Vol. I [137]) with the sulfuric acid] without great loss in sensitivity, but with advantageous reproducibility of conditions. Furthermore the sensitivity of the orcinol reagent to hexuronic acids (λ_{max} 518 nm but also measurable at 425 and 440 nm) allows their fractionation and estimation by the machine using the appropriate resin packing.

Instruments could be designed to utilize the reactions of carbohydrates with 4-anisyltetrazolium chloride (Tetrazolium Blue) (4) or with periodic acid followed by determination of the liberated formaldehyde or acetaldehyde (see refs. 5 and 6). Such systems would be equally satisfactory, and it would avoid the use of concentrated sulfuric acid. The high viscosity of this reagent causes considerable peak broadening in the analytical system. Fluorimetric assays give greater sensitivity. Although fluorimetric carbohydrate assays have received little attention, the chromophore of the periodate–formaldehyde assay is fluorescent and an automated version has been developed (5, 6) and is very useful on account of its wide applicability.

References

(*1*) J. F. Kennedy, R. J. Sturgeon, B. J. Catley, and R. D. Marshall, *in* "Specialist Periodical Reports, Carbohydrate Chemistry," Vol. 4–11, Parts II, 1971–1979.

(*2*) J. F. Kennedy, *Biochem. Soc. Trans.*, **2**, 54 (1974).

(*3*) J. F. Kennedy and J. E. Fox, *Carbohydr. Res.*, **54**, 13 (1977).

(*4*) K. Mopper and E. T. Degens, *Anal. Biochem.*, **45**, 147 (1972).

(*5*) H. Cho Tun, J. F. Kennedy, M. Stacey, and R. R. Woodbury, *Carbohydr. Res.*, **11**, 225 (1969).

(*6*) J. F. Kennedy, *in* "Automation in Analytical Chemistry," pp. 532–541. Technicon, Basingstoke, England, 1974.

[2] Fully Automatic Gel-Permeation Chromatographic Analysis of Neutral Oligosaccharides

By John F. Kennedy and John E. Fox

The University of Birmingham Macromolecular Analysis Service, Department of Chemistry, University of Birmingham, Birmingham, England

Introduction

The fractionation of series of homologous oligosaccharides has been afforded considerably less attention than the corresponding fractionation of monosaccharides, partly through lack of demand, but also through lack of methodology. Gas phase chromatographic analysis cannot be used because of the decrease in volatility of the appropriately derivatized oligosaccharides with increase in oligosaccharide chain length. Whereas gel permeation (gel filtration, molecular weight sieving) is the obvious method of choice, normal bench equipment is not capable of separating oligosaccharides of DP 8 and above.

The gel permeation chromatography mode of operation of the analyzer described in the previous article (Vol. VIII [1]) provides good separation of oligosaccharides in the range DP 1–15 (*1*).

Equipment

The equipment described in the foregoing article [Vol. VIII [1]) is further modified by replacement of the dual short columns with a single long, insulated, jacketed column. For a number of reasons, it is impractical to use two columns running 180° out of phase, and so the relevant modification to the equipment requires breaking the pipe connection between F_1 and G_1 and linking it into the pipe between F_2 and G_2. (G_2 is the column location being used.) In this mode, C_1 and C_2 stop and start together (avoid spurs on the chart) and act as nonreturn valves when not actually operating, and the sample loader automatically becomes more versatile since samples can be loaded at will by selecting F_1, F_2, etc., whichever is not being pumped to the column. (Samples cannot normally be introduced when column elution is in progress.)

13

METHODS IN CARBOHYDRATE
CHEMISTRY, VOL. VIII

Gel-Permeation Chromatography

The jacketed column (1.5-m × 2-cm id) is packed at 20° with an aqueous slurry of BioGel P2, 400-mesh, pre-swollen in water and vacuum deaerated under water at 20° to remove entrapped air, to give a bed height of 1.5 m. When this bed height is reached, the column is fitted at the top with an adapter fitting. An adjustable column fitting is used to allow compensation for any compaction of the gel bed during operation. The column is slowly heated by the jacket to 65° while being eluted in ascending fashion with water treated as above at a flow rate of 32 ml/h.

Samples of carbohydrate (1–1000 μg of each component) solutions are centrifuged (if necessary), loaded into the storage loops, and automatically loaded onto the column as before (Vol. VIII [1]). The column is eluted in ascending fashion at 65° with treated water at a flow rate of 32 ml/h to give separations on a basis of molecular size. A typical elution profile for a starch hydrolysate is shown in Figure 1. The orcinol reagent, the reagent and column sampling rates, and the heating of assay mixture are described in Vol. VIII [1]). The analysis time is 9.5–10 h for a 32 ml/h flow rate.

Sample Preparation

For gel-permeation chromatography, samples can be accepted in virtually

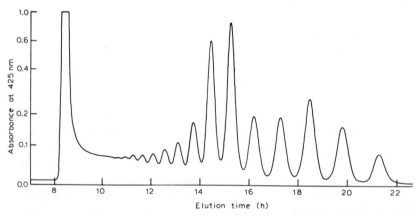

FIG. 1.—Chromatography of starch and oligosaccharides therefrom on BioGel P2 using a modified Jeolco JLC-6AH automatic carbohydrate analyzer (column flow rate 16 ml/h). From Kennedy and Fox (*1*). Reproduced with the kind permission of Elsevier Scientific Publishing Company, Amsterdam.

any form provided they are stable and not complexed, but it is desirable to bring the pH to 5–8 before loading.

Reagent and Sample Priorities

The recommendations given in Vol. VIII [1] apply equally well to gel permeation. Gel permeation columns should be flushed from time to time with 0.02 w/v azide solution to avoid bacterial contamination.

Standardization

The molecular sizes of unknown components of the sample are identified by their elution position compared with a chart for standards run under the same conditions (Vol. VIII [1]). Since over long periods of use the column flow rate and the length and characteristics of the gel may alter, and since the linkage type of the oligosaccharide affects its elution position, it is advisable to run a standard mixture once every six runs.

Components of the sample are quantitated by estimation of the area under the peak and comparison with standards (Vol. VIII [1]).

However, for most series of oligosaccharides, pure standards are not readily available for the higher members; but for many applications, it is sufficient to express the amount of each component as a percentage of the total carbohydrate present. Comparative quantitation is achieved by expression of the area under each peak as a percentage of the total area under all peaks. For a series of oligosaccharides of one component monosaccharide, the response per mole of component unit in the orcinol assay is independent of the DP of the oligosaccharide.

Sensitivity

Sensitivity aspects described for ion-exchange chromatography (Vol. VIII [1]) apply equally well to gel permeation. Using the standard 2-mm path length flow cells, as little as 1 μg of carbohydrate can be detected. All figures given in this paper are based on use of cells of 2-mm path length.

Column Elution Programming

No buffer changes, regeneration, or equilibrium steps are necessary in the running of the gel-permeation column. Time saving and run sharing can be achieved by programming the equipment to use the minimum column residence time of 5 h (elution of the void volume) as a means of overlapping runs, that is, the next sample is introduced to the column not more than 7–8 h before the completion of the previous run, that is, the runs are 240° rather than 360° out of phase. This gives a final turnaround time of 7–8 h per sample.

Column Flow Rate Changes

Marginally improved separations may be achieved if a slower flow rate is used. However, the minimum flow rate of the pumps C1 and C2 are both 32 ml/h. However each pump is of the double action type and consists of two piston (for example, C_{1a} and C_{1b}) pumps working 180° out of phase to even out the pulse effect of a piston pump; and for normal operation, the outputs from C_{1a} and C_{1b} are factory-jointed to give the final output from C_1. Although C_{1a} and C_{1b} cannot be decoupled to leave one stationary, by altering the piping only, one side of the pump can be used, thus giving a flow rate of 16 ml/h. It is not necessary to introduce any additional damping device to compensate for the fact that the column is now being pumped by a single action pump. It should, however, be stressed that the slower flow rate modification is not necessary if a homologous series of oligosaccharides is being analyzed for lower molecular weight components.

Column Packing Stability

Deaeration of the input water is essential (Fig. 2). "Cracking" of the gel bed may occur if the column is pumped in descending fashion, since the gravity enforced flow rate of the column is faster than the pump rate. To avoid this, the column is pumped in ascending fashion at all times. Normally, the flow in the column must be maintained even when the column is not in use.

Regular washing of the column with azide prevents microbial growth, and no decrease in performance level occurs within 1 year's operation and maintenance at 65°.

Comments

In the maltooligosaccharide series, plots of molecular weight against elution position give curves (Fig. 3), the shapes of which are commensurate with the accepted theories of gel filtration techniques. Similar fractionations, but in higher molecular weight ranges, should be achievable using larger gel pore sizes; for example, BioGel P6 extends the range upwards from the ~2000 molecular weight maximum of the BioGel P2 column.

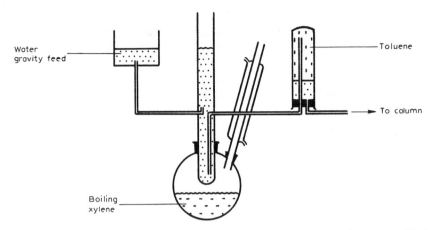

FIG. 2.—Deaeration and toluene impregnation system for column water input to modified Jeolco JLC-6AH automatic carbohydrate analyzer. From Kennedy and Fox (*1*). Reproduced with the kind permission of Elsevier Scientific Publishing Company, Amsterdam.

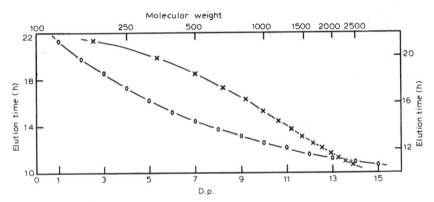

FIG. 3.—Relationships between the molecular weight and elution position for a series of maltooligosaccharides fractionated on BioGel P2 using a modified Jeolco JLC-6AH automatic carbohydrate analyzer: ×———×, log scale; O———O, linear scale. From Kennedy and Fox (*1*). Reproduced with the kind permission of Elsevier Scientific Publishing Company, Amsterdam.

Isomaltose is eluted from BioGel P2 in a position slightly different from maltose (*1, 2*) and runs faster than maltose (Table I). Similarly, panose is not eluted in a position coincident with that of maltotriose.

It would appear from our various data for the analysis of hydrolyzates of amylose, amylopectin, and starch that the presence of one or two α-D-(1→6)-linkages in maltooligosaccharides does not affect their elution positions.

TABLE I

Elution Positions of Maltooligosaccharides on Gel-Permeation Chromatography on a Jeolco JLC-6AH Analyzer

Carbohydrate	DP of carbohydrate	Elution times,[a,b] min	Relative elution times,[b] (Glucose = 100)
Glucose	1	544	100.00
Maltose	2	499	97.7
Isomaltose	2	491	90.3
Maltotriose	3	461	84.7
Panose	3	450	82.7
Maltotetraose	4	427	78.5
Maltopentaose	5	397	73.0
Maltohexaose	6	371	68.2
Maltoheptaose	7	348	64.0
Maltooctaose	8	330	60.7
Maltononaose	9	313	57.5
Maltodecaose	10	298	54.8
Maltoundecaose	11	285	52.4
Maltoduodecaose	12	274	50.4
Maltotredecaose	13	265	48.7
Maltquatuordecaose	14	256	47.1
Maltoquindecaose	15	249	45.8
Intermediate molecular weight (slowest)	16–25[c]	245	45.0
Intermediate molecular weight (fastest)	16–25[c]	205	37.7
High molecular weight	>25[c]	192	35.3

[a] Using column described herein at 32 ml/h flow rate.

[b] The values may vary slightly over a period according to differences in the length, state, and age of the resin bed.

[c] Approximate.

This is advantageous since separations of starch oligosaccharides, and calculations for quantitation, would be marred by overlapping/intermediate peaks. However, it must be borne in mind that in a starch hydrolyzate, the series of peaks cannot be specified as a homologous series of maltooligosaccharides, whereas in the case of amylose hydrolyzates they can. The technique is applicable to the separation of series of oligosaccharides other than those of amylosic origin although the relative elution positions will vary according to structure (3).

The Jeolco JLC-6AH analyzer may also be adapted to the separation, on a basis of molecular size, of orcinol–sulfuric acid positive carbohydrates by the ion-exclusion phenomena using ion-exchange resin as the column packing (4).

References

(*1*) J. F. Kennedy and J. E. Fox, *Carbohydr. Res.*, **54**, 13 (1977).
(*2*) M. John, G. Trenel, and H. Dellweg, *J. Chromatogr.*, **42**, 476 (1969).
(*3*) J. F. Kennedy and J. E. Fox, unpublished results.
(*4*) S. A. Barker, B. W. Hatt, J. F. Kennedy, and P. J. Somers, *Carbohydr. Res.*, **9**, 327 (1969).

[3] Separation and Automated Analysis of Phosphorylated Intermediates

By Paul J. Geiger, Shunwoo Ahn, and Samuel P. Bessman

Departments of Pharmacology and Nutrition, School of Medicine, University of Southern California, Los Angeles, California

Introduction

Within the past few years, automated analysis of phosphorylated metabolic intermediates as well as nucleotides and other phosphate-containing compounds involved in cell structure and function has become practicable and available (*1, 2*). With the development of automatic dry ashing of samples coupled to modern high performance liquid chromatography (HPLC) (Vol. VIII [4]), even compounds with carbon–phosphorus bonds are easily determined without lengthy wet ashing or inconvenient enzymic procedures. In addition to the system outlined in this article, Jellinek and Schnitger (*3, 4*) have also described automated analyzers, but the time to complete a single run is longer and the sensitivity somewhat lower than with the methods described here.

Separation of phosphorylated intermediates is accomplished by anion exchange on quaternary ammonium resins with borate-containing ammonium chloride buffers of pH 8.5–9.0. Ammonium chloride is used because of the ease with which it sublimes and is removed during ashing; separation of the phosphorylated sugar intermediates is improved by the use of borate for complexation, a well-known technique in the chromatography of carbohydrates (for example, *5*, p. 154).

Phosphate compounds in a typical sample applied to the HPLC column are allowed to bind to the water-washed resin and washing is continued to remove excess quantities of nonbound compounds and cations. After washing for a period dependent upon the amount of impurities to be removed, elution is started. This may be with solvent of constant composition or with gradient elution, depending upon the complexity of the sample to be analyzed. Gradient elution is recommended for physiological samples, while buffers of constant composition may be used under special circumstances (see "Results" below).

The method presented here describe improvements made since earlier descriptions (*2*); two major improvements are the use of resin in the precolumn and the use of constant borate concentration for routine sample

21

METHODS IN CARBOHYDRATE
CHEMISTRY, VOL. VIII

analysis. The latter permits an increase or decrease in column sizes and, concomitantly, volumes of buffer in a straightforward, linear manner depending on time and resolution requirements. For solving individual problems associated with these chromatographic variables of speed and resolution, for instance, the very useful papers by Snyder should be consulted (6, 7).

Although many HPLC systems now in use in many laboratories utilize very fine resins (< 10 μm bead diameters) and thus require high pressures (> 1000 psi) and stainless steel fittings to contain them, we have found glass columns with plastic fittings and 20-μm resins less expensive and easier to handle. Moreover, these systems have provided adequate resolving power for the metabolites of interest. Additionally, stainless steel systems (generally 316 ss) are subject to attack by halide salts. Such systems are, thus, harder to maintain and tend to introduce impurities into the ashing unit and detector components of the phosphate analyzer.

Procedures[1]

Apparatus

The design of the instrument for dry ashing and its construction and operation have been detailed in previous publications (1, 2). The equipment is available currently from Alsab Scientific Products, Inc. (Los Angeles, CA). Briefly, the unit accepts column effluent or samples placed into collecting cups by hand at a position just preceding the electric furnace. The furnace dries and ashes samples in the silica cups as they are passed through it by means of a 40-place, rotating turntable. The turntable moves one cup position every 30 sec. Cups emerging from the furnace with the ashed samples are permitted to cool for approximately 2 min, and the color developing reagents are added at this point once every 30 sec. The color is allowed to develop for approximately 2 min before the position of the turntable reaches the point where the colored solution is withdrawn from the cup and passed through the colorimeter. A washing sequence next washes each cup thoroughly before it is returned to the starting position where it receives a small amount of oxidizing agent just before it receives more eluate.

---

[1] Abbreviations used in this article: PC, phosphocreatine; G1P, D-glucose 1-phosphate; P$_i$, inorganic phosphate; αGP, glycerol phosphate; DHAP, dihydroxyacetone phosphate; M6P, D-mannose 6-phosphate; GAP, D-glyceraldehyde 3-phosphate; F6P, D-fructose 6-phosphate; G6P, D-glucose 6-phosphate; NAD, nicotinamide adenine dinucleotide; 3-PGA, 3-phospho-D-glyceric acid; AMP, adenosine monophosphate; PEP, phosphoenolpyruvate, FDP, D-fructose 1,6-diphosphate; DPG, 2,3-diphospho-D-glyceric acid; IMP, inosine monophosphate; ADP, adenosine diphosphate; ATP, adenosine triphosphate; GDP, guanosine diphosphate; GTP, guanosine triphosphate; nmole, nanomole.

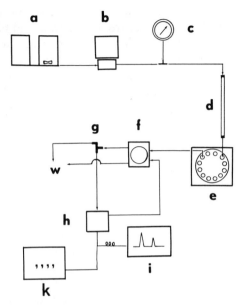

FIG. 1.—Schematic diagram of the chromatographic system. a, gradient generator; b, high pressure pump; c, pressure gauge; d, high performance liquid chromatography column; e, ashing unit; f, peristalic pump; g, debubbler; h, colorimeter set at 820 nm; i, strip chart recorder; w, waste; k, computer.

Figure 1 is a block diagram of the system. Starting with the simple gradient device, two cylindrical chambers produce a linear ammonium chloride–potassium (or ammonium) tetraborate gradient. Stirring is accomplished magnetically. Gradient solution is pumped to the column with a Milton Roy Instrument mini-pump (LDC Division, Milton Roy Co., Riviera Beach, Fl). Solutions must be carefully filtered and degassed before being allowed to enter the pump. Applied pressures up to 500 psi are read from a stainless steel gauge (Ametek/U.S. Gauge, Sellersville, PA).

The columns most frequently used are glass, 0.3-cm × 50-cm and 0.3-cm × 15-cm, (Microbore; Altex, Berkeley, CA) with polypropylene end fittings which support the resin by means of fritted discs. For most high resolution work with physiological samples, the 50-cm analytical column is packed with a 1:1 w/w mixture of strong anion exchange resins Aminex A-25 (Bio-Rad Laboratories, Richmond, CA) and DA-X4 (Durrum Chemical Co., Palo Alto, CA). These resins are high efficiency resins with about 8% and 4% crosslinking, respectively. A precolumn is used for sample loading and may be from 15-cm to 50-cm long, depending upon the volume of sample required. The precolumn contains another strong anion exchanger, AG MP-1, 200–400 mesh (Bio-Rad) which serves to accept the sample and protect the analytical column from impurities in the sample and in the elution buffers. (A yellow impurity arises from the borate and binds tightly to the resins.) The precolumn resin is, therefore, discarded after each run or pooled for subsequent regeneration while the analytical column may be used for 3 to 5 runs or until the resolution has degraded to an unacceptable level. The analytical resin is then unpacked and regenerated (see "Resins" below).

Columns are connected to the high pressure pump and gradient solution reservoir by 0.023-in Teflon tubing. This and other plastic tubing and fittings are obtainable from a number of suppliers. An excellent description of their handling and use is available (5). The 0.023-in Teflon tubing is also used to introduce the column effluent into the ashing unit and to connect the ashing unit to the colorimeter through the peristaltic pump. Connections are made as short as possible from the exit connection of the column throughout the rest of the system in order to minimize peak width (band-spreading). Peak width as observed in a chromatogram is the sum of intracolumn and extracolumn variances. The intracolumn variance is more or less fixed once the choice of column and resin have been made and the packing has been accomplished; the extracolumn variance can be significantly controlled by reducing tubing lengths and diameters, controlling dead volumes, etc.

It is helpful if the pump tubing for the peristaltic pump is color coded: white (0.040 in id) for the supply to the colorimeter and orange (0.035 in id) for the exit from the colorimeter. These pump tubings are available from many suppliers. Most satisfactory are the Desaga (Brinkman Instruments, NY) and the Minipulse II pumps (Gilson Electronics, Middletown, WI).

The colorimeter is a Glenco model 57V (Glenco Scientific, Houston, TX) to which is attached an integrating recorder. The micro flow cell most suitable with the colorimeter has a volume of 25 μl and an optical path length of 12 mm. The colorimeter output is also fed to a computer programmed to resolve curves and calculate peak areas (8).

Solutions

All solutions are made with analytical reagent-grade chemicals or the best grade available and deionized or glass-distilled water. They are filtered with a 1.2-μm filter (Millipore or other suitable) to remove particulate matter which can contaminate the valves and result in irregular pumping or pressure loss to the column. Such contamination would cause irreproducible retention times and non-Gaussian peaks in the chromatogram.

The color reagents used to develop the molybdenum blue reduction product with inorganic phosphate are modified from the recommendations of Ames (9). The first is 2% ascorbic acid in 2.5% sulfuric acid and the second, 0.5% ammonium molybdate in 2.5% sulfuric acid. Each of these solutions contains 1 ml of Ultrawet 60L (Sigma Chemical Co., St. Louis, MO) per liter to promote better flow in the small-bore tubing and to aid wash-out from sample to sample in the tubing as well as in the flow cell. The color reagents are mixed 1:1 v/v on the ashing unit itself (*1*) by means of automatic, dispensing pipets.

The oxidizing reagent is 3 N nitric acid containing 0.01 M potassium or ammonium tetraborate and is also added with an automatic, dispensing pipet. It should be noted that borate acts as a flux and is essential for smooth oxidation and color development in the ashed phosphate samples.

Solvents for elution are made from stock 2 M ammonium chloride and 0.2 M borate solutions. For analysis with the use of 50-cm columns, two gradient solution combinations have been developed: 0.1 M or 0.07 M NH$_4$Cl in the mixing chamber of the gradient maker and 0.6 M NH$_4$Cl in the other chamber. These solutions both contain 0.05 M borate; total volume is 160 ml. The 0.07 M buffer has been most useful in separating the very large PC peaks from G1P and P$_i$ in muscle samples; nevertheless, separation in the "G6P-NAD region" of the chromatogram is better with the 0.1 M starting buffer.

Resins

Resins (A-25 and DA-X4) obtained from the manufacturer are slurried with 1 N hydrochloric acid (HCl) and washed with glass distilled water, then washed with 0.5 or 1 N sodium hydroxide (NaOH) and again with water. The process is repeated and the final wash is made with 1 N HCl in which the resin is kept as a 1:1 v/v slurry after degassing under vacuum.

This washing procedure is usually satisfactory, but if the resin requires more thorough washing (unsatisfactory resolution), then the resin is converted to the hydroxide form with 25 volumes of 1 N NaOH and then reconverted by neutralizing with HCl (1 volume of 1 N). Other methods of washing are (*a*) conversion to the acetate form with glacial acetic acid, then reconverting to the chloride form and (*b*) boiling for 30 min with 6 N HCl; but these procedures are not necessary as long as the resins are handled with sufficient attention to cleanliness.

Degassing the AG MP-1 resin is usually not necessary. It can be used by merely slurrying with 1 N HCl (1:1 v/v) in which it is stored.

Columns

Into an 0.3-cm × 50-cm column containing about 2 cm of 0.6 M ammonium chloride (NH$_4$Cl) resin is introduced by means of a syringe connected to a length of polyethylene or Teflon tubing. Sufficient resin is used to pack the column only to a level of about one-sixth or less of its length. Exclusion of all air bubbles is necessary. The column, aligned in a vertical position, is then pumped with 0.6 M NH_4Cl (no borate). A precolumn containing about 2 cm of AG MP-1 (or Dowex 1-X4, 200–400 mesh) to act as a filter is used ahead of the analytical column being packed. Pumping is continued at or below maximum operating pressure, about 500 psi, until a sharply defined top of the resin bed is observed. Pumping is stopped; the solution above the bed is carefully removed, and more resin is introduced.

The above process is repeated until the column is filled and the resin is trapped between the fritted discs of the column ends. Alternatively, a small amount of a material like Celite 545 AW (Supelco, Inc., PA) may be used at the top to help trap the resin bed and remove any dead space. Water is then used to wash out NH_4Cl to an end point determined with silver nitrate (a faintly opalescent solution is approximately equivalent to 0.001 M NH_4Cl or less and is acceptable).

Subsequent to each use, columns are regenerated with a slug of 1–3 N HCl (1–3 ml) conveniently applied by means of the precolumn-filter arrangement used in packing.

Sample Preparation

The most important first step in metabolite analysis is care in sample preparation. Thus, loose cells (from culture or blood samples) can be ejected from a syringe or otherwise rapidly mixed with an equal volume or twice the volume of ice-cold trichloroacetic acid (TCA) or perchloric acid (PCA) so that the final concentration of acid is about 3% for the TCA and 0.6 M for the PCA. Precipitated protein is removed by centrifugation (in the cold) and washed once with a small amount of ice water; the washings are then combined with the supernatant extract. The following operations are all done at 0–4°.

TCA is removed by extraction with water-saturated ether (3–4 times with 4–5 volumes). Neutralization may be completed with a small amount of potassium hydrogen carbonate ($KHCO_3$) using a glass electrode or bromthymol blue as indicator, and the extract is then brought to a known volume.

PCA is removed by simultaneous neutralization and precipitation with a small amount of a saturated solution of potassium carbonate (K_2CO_3) or 10 M KOH. The solution is swirled during addition to avoid local concentration build-up, and the mixture is brought to pH 7.0 (glass electrode or bromthymol blue added directly to the sample). The precipitate of potassium perchlorate is packed by centrifugation, and the supernatant is decanted; the precipitate is washed once with 3 volumes of ice water; the mixture is centrifuged, and the supernatant is combined with the previous one and made up to a known volume. The precipitate ($KClO_4$) is discarded.

Unlike loose cells, tissue samples require very rapid cooling as a first preparative step to stop enzymic activity. Thus, freeze clamping (*10*) is followed by pulverizing in a mortar, all at liquid nitrogen temperature. The powdered tissue is then extracted by slurrying with 3 *M* PCA at $-10°$ to allow complete penetration and fixation of the powder by the acid before dilution to a final concentration of 0.6 *M*. Subsequent operations are done as described above. The procedure is about the same as has been detailed by Lowry and Passonneau (*10*).

All extracts as well as frozen tissue powders may be stored indefinitely at $-50°$ or below.

Loading and Running a Sample

Most routine chromatographic runs are done at room temperature in our laboratory. The system will accept from 500 to 1000 nmole of total phosphate. The best sample size may be determined by placing aliquots of the sample extract in alternate cups of the analyzer to permit ashing and color development. Comparison is then made with aliquots of a standard inorganic phosphate (KH_2PO_4) solution adjusted to convenient concentrations of from 0.1 to 1.0 m*M* depending upon the micropipet or syringe available. A pushbutton device together with a precision 250-μl syringe (Hamilton Co., Reno, NV) is useful for delivering 5-μl quantities in a very reproducible manner. Then, depending upon the concentration of the total phosphate that has been determined in the sample extract, a volume of from 100 to 1000 μl is loaded on a precolumn of suitable length also by means of a precision syringe (for example, Hamilton or Glenco). The precolumn contains 2 cm of AGMP-1 resin topped by a few millimeters of Celite 545 (AW) and is prepared with care to insure a flat resin bed surface at 90° to the long axis of the column. Water is removed and sample is added, followed by additional water if necessary to fill the precolumn and exclude any air bubbles. The precolumn is then closed, and a flow of water is started with the high pressure pump.

The flow rate is measured by connecting the tube from the column exit to a 0.1-ml graduated pipet with a short length of PVC tube sleeving. For the 50-cm column, a flow rate of 0.1 ml in 28–30 sec secures adequate resolution of a complex physiological sample in about 7 hr. Shorter columns and running times may be adequate, depending upon the particular metabolites being examined. Thus, almost all glycolytic intermediates elute sooner than most nucleotides. Once the flow rate has been established, the gradient is started, and an indexing mark or signal is made on the recorder chart at the same time.

Runs are routinely made at 20–25° to avoid destroying GAP or DHAP in the extracts, as these compounds are sensitive to alkali and are hydrolyzed rapidly at elevated temperatures.

If preparative columns (0.9-cm × 30-cm) are to be used (available from Altex or Glenco), a stream splitting valve (Glenco) is used to permit only 0.1–0.2 ml of effluent to enter the ashing unit in each 30-sec cycle of collection; the remainder of the stream is used in appropriate ways for further analysis. The scaling up to larger columns is first made in direct proportion to that used for analytical work. This depends somewhat upon the peak size and separation. Nevertheless, if the 0.9-cm id column contains 10 times as much resin as the 0.3-cm unit, the gradient volume is increased proportionately to 1600 ml total, and the flow is set to 1–2 ml/min. In practice, we have found that 800 ml is often satisfactory, perhaps as the result of increased efficiency with larger columns as pointed out by Wolf (11).

Interpretation of Results

By means of relative retention times, peaks obtained in a chromatogram of a biological sample are compared to those obtained using authentic standards. Relative retention time (t_R) is defined as the ratio of the time (or distance) on the chart for the peak of interest to that of an arbitrarily chosen reference peak, inorganic phosphate, for example. Closely similar numbers are presumptive evidence for the metabolite of interest, but cochromatography with an authentic compound is then necessary to insure identification of at least the compound of interest.

Other methods of identification involve simplifying the sample before placing it on the column, for example, removal of nucleotides with charcoal, partial hydrolysis (1 M $HClO_4$, 100°, 10 min, conditions that permit only true phosphate esters to remain), and extraction of the inorganic phosphate (12). The extraction must be applied judiciously to prevent catalytic destruction of phosphocreatine, for example.

Enzymic methods are powerful tools for further identification of specific metabolites, isolated or not. A suspected metabolite converted to another, for example, DHAP to αGP or FDP to GAP and DHAP, and again chromatographed, affords virtually conclusive evidence for identification. Suitable conditions for many enzymic assays are found in reference 10.

Typical Results

In Table I are found t_R values measured from chromatograms of five different aliquots of one extract from rat gastrocnemius muscle run on different

TABLE I

Relative Retention Times and Quantities of Metabolites
Determined in Five Different Samples of
Rat Gastrocnemius Muscle[a,b]

	$t_R{}^c$	SD	Nanomoles[e]	SD
PC	0.84	0.01	off	
G1P	0.94	0.00	6.91	0.30
P_i	1.00	—	off	
αGP	1.06	0.01	23.3	1.4
DHAP	1.29	0.01	9.79	1.7
M6P	1.33	0.01	—[f]	
GAP	1.42	0.01	0.21	0.17
F6P	1.48	0.02	10.3	0.61
G6P	1.64	0.01	45.6	2.6
NAD	1.72	0.01	—[f]	
3PGA	2.07	0.03	11.8	0.72
X_1	2.21	0.03	0.58	0.09
AMP	2.38	0.03	5.40	0.28
PEP	2.42	0.03	—[f]	
FDP	2.58	0.04	29.3	1.5
DPG	2.66	0.04	1.17	0.18
IMP	2.86	0.05	51.0	2.7
X_{11}	2.98	0.05	0.56	0.16
ADP	3.23	0.05	16.4	0.97
NADH	3.60	0.05	0.69	0.17
ATP	3.80	0.05	98.4	5.1
GTP	3.93	0.06	—[f]	

[a] Average of five samples ± standard deviation.
[b] X values are unknowns in this particular tissue.
[c] Based upon inorganic phosphate (P_i) assigned the value of 1.00.
[d] Standard deviation.
[e] Nanomoles from the equivalent of 15 mg of tissue, wet weight.
Off indicates peak too large to be determined.
[f] Inseparable from the previous peak.

days. Each complete run required 6 h; buffers were 0.07 M NH_4Cl (80 ml) in the mixing chamber of the gradient maker and 0.6 M NH_4Cl (80 ml) in the other chamber together with 0.05 M $K_2B_4O_7$ in both. They were run at room temperature with mixed resin (A-25 + DA-X4) in 0.3-cm × 50-cm column, 0.22 ml/min flow rate, colorimeter set at 820 nm with 0.5 absorbance units full scale sensitivity. Also illustrated in Table I is the precision achieved in quantifying metabolites from the same five analyses.

Figure 2 is typical of the chromatograms obtained from this series of analyzes. The two "square-topped" peaks following GTP are the P_i standards (20 nmoles) included in every run. Chromatograms from other tissues such as liver and brain are more complex and resolution frequently is not

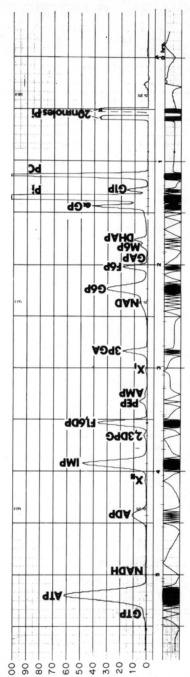

FIG. 2.—Chromatogram of rat gastrocnemius muscle metabolites. Extract on the column was equivalent to 15 mg wet weight. Muscle was aged a few minutes before fixation with perchloric acid.

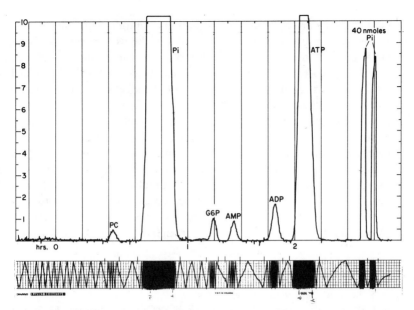

Fig. 3.—Chromatogram of metabolites generated by a mitochondrial preparation. Extract on the column contained metabolites resulting from a 5-sec incubation with the equivalent of 50 μg of mitochondrial protein.

as good as shown here since the pentose pathway intermediates as well as UMP and CMP and other unknowns appear in the region that includes G6P, NAD, and 3-PGA.

Figure 3 illustrates the application of the system to bound creatine kinase in heart mitochondria. Chromatography was conducted with AG MP-1 resin packed in an 0.3-cm × 15-cm column; 200-μl samples routinely were loaded by means of an empty precolumn. The less expensive AG MP-1 resin was used and discarded after each run owing to the use of relatively large amounts of radioactive phosphorus in the experiments and the desire to prevent contamination from sample to sample.

To separate relatively large amounts of P_i from the PC for the quantitative measurement of both the phosphorus and radioactivity in each peak, the mixing chamber can be filled with 60 ml of 0.05 M NH_4Cl (in 0.05 M $K_2B_4O_7$) while the other chamber contains 50 ml of the usual 0.6 M NH_4Cl (in 0.05 M $K_2B_4O_7$). After a 15-min water wash to remove excess mannitol (necessary for preparation of mitochondria), the solution in the mixing chamber is connected to the column and elution is allowed to proceed ($\sim$30 min) until the depths of the buffers are the same in both chambers. Without interrupting the flow to the column, the valve between chambers is then opened and the gradient elution started. Thus the first period of elution with solvent of constant composition provides excellent resolution between a very small peak, PC (1–5 nmole and about 5000 cpm) and a very large one, P_i (250 nmole and about 1,000,000 cpm). The separation can be conducted at 25° at a flow rate of 0.4 ml/min; colorimeter, 820 nm with 0.5 absorbance units full scale sensitivity. A fraction collector (Büchler, alpha 200) collects effluent from the colorimeter in 1-min fractions, about 0.5 ml each, that are counted with the use of 5.5 ml of ACS scintillation solvent (Amersham-Searle) per vial.

The method provides a tool for investigation of problems in metabolic regulation, such as the study of control points in intermediary metabolism, perturbed, for instance, by various drugs or chemical probes. For such studies, two things are required which the system and methods supply: the simultaneous measurement of numerous phosphorylated metabolic intermediates and (using ^{33}P or ^{32}P) the flux through the metabolic pathways (13).

References[2]

(1)   S. P. Bessman, *Anal. Biochem.*, **59**, 524 (1974).

(2)   S. P. Bessman, P. J. Geiger, T. C. Lu, and E. R. B. McCabe, *Anal. Biochem.*, **59**, 533 (1974).

(3)   M. Jellinek, H. Amako, and V. Willman, in "Advances in Automated Analysis," 5th Technicon Intern. Cong., 1970, Vol. I, Thurman Associates, Miami, Florida, 1971, p. 587.

(4)   H. Schnitger, K. Papenberg, E. Ganse, R. Czok, Th. Bucher, and H. Adam, *Biochem. Z.*, **332**, 167 (1959).

(5)   J. X. Khym, "Analytical Ion-Exchange Procedures in Chemistry and Biology: Theory, Equipment, Techniques," Prentice Hall, New Jersey, 1974.

(6)   L. R. Snyder, *J. Chromatogr. Sci.*, **10**, 200 (1972).

(7)   L. R. Snyder, *J. Chromatogr. Sci.*, **10**, 369 (1972).

(8)   E. C. Layne, unpublished.

(9)   B. N. Ames, in "Methods in Enzymology," S. P. Colowick and N. O. Kaplan, eds., Vol. VII. Academic Press, New York, N.Y., 1966, p. 115.

(10) O. H. Lowry and J. V. Passonneau, "A Flexible System of Enzymatic Analysis," Academic Press, New York, N.Y., 1972, p. 123 ff.

(11) J. P. Wolf, III, *Anal. Chem.*, **45**, 1248 (1973).

(12) O. Lindberg and L. Ernster, in "Methods of Biochemical Analysis," David Glick, ed., Vol. III, 1956, p. 7.

(13) E. A. Newsholme and C. Start, "Regulation in Metabolism," John Wiley, New York, N.Y., 1976, p. 16 ff.

---

[2] References are intended as a useful guide and are not meant to be all inclusive of published work available on phosphorylated metabolic intermediates and HPLC.

[4] High-Performance Liquid Chromatography

By G. D. McGinnis and P. Fang

Forest Products Utilization Laboratory, Mississippi State University, Mississippi State, Mississippi

Introduction

High-performance liquid chromatography (HPLC) was defined by Snyder and Kirkland as "an automated, high-pressure liquid chromatography in columns, with a capability for the high-resolution separation of a wide range of sample types, within times of a few minutes to perhaps an hour" (*1*). More detailed descriptions of the theory, equipment, and the general application of this technique can be found in several books (*1–4*), review articles (*5–7*), and bibliographies (*8–9*).

High-performance liquid chromatography has many advantages over the other methods of carbohydrate analysis. As compared to paper chromatography and open-column chromatography, it is much faster, more sensitive, and more suitable for routine quantitative analysis. The gas–liquid chromatographic (glc) methods for carbohydrate analysis (Vol. VI [1]–[5]) are 50–100 times more sensitive than the HPLC detector system (differential refractometer detector). Both the HPLC and glc procedures are approximately equivalent in terms of actual separation of monosaccharides, in precision and accuracy, and in their potential for automation (*10*). However, the HPLC procedure for carbohydrate analysis is much faster and more suitable for routine analysis since the carbohydrates in aqueous solutions can be determined without derivatization or potential sample loss and with a minimum of sample preparation. Also HPLC can be used to analyze a much wider range of carbohydrates, including the thermally labile carbohydrates and oligo- and polysaccharides which are not volatile enough to be analyzed by gas–liquid chromatography.

Many of the early high-performance separations of carbohydrates were done using ion-exchange columns. Two of the more widely used methods involved the separation of the carbohydrates as borate complexes using anion-exchange resins and the separation of carbohydrates using anion- and cation-exchange resins with mixed solvents (*9, 11*). Both of these methods are capable of resolving a wide range of monosaccharides and disaccharides, can be automated, and will detect monosaccharides down to the microgram level. However, they are relatively slow and generally require 1–4 h for a normal monosaccharide analysis.

METHODS IN CARBOHYDRATE
CHEMISTRY, VOL. VIII

In the last 3 to 4 years, major advances have been made in column technology, including the development and commercial availability of uniformly sized packing materials in the 5 to 10 μm regions and the development of chemically bonded stationary phases for carbohydrate analysis. The new types of columns have made it possible to separate a wide range of carbohydrates with high resolution in relatively short periods of time (15–60 min). This article will describe the techniques used for separating carbohydrates by high-performance liquid chromatography with special emphasis on the use and capability of the newer types of column-packing materials.

Procedure

Column-Packing Materials

Some of the newer packing materials which can be used in separating unsubstituted carbohydrates are chemically bonded silicas, such as the μBondapak/carbohydrate from Waters Associates (Milford, MA) (12–14), Partisil-10 PAC from Whatman, Inc. (Clifton, NJ) (15), the Amino Sil-x-1 from the Perkin-Elmer Corporation (Norwalk, CT), and LiChrosorb-NH$_2$ from E. Merck & Co. (West Germany). Cation-exchange resin, such as the Aminex-A-5 (Bio-Rad Laboratories, Richmond, CA), has also been used for the high-performance separation of unsubstituted monosaccharides (16–17). The substituted carbohydrates containing acidic or basic groups, such as the nucleosides or nucleotides, can be separated on high-efficiency, ion-exchange columns (9, 18, Vol. VIII [3]), while many of the neutral, substituted carbohydrates are separated on microparticulate silica gel columns (19–20).

Column Preparation

The various methods for packing high-performance columns are described in detail by Kirkland (3) and Snyder and Kirkland (1) and will not be reviewed here. Packing columns containing fine particles (5–10 μm) require special procedures and equipment to obtain maximum column efficiency. It is generally recommended for routine laboratory analysis to purchase these columns prepacked directly from the manufacturers.

Solvent and Sample Purification

If "distilled-in-glass" or "spectro" grade solvents are being used, the only solvent pretreatment necessary is to filter the solvent through a 0.5 μm filter

(Millipore Corp., Bedford, MA) in order to remove particulate material. In many cases, however, the less expensive "reagent-grade" solvents can be used if adequate precaution is taken. The necessity for further purification of the solvents depends on many factors, such as the type of detector, the type of column, and the particular solvent being used. One widely used method for purifying solvents for liquid chromatography is by passing the solvent through a column of dry, porous activated silica, such as Davison Grade 35 silica (W. R. Grace, 12–42 mesh) just prior to use (*1, 3*). The silica is activated at 180° for 3 h and cooled. Further purification of the solvent can also be achieved with the HPLC unit by placing a 5–10-cm precolumn between the pump and the sample injection unit. The precolumn should be packed with the same deactivated absorbent used in the separation column, but can be larger and need not be well packed.

The solvent should also be degassed just prior to being used. This is especially important if water or water-containing solvents are being used. Removal of the dissolved gases can be done by means of heat, vacuum, or high-intensity ultrasonic waves (*21–23*).

The carbohydrate samples in aqueous or organic solutions can generally be analyzed directly without pretreatment except for filtration through a 0.5-μm pore-diameter filter to remove the particulates. However, if the samples contain large amounts of salts or proteins, it is best to remove the salts or proteins by deionization through a mixed-bed resin such as Amberlite MB-1. This deionization can be achieved by simply adding the washed resin to the sample in a flask and shaking on a wrist-action shaker for 10 min and filtering (*16*) or by passing the sample through a column of the mixed-bed resin (*24*). This type of pretreatment will increase the lifetime of the analytical HPLC column considerably.

Detectors

The most widely used detector for carbohydrate analysis is the differential refractometer. This detector has moderate sensitivity for carbohydrates and has a wide linear range. The lower limit of sensitivity for this detector is around 20 μg which means that samples containing 0.01% of an individual carbohydrate can be detected when injecting 200 μl of the sample (*13*).

Most of the manual and electronic techniques that have been developed for integrating gas chromatography signals can also be used with the refractometer. Palmer used peak heights to determine the concentration of D-glucose, D-xylose and cellobiose from a cellulose hydrolyzate on a μBonda-pak/carbohydrate column (*13*). He found that the detector response was linear from 20–400 μg, and the precision coefficient of variation was 1–2% for D-glucose and cellobiose and about 3% for the minor component,

D-xylose. Linden and Lawhead (12) using the same type of column with a digital integrator found a coefficient of variation of 0.37–1.8 for three different carbohydrates. Recently, Scobell and co-workers (16) using an Aminex A-5 column and an electronic integrator reported accuracy on the order of ±0.2% absolute could be obtained for a wide variety of carbohydrates. They also found that the calibration curves (peak area versus amount injected) for D-fructose, D-glucose, maltose, maltotriose, and maltohexose were almost identical.

Discussion

The types of separations achieved by high-performance liquid chromatography are shown in a series of chromatograms. The first group of chromato-

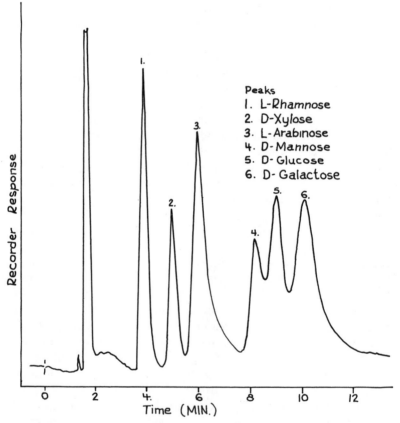

FIG. 1.—Separation of monosaccharides on a μBondapak/carbohydrate column. Column 0.4-cm id × 30-cm, mobile phase, 15:85 v/v water–acetonitrile, 2.5 ml/min; temperature 25°; refractive index detector (13).

grams (Figs. 1–4) was done using two commercially available, chemically bonded, stationary-phase columns. The second group of chromatograms (Figs. 5 and 6) consists of separations of carbohydrates on Aminex A-5, a cation-exchange resin, while the final group of chromatograms (Figs. 7 and 8) was done using Partisil-10, a commercially available microparticulate silica gel.

The initial chromatogram (Fig. 1) shows the separation of a group of monosaccharides 85:15 v/v using acetonitrile–water as the eluting solvent

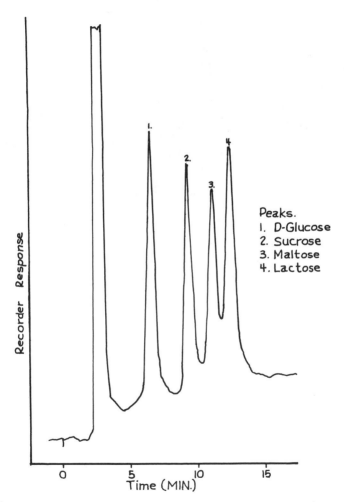

Peaks.
1. D-Glucose
2. Sucrose
3. Maltose
4. Lactose

FIG. 2.—Separation of disaccharides on a μBondapak/carbohydrate column. Column 0.4-cm id × 30-cm, mobile phase, 20:80 v/v water–acetonitrile, 1.5 ml/min; temperature 25°; refractive index detector (13).

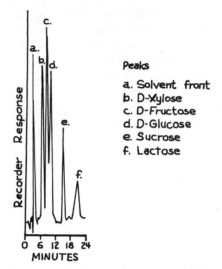

Peaks

a. Solvent front
b. D-Xylose
c. D-Fructose
d. D-Glucose
e. Sucrose
f. Lactose

FIG. 3.—Separation of mono- and disaccharides on a Partisil-10 PAC column. Column 0.46-cm id × 25-cm, mobile phase, 20:80 v/v water–acetonitrile, pH 5.0 with H_3PO_4, 1.3 ml/min; room temperature; refractive index detector (15).

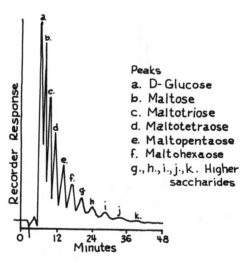

Peaks

a. D-Glucose
b. Maltose
c. Maltotriose
d. Maltotetraose
e. Maltopentaose
f. Maltohexaose
g., h., i., j., k. Higher
     saccharides

FIG. 4.—Separation of maltodextrins on a Partisil-10 PAC column. Column 0.46-cm id × 25-cm, mobile phase, 65:35 v/v acetonitrile–0.0025 M Na $C_2H_3O_2$, pH 5.0 with acetic acid, 65 ml/h; room temperature; refractive index detector (15).

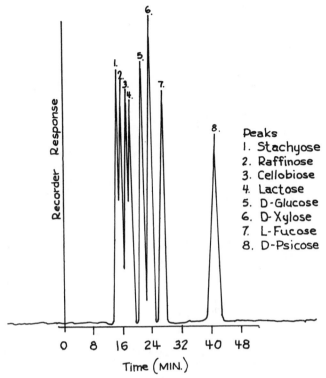

Peaks
1. Stachyose
2. Raffinose
3. Cellobiose
4. Lactose
5. D-Glucose
6. D-Xylose
7. L-Fucose
8. D-Psicose

Time (MIN.)

FIG. 5.—The separation of carbohydrates on an Aminex A-5(Ca^{+2}) column. Column 0.305-in id × 10-in, mobile phase water, 0.3 ml/min; column temperature 85°; detector temperature 45°; refractive index detector (16).

with a μBondapak/carbohydrate column (13). The same column can be used to separate disaccharides and higher saccharides by simply increasing the amount of water in the solvent. This is illustrated in Figure 2 which shows the results when a series of mono- and disaccharides were separated 80:20 v/v using acetonitrile–water. More complex mixtures of carbohydrates can be separated using gradient elution devices, solvent programming, and by increasing the length of the column.

All carbohydrate separations that have been done with the chemically bonded stationary phase have used mixtures of acetonitrile–water as the eluting solvents. With some of the chemically bonded columns, the separation of the carbohydrates can be improved if small amounts of acid are added to the acetonitrile–water eluting solvent. The type of separation with a

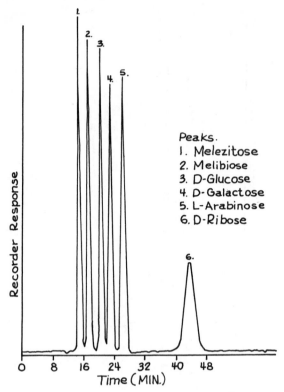

FIG. 6.—The separation of carbohydrates on an Aminex A-5(Ca^{+2}) column. Column 0.305-in id × 10-in, mobile phase water, 0.3 ml/min; column temperature 85°, detector temperature 45°; refractive index detector (*16*).

Partisil-10 PAC column, using the acidified acetonitrile–water mixture, is shown in Figures 3–4 (*15*). Chemically bonded columns are quite versatile and have been used to separate not only hexoses but also a variety of other types of carbohydrates, including pentoses (*14, 25, 26*), deoxysugars (*14*), ketoses (*12*), amino sugars (*12, 25*), and sugar alcohols (*12, 25*).

High-performance liquid-chromatographic separations of unsubstituted carbohydrates can also be done using cation-exchange columns. Generally, the separation on ion-exchange columns is done at an elevated temperature, which is a disadvantage. However, these columns use water as the eluting solvent and, in general, are less expensive than the chemically bonded stationary-phase columns. Recent work by Scobell and co-workers (*16*) and by Goulding (*17*) indicate that these columns are capable of separating a wide variety of carbohydrates in relatively short times. This is illustrated

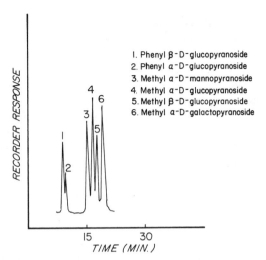

FIG. 7.—The separation of glycosides on a silica gel column (Partisil-10). Column 0.46-cm id × 25-cm; mobile phase 1:9 v/v water–acetonitrile, 0.54 ml/min; temperature 25°; refractive index detector (*19*).

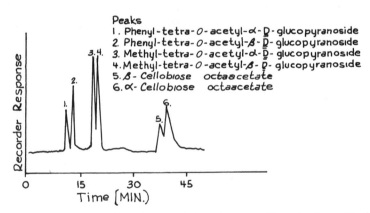

FIG. 8.—The separation of acetylated carbohydrates on a silica gel column (Partisil-10). Column 0.46-cm id × 25-cm; mobile phase 1:1 v/v ethyl acetate–hexane, 0.54 ml/min; temperature 25°; refractive index detector (*19*).

in Figures 5 and 6, which show the results of a separation of a series of mono- and oligosaccharides that are commonly found in the sweetner industry (*16*).

High-performance liquid chromatography can also be used to separate substituted carbohydrates with acid or base functions. Most of the studies in this area have dealt with the separation of the nucleosides and nucleotides using ion-exchange columns. More detailed information on the types of columns, extraction procedures, and chromatographic conditions can be found in recent reviews by Brown (*18*) and Zadrazil (*8*).

Carbohydrates substituted with neutral groups can also be separated using high-performance liquid chromatography. Some examples of this type of separation are shown in Figures 7 and 8. The first chromatogram (Fig. 7) shows the separation of a series of glycosides on a column of microparticulate silica gel (Partisil-10, Whatman, Inc.) using 9:1 v/v acetonitrile–water as the eluting solvent. The same type of column can be used to separate a series of completely substituted carbohydrates by reducing the polarity of the eluting solvent. Figure 8 illustrates the separation of a group of completely substituted carbohydrates using 1:1 v/v ethyl acetate–hexane as the eluting solvent (*19*).

The microparticulate silica columns are extremely versatile columns for carbohydrate separations. These types of columns can be used to separate the unsubstituted as well as the partially and completely substituted carbohydrates (*19–20, 26*). In most cases, solvent selection is relatively easy, because many of the solvent systems which have been developed for thin-layer chromatography on silica gel can also be used for HPLC on silica gel. Another advantage is that microparticulate silica gel is available in large preparative columns at a reasonable cost while most of the chemically bonded columns are only available in analytical sizes.

References

(*1*) L. R. Snyder and J. J. Kirkland, "Introduction to Modern Liquid Chromatography," John Wiley, New York, 1974.

(*2*) P. R. Brown, "High-Pressure Liquid Chromatography," Academic Press, New York, 1973.

(*3*) J. J. Kirkland, "Modern Practice of Liquid Chromatography," Wiley-Interscience, New York, 1971.

(*4*) J. N. Done, J. H. Knox, and J. Loheac, "Applications of High-Speed Liquid Chromatography," John Wiley, London, 1974.

(*5*) H. F. Walton, *Anal. Chem.*, **48**, No. 5, 52R (1976).

(*6*) J. J. DeStefano and J. J. Kirkland, *Anal. Chem.*, **47**, No. 12, 1103A (1975).

(*7*) P. Jandera and J. Churacek, *J. Chromatogr.*, **98**, 55 (1974).

(*8*) "Bibliography of Liquid-Column Chromatography, 1971–73," *J. Chromatogr.*, *Suppl. Vol.*, **6**, Z. Deyl and J. Kopecky, eds., Elsevier, Amsterdam, 1976.

(*9*) K. Capek and J. Stanek, Jr., "Carbohydrates in Liquid-Column Chromatography," Z. Deyl, K. Macek, and J. Janak, eds., Elsevier, Amsterdam, Vol. 3, 1975, pp. 465–522.

(*10*) J. K. Palmer, "Gas Liquid Chromatography and High-pressure Liquid Chromatography of Drugs," Marcel Dekker, New York, Vol. II, Chapter 38 (in press).

(*11*) O. Samuelson, "Partition Chromatography on Ion-Exchange Resins in Methods in Carbohydrate Chemistry," Academic Press, New York, Vol. 6, 1972, p. 65.

(*12*) J. C. Linden and C. L. Lawhead, *J. Chromatogr.*, **105**, 125 (1975).

(*13*) J. K. Palmer, *Anal. Lett.*, **8(3)**, 215 (1975).

(*14*) R. B. Meagher, S. J. and A. Furst, *J. Chromatogr.*, **117**, 211 (1976).

(*15*) F. M. Rabel, A. G. Caputo, and E. T. Butts, *J. Chromatogr.*, **126**, 731 (1976).

(*16*) H. D. Scobell, K. M. Brobst, and E. M. Steele, *Cereal Chem.*, **54**, 905 (1977).

(*17*) R. W. Goulding, *J. Chromatogr.*, **103**, 229 (1975).

(*18*) P. R. Brown, "The Use of High-Pressure Liquid Chromatography in Pharmacology and Toxicology in Advances in Chromatography," Marcel Dekker, New York, Vol. 12, 1975, pp. 1–30.

(*19*) G. D. McGinnis and P. Fang, *Carbohydr. Res.*, 1977 (accepted for publication).

(*20*) J. L. Rocca and A. Rouchouse, *J. Chromatogr.*, **117**, 216 (1976).

(*21*) R. Battino, D. F. Evans, and M. Bogan, *Anal. Chim. Acta* **43**, 518 (1968).

(*22*) R. Battino, M. Banzhof, M. Bogan, and E. Wilheim, *Anal. Chem.*, **43**, 806 (1971).

(*23*) V. E. Dell'Ova, M. B. Denton, and M. F. Burke, *Anal. Chem.*, **46**, 1365 (1974).

(*24*) J. K. Palmer and W. B. Brandes, *J. Agr. Food Chem.*, **22**, 709 (1974).

(*25*) R. Schwarzenbach, *J. Chromatogr.*, **117**, 206 (1976).

[5] Gel Permeation Chromatography

By Roy L. Whistler and Abul Kashem M. Anisuzzaman

Department of Biochemistry, Purdue University, Lafayette, Indiana

Introduction

Gel permeation chromatography (gpc), or exclusion chromatography, is a technique for separating molecules according to differences in their sizes. The stationary phase is a bed of swollen gel which is composed of spongelike, porous particles. The sample to be separated into constituent components is allowed to pass through the stationary phase with the aid of a solvent. Molecules larger than the pores of the gel bed cannot enter the gel matrix and are eluted rapidly. In contrast, smaller molecules enter the gel-pores to a varying extent depending on their size and shape. Large molecules travel through the gel at a faster rate than small molecules, and as a consequence, components are eluted sequentially in order of decreasing molecular size (Fig. 1).

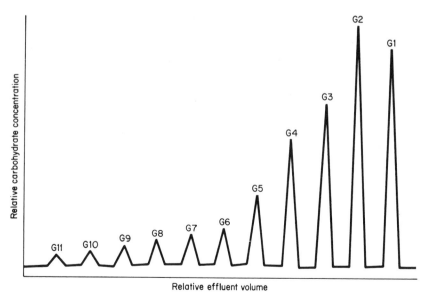

FIG. 1.—Glucose oligosaccharides chromatographed on BioGel P-2, −400 mesh. G1 through G11 = glucose to maltoundecaose [John and coworkers (65)].

45

METHODS IN CARBOHYDRATE
CHEMISTRY, VOL. VIII

After the pioneering research of Porath and Flodin (*1*) on cross-linked dextran gel, gpc has become in increasingly useful method for the separation and characterization of proteins, peptides, enzymes, nucleic acids, and carbohydrates. Several books (*2–4*) and articles (*5–15*) describing the theory and applications of this technique have appeared. In the area of carbohydrates, earlier applications were reviewed (*16–18*). Since then, other reports (*19–60*) on the application of gpc in this and related fields have appeared.

Gel permeation chromatography is an economical method of separation because the same gel column can be used repeatedly, and since the separation is based on differences in molecular size, the technique can also be used for the determination of molecular weight (*61–64*). Another advantage is that the process is gentle and no structural modification of components takes place during the operation. John and co-workers (*65*) have shown that the method can be used for quantitative estimation of carbohydrates.

Materials and Methods

Gels

Gels used for the separation of carbohydrates include cross-linked dextran, cross-linked polyacrylamide, cross-linked poly(acryloylmorpholines), and agaran (agarose).

Cross-Linked Dextran.—A water-insoluble gel is obtained by reacting an aqueous solution of linear dextran with epichlorohydrin under alkaline conditions. The extent of cross-linking and, hence, the swelling properties and average pore size within the matrix are regulated by varying the molecular weight of the dextran and the proportion of epichlorohydrin (*66*). Some eight different commercial dextran gels designated Sephadex G-10 through G-200 are available (*67*). Their properties are summarized in Table I.

Sephadex LH-20 prepared by hydroxypropylation of Sephadex G-25 (*67*) swells both in water and in polar organic solvents and is useful for a variety of solvents (*68–71*).

Cross-Linked Polyacrylamide.—The gel is prepared (*72*) by the co-polymerization of acrylamide with N,N'-methylenebisacrylamide. As with dextran gels, polyacrylamide gels having different swelling properties are obtained by varying the proportions of the reactants (*72*). Polyacrylamide gels are not subject to attack by microorganisms and some ten commercial gels are available (*73*). Their properties are given in Table II.

Cross-Linked Poly(acryloylmorpholine).—Gels are prepared by free-radical copolymerization of N-acryloylmorpholine and N,N'-methylenebisacryl-

TABLE I

Properties of Commercial Dextran Gels (Sephadex)[a]

Sephadex type	Particle diameter dry, μ	Water regain, ml/g	Bed volume, ml/2	Fractionation range, M.W. of dextran	Swelling time at 98°, h	Maximum pressure, cm H$_2$O
G-10	40–120	1.0 ± 0.1	2–3	–700	1	>200
G-15	40–120	1.5 ± 0.1	2.5–3.5	–1500	1	>200
G-25 course	100–300	2.5 ± 0.2	4–6	100–5000	1	>200
medium	50–150					
fine	20–80					
superfine	10–40					
G-50 course	100–300	5.0 ± 0.3	9–11	500–10000	1	>200
medium	50–150					
fine	20–80					
superfine	10–40					
G-75	40–120	7.5 ± 0.5	12–15	1000–50000	3	160
superfine	10–40					
G-100	40–120	10 ± 1.0	15–20	1000–100000	5	96
superfine	10–40					
G-175	40–120	15 ± 1.5	20–30	1000–150000	5	36
superfine	10–40					
G-200	40–120	20 ± 2.0	20–40	1000–200000	5	16
superfine	10–40					

[a] According to data from manufacturer: Pharmacia Fine Chemicals, Uppsala, Sweden.

TABLE II

Properties of Commercial Polyacrylamide Gels (BioGel)[a]

BioGel type[b]	Water regain, ml/g	Bed volume, ml/g	Fractionation range, (M.W.)[c]	Swelling time at 100°, h	Maximum pressure, cm H$_2$O
P-2	1.5	3.0	100–1800	2	>100
P-4	2.4	4.8	800–4000	2	>100
P-6	3.7	7.4	100–6000	2	>100
P-10	4.5	9.0	1500–20000	2	>100
P-30	5.7	11.4	2500–40000	3	>100
P-60	7.2	14.4	3000–60000	3	100
P-100	7.5	15.0	5000–100000	5	60
P-150	9.2	18.4	15000–150000	5	30
P-200	14.7	29.4	30000–200000	5	20
P-300	18.0	36	60000–400000	5	15

[a] According to the manufacturer: Bio-Rad Laboratories, Richmond, California.
[b] All types are available in particle sizes (mesh, wet state): 50–100, 100–200, minus 400. The grades P-2 through P-10 are also available in fractions 200–400.
[c] Fractionation range for globular proteins or peptides; values for molecules of other geometric configuration will be somewhat lower.

TABLE III

Properties of Commercial Agaran Gels[a]

Type	Agaran concentration, % in gel	Particle diameter, hydrated	Fractionation of globular protein	Range dextran	Maximum pressure, cm H$_2$O
Sepharose 2B	2	60–250 μ	10^4–4 × 10^7	10^4–2 × 10^7	40
Sepharose 4B	4	40–190 μ	10^4–2 × 10^7	10^4–5 × 10^6	80
Sepharose 6B	6	40–210 μ	10^4–4 × 10^6	10^4–10^6	> 200
BioGel A-0.5m	10	50–100 mesh 100–200 mesh 200–400 mesh	> 10^4–5 × 10^5	b	> 100
BioGel A-1m	6	50–100 mesh 100–200 mesh 200–400 mesh	10^4–5 × 10^6	b	> 100
BioGel A-15m	4	50–100 mesh 100–200 mesh 200–400 mesh	4 × 10^4–1.5 × 10^7	b	90
BioGel A-50m	2	50–100 mesh 100–200 mesh	10^5–5 × 10^7	b	50
BioGel A-150m	1	50–100 mesh 100–200 mesh	10^6–1.5 × 10^8	b	30

[a] According to manufacturer: Sepharose 2B through Sepharose 6B from Pharmacia Fine Chemicals, Uppsala, Sweden; BioGel A-0.5m through BioGel A-150m from Bio-Rad Laboratories. Richmond, California.

[b] Values for linear molecules are somewhat lower than those for globular protein.

amide in aqueous solution (74). Five commercial grades are available (75). These gels can be used in both aqueous and organic solvents (76), are stable to heat and extremes of pH (77), and are not susceptible to microbial attack.

Agaran (Agarose).—Gels containing 1–10% agaran (linear polysaccharide of alternating D-glucosyl and 3,6-anhydro-L-galactosyl units) are commercially available. Some of their properties are summarized in Table III.

Columns

The success and reproducibility of gpc depends to a large extent on the design of the column and its accessories. Columns made of glass tubing (2–250 cm long; 5–50 mm diameter) are commonly used. Tubings having diameters less than 2 cm tend to give "wall effects"; sometimes these effects can be eliminated by rinsing the column with a 1% solution of dichloromethylsilane in benzene (65, 78). Use of a sintered glass as bed support should be avoided because it becomes easily clogged and hampers even flow of the eluent. However, a support of nylon or poly(tetrafluoroethylene)

(Teflon) fabric gives satisfactory results. The "dead volume" (volume of the column beneath the bed support) should be small, not exceeding 0.1% of the total column volume. Columns filled with plunger-type flow adaptors (77) are useful, particularly for recycling techniques or upward eluent flow. Flow adaptors at the top and bottom of the column permit automatic sample application (65, 76, 79) or connection of two or more columns in series (79, 80). Use of columns in series or in recycling (80, 81) gives better separations of substances having small molecular weight differences. Generally elution characteristics of the sample are temperature dependent (81) and thermostating of the column is recommended for reproducible results (65).

Slurry Preparations

Commercial dextran, polyacrylamide and poly(acryloylmorpholine) gels are supplied in the dry state and require swelling in a suitable solvent before their use. Swelling is done by dispersing the dry gel in an excess of the solvent at ambient or higher temperatures (90°–100°). After allowing the gel to equilibrate with the solvent and when it has become completely swollen, fine particles are removed by elutriation, and trapped air bubbles are removed either by boiling or by aspiration.

Agaran gels are delivered in the swollen state and are used directly after degassing by aspiration. The boiling procedure should be avoided for agaran gels because of the solubility of agaran at temperatures above 40°. After swelling and deaeration, the gel is allowed to settle and excess solvent is removed until a slurry of desired density is obtained. A thick slurry has a tendency to retain air bubbles, while a dilute slurry may lead to an unevenly packed column.

Column Packing

The column with bed support is mounted vertically; and the bottom flow adaptor, if used, is installed. The column is then filled with the solvent. Air bubbles from the dead space beneath the column are flushed off by opening the outlet. The outlet is closed to retain solvent to a quarter of the column volume. To obtain uniform packing, it is convenient to add all the gel at one time and not at intervals. This can be facilitated by attaching an extension tube or a gel reservoir to the top of the column. After pouring all the gel needed into the column, sedimentation is immediately started by opening the outlet. When the solvent level drops to within 3 cm of the column top, the outlet is closed; and the extension tube or gel reservoir is removed. If required, an upper flow adaptor is fitted. While flow rates can be controlled by means of the bottom outlet valve, it is desirable to prevent excessive head pressure which can easily compress and deform the gel particles causing slow, uneven flow and poor separations. Consequently, low head pressure should be maintained. It is recommended to control the hydrostatic pressure difference through adjustment of the levels of the effluent tube opening in relation to the level of eluent in its nonpressurized vessel. An upward reaching effluent tube from the bottom of the column may be arranged so that its external outlet is at the appropriate level below the eluent level in the reservoir.

Next the outlet is opened, and 2–3 column volumes of the eluent is allowed to pass through the gel bed to stabilize it. To protect the upper surface of the gel bed, it is convenient to keep it covered with a piece of filter paper or nylon net. The quality of packing can be checked by observing the migration of a colored zone through the column. Such a zone can be formed with a 0.2% solution of a colored compound such as Blue Dextran 2000 (67).

Sample Preparation

For good separation, the volume of the sample solution should be 1–2% of the bed volume. On a preparative scale, however, the sample volume can be raised to 10–25% of the bed volume. Care should be exercised to keep the viscosity low so as to prevent broadening of zones (83).

Sample Application

On a Drained-Bed Surface.—The eluent level is lowered to the gel surface, and the sample solution is carefully layered on top of the gel bed. By opening the column outlet, the sample is then allowed to enter the column packing. Then the gel surface is rinsed 2–3 times with a small amount of the eluent, and the column is filled with eluent and connected to the solvent reservoir.

Injection of Sample

Without removing the excess of eluent above the surface of the column, the sample solution is layered on top of the gel bed with the aid of a hypodermic syringe. If the density of the sample solution is not sufficiently high, an inert substance such as sodium chloride can be added to raise the density. After application, elution may be started immediately.

Sample Addition by Means of a Flow Adaptor.—Conveniently, sample may be added to the column through the use of a flow adaptor. This method allows the sample to be applied either at the top of the column, or at the bottom as required for upward flow elution. The adaptor is pressed until it is in contact with the gel bed surface. The excess eluent leaves the column through the adaptor. The adaptor tubing is then connected to the sample solution reservoir through a three-way valve, and a measured volume of the sample solution is introduced to the column by gravitational flow.

Eluant

Deionized water is most commonly used as eluant for underivatized carbohydrates. Aqueous solutions containing buffer salts can also be used, but if the product is isolated by lyophilization, the use of a volatile buffer such as ammonium acetate is recommended. For derivatized sugars, organic solvents such as acetone, methanol, and ethanol can be used.

Sample Detection

Eluant portions can be obtained from a fraction collector and individually tested for product, or the eluate can be continuously monitored (65, 76, 79, 84).

Instruments are available for automatic analysis of the eluate. An automatic differential refractometer with a recorder (Water Associates Inc., Framingham, MA) is conveneint. Solvent from the solvent reservoir can be pumped to a reference cell of the refractometer while eluate is automatically introduced into the sample cell. The differential refractive index is measured and recorded automatically. The eluate then passes to the fraction collector through a siphon.

Alternatively, the eluate can be analyzed by using an Auto-Analyzer (Technicon Corporation, Tarrytown, NY). A fraction of the eluate is introduced into the analyzer while the rest goes to the fraction collector. In the analyzer, the eluate is mixed with orcinol–sulfuric acid (62–64, 79; Vol. I [134]) or cysteine–sulfuric acid (76; Vol. I [131], [134], [135]). The color developed by heating is determined spectrophometrically and recorded.

Prevention of Microbial Growth

Microbial growth seldom occurs during chromatographic operations. However, microbial infection which may occur during storage of a packed column or slurries can be prevented by the addition of an antimicrobial agent. Generally a 0.02% sodium azide, 0.05% trichlorobutanol, or 0.01% phenyl mercuric acetate is used for this purpose. These may be eluted from columns before chromatographic operations.

References

(1)   J. Porath and P. Flodin, *Nature*, **183**, 1657 (1959).

(2)   H. Determann, "Gel Chromatography," Springer–Verlag, New York, 2nd Ed., 1969.

(3)   L. Fisher, "Introduction to Gel Chromatography," Elsevier, Amsterdam, 1969.

(4)   B. Goelette and J. Porath, *Gel Filtration*, *in* "Chromatography," E. Heftmann, ed., Reinhold, New York, 2nd Ed., 1967, pp. .

(5)   J. Reiland, *Methods Enzymol.*, **22**, 287 (1971).

(6)   S. C. Churms, *Advan. Carbohydr. Chem. Biochem.*, **25**, 13 (1970).

(7)   K. H. Altgelt, *Advan. Chromatogr.*, **7**, 3 (1968).

(8)   H. Vink, *J. Chromatogr.*, **52**, 205 (1970).

(9)   J. M. Curling, *Exp. Physiol. Biochem.*, **3**, 417 (1970).

(10) J. C. Giddings and K. L. Mallik, *Anal. Chem.*, **38**, 997 (1966).

(11) N. V. B. Marsden, *Ann. N.Y. Acad. Sci.*, **125**, 428 (1965).

(12) T. C. Laurent and J. Killander, *J. Chromatogr.*, **14**, 317 (1965).

(13) G. K. Ackers, *Biochemistry*, **3**, 723 (1964).

(14) P. G. Squire, *Arch. Biochem. Biophys.*, **107**, 471 (1964).

(15) T. L. Chang, *Anal. Chem. Acta*, **42**, 51 (1968).

(16) K. Granath, *in* "New Biochemical Separations," A. T. James and L. J. Morris, ed., Van Nostrand, London, 1964, p. 93.

(17) "Bibliography of Column Chromatography (1967–1970) and Survey of Applications," *J. Chromatogr.*, *Suppl.* **Vol. 3**, 107–158, 1973.

(18) "Bibliography of Column Chromatography (1971–1973) and Survey of Applications," *J. Chromatogr.*, *Suppl.* **Vol. 6**, 115–154 (1976).

(19) F. Shafizadeh and Y. Z. Lai, *Carbohydr. Res.*, **31**, 57 (1973).

(20) J. J. Marshall, *Carbohydr. Res.*, **34**, 289 (1974).

(21) L. A. Elyakova and T. N. Zvyagintseva, *Carbohydr. Res.*, **34**, 241 (1974).

(22) G. P. Belue and G. D. McGinnis, *J. Chromatogr.*, **97**, 25 (1974).

(23) S. C. Churms and A. M. Stephen, *Carbohydr. Res.*, **35**, 73 (1974).

(24) R. Girard and A-M. Staub, *Carbohydr. Res.*, **37**, 127 (1974).

(25) J. I. Javid and R. J. Winzler, *Biochemistry*, **13**, 3635 (1974).

(26) J. A. Cifonelli, *Carbohydr. Res.*, **37**, 145 (1974).

(27) M. C. Wang and S. Bartnicki-Garcia, *Carbohydr. Res.*, **37**, 331 (1974).

(28) J. Schrager and M. D. G. Oates, *Biochim. Biophys. Acta*, **372**, 183 (1974).

(29) J. A. Thoma, G. V. K. Rao, A. Bowanko, A. Jennings, and C. Crook, *Carbohydr. Res.*, **38**, 279 (1974).

(30) R. G. Brown and B. Lindberg, *Carbohydr. Res.*, **38**, 369 (1974).

(31) F. Andre and F. Descos, *Biochim. Biophys. Acta*, **386**, 129 (1975).

(32) H. Carchon and C. K. De Bryne, *Carbohydr. Res.*, **41**, 175 (1975).

(33) B. D. E. Gaillard and G. N. Richards, *Carbohydr. Res.*, **42**, 135 (1974).

(34) K-G. Rosell and S. Stevensson, *Carbohydr. Res.*, **42**, 297 (1975).

(35) W. G. Carter and M. E. Etzler, *Biochemistry*, **14**, 5118 (1975).

(36) Y. Yoshiok, M. Emori, T. Ikekawa and F. Fukuoka, *Carbohydr. Res.*, **43**, 305 (1975).

(37) I. R. Chester and P. M. Meadow, *Eur. J. Biochem.*, **58**, 273 (1975).

(38) S. C. Chums and A. M. Stephen, *Carbohydr. Res.*, **45**, 291 (1975).

(39) D. Fialova, M. Ticha and J. Kocourek, *Biochim. Biophys. Acta*, **393**, 170 (1975).

(40) T. P. Nowak and S. H. Barondes, *Biochim. Biophys. Acta*, **393**, 115 (1975).

(41) C. Ouyang and F-Y. Yang, *Biochim. Biophys. Acta*, **386**, 479 (1975).

(42) D. J. Manners and G. Wilson, *Carbohydr. Res.*, **48**, 255 (1976).

(43) Y. Inoue and K. Nagasawa, *Carbohydr. Res.*, **46**, 87 (1976).

(44) L. Gatteno, D. Bladier, M. Garnier, and P. Cornillot, *Carbohydr. Res.*, **52**, 197 (1976).

(45) E. Moczar and G. Vass, *Carbohydr. Res.*, **50**, 133 (1976).

(46) G. R. Wollard, E. B. Rathbone and L. Novellie, *Carbohydr. Res.*, **51**, 249 (1976).

(47) C. Amar, J-M. Delaumeny, and E. Vilkas, *Biochim. Biophys. Acta*, **421**, 263 (1976).

(48) M. Cermakova, G. Entlicher, and J. Kocourek, *Biochim. Biophys. Acta*, **420**, 236 (1976).

(49) P. H. Fishman, R. O. Brady, and S. A. Anderson, *Biochemistry*, **15**, 201 (1976).

(50) W. A. Frazier, S. D. Rosen, R. W. Rietherman, and S. H. Barondes, *J. Biol. Chem.*, **250**, 7714 (1976).

(51) Y. Fujita, K. Oishi, K. Suzuki, and K. Imahari, *Biochemistry*, **14**, 4465 (1975).

(52) L. Höglund, *Comp. Biochem. Physiol.*, **53B**, 9 (1976).

(53) M. Moczar and E. Moczar, *Comp. Biochem. Physiol.*, **53B**, 255 (1976).

(54) H. Moulki, R. Bonaly, B. Fournet, and J. Montreuil, *Biochim. Biophys. Acta*, **420**, 279 (1976).

(55) M. J. Prigent and R. Bourrillion, *Biochim. Biophys. Acta*, **420**, 112 (1976).

(56) P-E. Jansson and B. Lindberg, *Carbohydr. Res.*, **54**, 261 (1977).

(57) P. J. Cashion, H. J. Notman, T. Cadgar, and G. M. Sathe, *J. Chromatogr.*, **135**, 189 (1977).

(58) S. C. Churms, E. H. Merrifield, and A. M. Stephen, *Carbohydr. Res.*, **55**, 3 (1977).

(59) H. J. Jennings and A. K. Bhattachrjee, *Carbohydr. Res.*, **55**, 105 (1977).

(60) H. Yamada, Y. Aramaki, and T. Miyazaki, *Biochim. Biophys. Acta*, **497**, 396 (1977).

(61) G. N. Bathgate, *J. Chromatogr.*, **47**, 92 (1970).

(62) P. I. Bekker, S. C. Chums, A. M. Stephen, and G. R. Woolard, *Tetrahedron*, **25**, 3359 (1969).

(63) K. A. Granath and B. E. Kvist, *J. Chromatogr.*, **28**, 69 (1967).

(64) M. Van Lancker and E. Veirman, *Ann. Sci. Text. Belg.*, **20**, 98 (1972).

(65) M. John, G. Trenel, and H. Dellweg, *J. Chromatogr.*, **42**, 476 (1969).

(66) P. Flodin, Dissertation, Uppsala University, 1962.

(67) Pharmacia Fine Chemicals, Uppsala, Sweden.

(68) J. H. Zwaving, *J. Chromatogr.*, **35**, 562 (1968).

(69) A. Repas, B. Nikolin, and K. Dursun, *J. Chromatogr.*, **44**, 184 (1969).

(70) A. Klemer and R. Kutz, *Tetrahedron Lett.*, 1693 (1969).

(71) A. N. De Belder and B. Norrman, *Carbohydr. Res.*, **8**, 1 (1968).

(72) S. Hjerten, *Arch. Biochem. Biophys.*, **Suppl. 1**, 147 (1962).

(73) Bio-Rad Laboratories, Richmond, California.

(74) R. Epton, C. Holloway, and J. V. McLaren, *J. Appl. Polymer Sci.*, **18**, 179 (1974).

(75) Koch-Light Laboratories, Colnbrook, Buckinghamshire, England.

(76) R. Epton, C. Holloway, and J. V. McLaren, *J. Chromatogr.*, **90**, 249 (1974).

(77) R. Epton, C. Holloway, and J. V. McLaren, *J. Chromatogr.*, **117**, 245 (1976).

(78) A. Schwartz and B. A. Zabin, *Anal. Biochem.*, **14**, 321 (1966).

(79) K. Kainuma, A. Nogami, and J. Mercier, *J. Chromatogr.*, **121**, 361 (1976).

(80) J. Porath and H. Bennich, *Arch. Biochem. Biophys.*, **Suppl. 1**, 125 (1962).

(81) J. Killander, *Biochem. Biophys. Acta*, **93**, 1 (1964).

(82) W. Brown, *J. Chromatogr.*, **52**, 273 (1970).

(83) P. Flodin, *J. Chromatogr.*, **5**, 103 (1961).

(84) J. M. Goodson and V. DiStefano, *J. Chromatogr.*, **45**, 139 (1969).

[6] Analysis and Structural Characterization of Amino Sugars by Gas–Liquid Chromatography and Mass Spectrometry

By Clifford G. Wong, Sun-Sang Joseph Sung,
and Charles C. Sweeley

*Department of Biochemistry, Michigan State University,
East Lansing, Michigan*

Introduction

Amino sugars are important components in the oligosaccharide structures of glycoproteins, glycosphingolipids, mucopolysaccharides, bacterial peptidoglycans, lipopolysaccharides, and antibiotic substances and in the free oligosaccharides of urine and milk. Within the past ten years, combined gas-liquid chromatography–mass spectrometry has become practical for the sub-microgram-scale identification and characterization of these complex carbohydrates. In this chapter are presented the retention behavior of several kinds of amino sugar derivatives on gas-liquid chromatography (glc) and the major ions produced from these substances by electron impact ionization mass spectrometry (ms).

The three most common types of derivatization for carbohydrates are acetylation, methylation, and trimethylsilylation. Although the preparation of acetyl derivatives of monosaccharides is a simple technique, there are a few complications. When dealing with alditols produced by borohydride reduction of sugars, borate complexes are formed and can interfere with the acetylation reaction (*1*). Thus, it is important to remove borate prior to the acetylation step. Another problem may arise in the possible decomposition of sugar acetates on the column, as reported by Bishop *et al.* (*2*), Perry (*3*) and Gunner *et al.* (*4*). Stellner *et al.* (*5*) have reported very poor recoveries of their partially methylated hexosaminitol acetates due to the inherent design of individual glc-ms models.

Trimethylsilylation of sugars, as reported by Sweeley *et al.* (*6*), is a simple and rapid method for derivatization. However, it must be kept in mind that the treatment of hexosamine hydrochlorides with trimethylchlorosilane and hexamethyldisilazane in pyridine will not yield silyl derivatives of the amino

55

METHODS IN CARBOHYDRATE
CHEMISTRY, VOL. VIII

groups. However, use of *N,O*-bis(trimethylsilyl)acetamide (BSA) (*7, 8*) or *N,O*-bis(trimethylsilyl)trifluoroacetamide (BSTFA) (*22, 23*) as the trimethylsilylating reagent has been shown to effectively trimethylsilylate all functional groups.

Methylation is an important method for structure elucidation of complex polysaccharides (*9*). A convenient method for methylation of carbohydrates, giving very high yields of permethylated derivatives, has been developed by Hakomori (*10*, Vol. VI [64]). Complete methylation of all accessible functional groups, including the *N*-acetamido groups, can be accomplished in one step. The oligosaccharide products themselves have been analyzed by mass spectrometry up to molecular weights approaching 2,000. Subsequent acetylation of the acid-hydrolyzed and reduced alditols yields the partially methylated alditol acetates. Separation of these substances on an ECNSS-M GC column and MS analysis of the eluted components gives structural information of a polysaccharide in regard to glycosidic linkages and carbohydrate composition (*9*). The individual partially methylated alditol acetates are identified not only by their glc retention times, but also by their characteristic fragmentation patterns on the mass spectrometer.

Procedures

Equipment

Relative retention times of the various sugars reported here were determined with a Hewlett-Packard F & M Model 402 gas chromatograph equipped with a hydrogen flame ionization detector. The carrier gas was nitrogen, with flow rates between 40 and 50 ml/min. Glass columns containing various packings were 6 ft × 2 mm (i.d.).

The GC–MS runs were performed on an LKB 9000 gas chromatograph–mass spectrometer, interfaced to a PDP 8/e minicomputer (Digital Equipment Co., Maynard, Mass.) for data compilation and analysis. The mass spectrometer was operated at 70 eV with an accelerating voltage of 3.5 kV and an ion source temperature of 290°. The helium carrier gas flow rate was approximately 30 ml/min. Coiled glass columns were 6 ft × 2 mm (i.d.).

2-Acetamido-2-deoxyalditols and 2-Amino-2-deoxyalditols

In separate one-dram vials with Teflon-lined screw-caps, about 2.5 mg of each *N*-acetylhexosamine and hexosamine hydrochloride (Sigma Chem. Co., St. Louis, Mo. and Pfanstiehl Laboratories, Waukegan, Ill.) are mixed with 15 mg of sodium borohydride in 1.5 ml of water. The reaction is allowed to proceed overnight (6 h) at 4° and is stopped by the dropwise addition of glacial acetic acid until the pH of the solution is acid (pH 2–3) and hydrogen gas can no longer be seen bubbling from the solution. The acidified solutions are then taken to dryness under a stream of nitrogen on a 50° water bath with successive additions of methanol (total volume approximately 20 ml) and evaporation to remove borate completely as the volatile trimethylborate ester. Finally, the dried residues are dissolved in 0.5 ml of water and used as stock solutions for derivatization.

Methyl 2-Acetamido-2-deoxyglycosides

Into a Teflon-lined, screw-capped test tube (10 × 1.3 cm) containing 5 mg of the N-acetylhexosamine, 3 ml of 0.75N anhydrous methanolic HCl (Vol. IV [21], Vol. VI [69], Vol. VII [34]) is added; and the mixture is heated at 80° for 3 h. Losses of solvent from leaky caps are minimized by momentarily loosening the cap after about 10 min heating to reduce the pressure. After methanolysis, powdered silver carbonate is added in small portions to neutralize the reaction mixture (pH 6 by litmus paper test).

For further conversion to N-acetyl derivatives, 0.3 ml of acetic anhydride is added to the tubes, and the reaction mixtures are kept at 20°–25° for 6 h. The mixtures are centrifuged; the supernatant fraction is transferred to a one-dram, Teflon-lined, screw-capped vial; and the solvent is removed by a stream of nitrogen. The silver chloride precipitate is washed twice with 2-ml portions of anhydrous methanol (Vol. VII [3]), and the combined supernatants are quantitatively transferred to one-dram vials and dried down under nitrogen. The methyl-2-acetamido-2-deoxyhexosides are redissolved in 1 ml of water and used as standards (5 mg/ml) for subsequent derivatization.

For some biological samples, an incubation time of 18–24 h is preferred for quantitative acid-catalyzed methanolysis.

Partially Methylated Alditol Acetates

Permethylation of carbohydrates is done under dry nitrogen by the method of Hakomori (10). Hexane is redistilled after refluxing with 20 g/l of barium oxide for 2 h and is stored over sodium. Dimethyl sulfoxide is dried by refluxing with 50 g/liter of barium oxide for 2 h, redistilled, and stored over molecular sieves (Vol. VI [64], Vol. VII [26]). All other solvents are redistilled. A sample (0.9 g of 57% oil emulsion) of sodium hydride (Alfa Inorganics, Beverly, Mass.) is washed 7 times with 15-ml portions of dry redistilled hexane. Dry redistilled dimethylsulfoxide (10 ml) is added and allowed to react at 65°–70° for about 90 min, until the bubbling of hydrogen ceased. The methylsulfinyl ion solution (0.5 ml) is added to a solution of 0.5 g of the sample in 0.5 ml of dimethylsulfoxide, and the mixture is allowed to react for 30 min with periodic sonication. Two ml of redistilled iodomethane (methyl iodide) (Pflatz and Bauer, Stamford, Conn.) is then slowly added, and the mixture is allowed to stand for 2 h at 20°–25°. The reaction mixtures are then mixed with 5 ml of chloroform and washed twice with 5 ml of water, once with 5 ml of a 20% solution of sodium thiosulfate ($Na_2S_2O_3$), and three times with water. The organic phases are evaporated to dryness under nitrogen with the aid of absolute ethanol to remove water by azeotropic distillation, and the residues are hydrolyzed in 0.5 ml of 0.5N H_2SO_4 in 95% acetic acid for 24 h at 85°. Water (0.5 ml) is then added, and heating is continued for an additional 5 h at 85°.

A small column containing 2 ml of Dowex 1X8 anion-exchange resin [acetate form] (50–100 mesh) is used to absorb the sulfate, the permethylated carbohydrates being eluted with 2–3 ml of acetic acid. The hydrolyzate is transferred to a 1-dram vial and evaporated to dryness under nitrogen. Reduction with 0.5 ml of sodium borohydride (10 mg/ml) for 2 h at 20–25° yields the partially methylated alditols. After the addition of several drops of glacial acetic acid, the solutions are dried under nitrogen. Borate is removed as its methyl ester, as described above, using 1–2 drops of acetic acid and 2 ml of methanol and heating in a 50° water bath for 5 min under a stream of nitrogen. Esterification is repeated three more times. The dried sample is acetylated in 0.5–1 ml of acetic anhydride for 60–90 min at 100°. After drying under nitrogen with the aid of toluene, the sample is dissolved in 2 ml of dichloromethane (methylene chloride), washed three times with 1–2 ml of water, redried under nitrogen, and redissolved in 0.5–1 ml of dichloromethane for GC and GC-MS analyses.

Partially methylated glucosaminitol (2-amino-2-deoxy-D-glucitol) acetates may be synthesized from D-glucosamine hydrochloride by the method of Tai et al. (11).

Trimethylsilyl Derivatives

Reagent I.—Pyridine (redistilled, stored over KOH), 10 volumes (Vol. II [43], [53], [63], [73]; Vol. IV [73]; Vol. VII [2]). Hexamethyldisilazane (commercial reagent), 4 volumes. Trimethylchlorosilane (commercial reagent), 2 volumes. The reagents are added to a 7-ml, screw-capped test tube with a Teflon-lined cap, mixed, and centrifuged. If moisture is excluded, the derivatizing solution can be used for 1 week.

Reagent II.—*N,O*-Bis(trimethylsilyl)trifluoroacetamide (BSTFA) containing 1% trimethylchlorosilane (Pierce Chemical Co., Rockford, Illinois).

2-Amino-2-deoxy-O-trimethylsilylhexosides.—Reagent I (100 μl) is pipeted into dry, 1-dram, Teflon-lined, screw-capped vials containing 125 μg of amino sugar. The mixture is allowed to stand at 20°–25° for 30 min. An appropriate aliquot (1–3 μl) is injected immediately into the gas chromatograph, for the N-trimethylsilyl hexosamine derivatives are present in appreciable amounts after 2 h at room temperature.

2-Deoxy-2-trimethylsilylamino-O-trimethylsilylhexosides.—Into 1-dram Teflon-lined screw-capped vials containing 125 μg of amino sugar is added 50 μl of dry pyridine, followed by 50 μl of BSTFA (Reagent II). The sealed vial is heated at 80° for 30 min, and an aliquot is injected into the GC. (Note: N-acetyl derivatives do not form any N-trimethylsilyl amide under these conditions).

Acetate Derivatives

Acetic anhydride (100 μl) and dry pyridine (100 μl) are added to dry, 1-dram, Teflon-lined, screw-capped vials containing 250 μg of amino sugar. The sealed vials are heated at 100° for 4 h, after which 2 ml of redistilled toluene is added and the mixture is dried by evaporation under a stream of nitrogen at 50°. This addition of toluene and subsequent evaporation are repeated once more to ensure the complete removal of acetic anhydride and pyridine. A solution of the acetylated sugar in 200 μl of dry, redistilled methylene chloride is used for GC analysis.

Results

Tables I–XII are a summary of the relative GC retention times and the major ions found in the mass spectra of each denoted amino sugar. Since stereoisomers and anomers of the carbohydrate derivatives give similar mass spectra, with small differences in peak intensity, the mass spectrum of only one stereoisomer is given.

TABLE I

Retention Times of the Peracetylated Amino Sugars

Compound	Relative retention time[a]	Other references
2-Acetamido-1,3,4,6-tetra-O-acetyl-2-deoxy-D-glucose[b]	0.36, 2.57	
2-Acetamido-1,3,4,6-tetra-O-acetyl-2-deoxy-D-galactose[b]	0.37, 0.43, 2.53, 2.77	
2-Acetamido-1,3,4,6-tetra-O-acetyl-2-deoxy-D-mannose[b]	0.46, 2.33, 2.48, 2.90	
2-Acetamido-1,3,4,5,6-penta-O-acetyl-2-deoxy-D-glucitol[c]	1.76	(12)
2-Acetamido-1,3,4,5,6-penta-O-acetyl-2-deoxy-D-galactitol[c]	2.34	(12)
2-Acetamido-1,3,4,5,6-penta-O-acetyl-2-deoxy-D-mannitol[c]	2.83	(12)

[a] Retention times are relative to 1,2,3,4,5,6,7-hepta-O-acetylperseitol on a column of 3% Poly A-103 on Gas Chrom Q 100/120 mesh (Applied Science Laboratories, Inc., State College, Pa.).
[b] Isothermal at 200°; internal standard retention time was 11 min.
[c] Isothermal at 210°; internal standard retention time was 6.8 min.

TABLE II

Retention Times of the 2-Acetamido-2-deoxy-O-trimethylsilyl Sugars

Compound[a]	Relative retention time[b]
2-Acetamido-2-deoxy-1,3,4,6-tetra-O-trimethylsilylglucopyranoside	0.93, 1.69
2-Acetamido-2-deoxy-1,3,4,6-tetra-O-trimethylsilylgalactopyranoside	1.35, 1.56
2-Acetamido-2-deoxy-1,3,4,6-tetra-O-trimethylsilylmannopyranoside	0.95, 1.27
Methyl 2-acetamido-2-deoxy-3,4,6-tri-O-trimethylsilylglucopyranoside	1.31, 1.41, 1.56
Methyl 2-acetamido-2-deoxy-3,4,6-tri-O-trimethylsilylgalactopyranoside	1.14, 1.37
Methyl 2-acetamido-2-deoxy-3,4,6-tri-O-trimethylsilylmannopyranoside	0.94, 1.60
2-Acetamido-2-deoxy-1,3,4,5,6-penta-O-trimethylsilylglucitol	1.69
2-Acetamido-2-deoxy-1,3,4,5,6-penta-O-trimethylsilylgalactitol	1.77
2-Acetamido-2-deoxy-1,3,4,5,6-penta-O-trimethylsilylmannitol	1.88

[a] Derivatized with either Reagent I or II.
[b] Retention times are relative to 1,2,3,4,5,6-hexa-O-trimethylsilylmannitol (10.5 min) on 3% SP 2100 on Supelcoport 80/100 mesh (Supelco, Inc., Bellefonte, Pa.) at column temperature of 180° isothermal.

TABLE III

Retention Times of the 2-Amino-2-deoxy-O-trimethylsily Alditols

Compound	Relative retention time[a]
2-Amino-2-deoxy-1,3,4,5,6-penta-O-trimethylsilylglucitol[b]	1.22
2-Amino-2-deoxy-1,3,4,5,6-penta-O-trimethylsilylgalactitol[b]	1.18
2-Amino-2-deoxy-1,3,4,5,6-penta-O-trimethylsilylmannitol[b]	1.23
2-Deoxy-2-trimethylsilylamino-1,3,4,5,6-penta-O-trimethylsilylglucitol[c]	0.92
2-Deoxy-2-trimethylsilylamino-1,3,4,5,6-penta-O-trimethylsilylgalactitol[c]	0.91
2-Deoxy-2-trimethylsilylamino-1,3,4,5,6-penta-O-trimethylsilylmannitol[c]	0.92

[a] Retention time relative to 1,2,3,4,5,6-hexa-O-trimethylsilylmannitol (10.4 min) on 3% SP 2100, Supelcoport 80/100 mesh (Supelco, Inc., Bellefonte, Pa.) at column temperature of 180° isothermal.
[b] Derivatized with trimethylsilylating Reagent I.
[c] Derivatized with trimethylsilylating Reagent II.

TABLE IV

Retention Times of the Partially O-Methylated
2-N-methylglucosaminitol Acetates

Position of O-CH$_3$ groups	Relative retention time[a]
3, 4, 6	1.00 (retention time = 8.3 min)
3, 6	1.68
3, 4	2.19
4, 6	2.51
3	2.91
4	3.64
6	4.51

[a] Isothermal at 190°, 3% OV-210 on Supelcoport 80/100 mesh (Supelco, Inc., Bellefonte, Pa.).

TABLE V

Major Fragment Ions Observed in
the Mass Spectrum of
2-Acetamido-1,3,4,6-tetra-O-acetyl-
2-deoxy-D-galactopyranose
(MW = 389)[a]

m/e	Relative Intensity
43	100.0 [CH$_3$CO]$^+$
72	10.9
84	10.4
97	6.9
108	3.3
110	0.9
114	48.9
126	9.1
139	13.6
144	1.3
150	4.7
156	19.7
168	6.6
181	7.2
198	5.0
199	6.7
210	1.8
241	14.3
330	2.6 (M$^+$-59)
346	1.5 (M$^+$-43)

[a] References 13 and 14 give detailed descriptions of fragmentation pathways and identifications of ions.

TABLE VI

Major Fragment Ions Observed in
the Mass Spectrum of
2-Acetamido-1,3,4,5,6-penta-O-
acetyl-2-deoxy-D-glucitol
(MW = 433)[a]

m/e	Relative intensity
43	100.0 [CH$_3$CO]$^+$
60	21.4
84	73.8
85	20.3
102	22.9
114	8.9
115	7.9
126	15.1
139	12.5
144	23.0 (M$^+$-289)
145	7.7
151	9.4
156	8.6
157	4.8
168	7.4
216	1.5
217	1.4
288	0.3
289	0.4
318	8.3 (M$^+$-73-42)
360	1.1 (M$^+$-73)
374	0.2 (M$^+$-59)
390	0.1 (M$^+$-43)

[a] References 15 and 16 give detailed descriptions of fragmentation pathways and identifications of ions.

TABLE VII

Major Fragment Ions Observed in the Mass Spectra of Partially
O-Methylated 2-N-Methylglucosaminitol Acetates

m/e	Position of CH_3O- groups						
	3,4,6	3,6	3,4	4,6	3	4	6
43	+	+	+	+	+	+	+
45	+	+	+	+	+	+	+
74	+	+	+	+	+	+	+
87	+	+	+	+	+	+	+
98	+	+	+	+	+	+	+
116	+	+	+	+	+	+	+
124		+			+		
128							+
129	+	+	+	+		+	+
142	+	+	+	+	+		+
145	+		+				
158	+	+	+	+	+	+	+
161	+		+	+			
170		+		+		+	+
173		+					
189			+			+	
202	+	+	+		+		
205	+						
230					+		
233		+					
261						+	
274					+		

[a] References 5, 9, 11, 16, 17, and 18 give detailed descriptions of fragmentation pathways and identifications of ions.

TABLE VIII

Major Fragment Ions Observed in
the Mass Spectrum of
2-Acetamido-2-deoxy-1,3,4,5,6-
penta-O-trimethylsilylmannitol
$(MW = 583)^a$

m/e	Relative intensity
73	100.0 $[(CH_3)_3Si]^+$
103	18.5
132	23.1
147	23.8
157	16.9
174	14.9
186	29.4
205	18.2
217	25.1
247	13.2
276	12.4
319	18.4
378	7.6 (M^+-205)
390	4.0 $(M^+-90-103)$
478	1.5 $(M^+-15-90)$
480	1.9 (M^+-103)
568	6.1 (M^+-15)

[a] References 8, 16, and 19 give detailed
descriptions of fragmentation pathways
and identifications of ions.

TABLE X

Major Fragment Ions Observed in
the Mass Spectrum of
2-Deoxy-2-trimethylsilylamino-
1,3,4,5,6-penta-O-trimethylsilyl-
D-mannitol $(MW = 613)^a$

m/e	Relative intensity
73	46.9 $[(CH_3)_3Si]^+$
103	6.5
204	100.0
205	21.1
217	5.0
307	1.4
420	2.0 $(M^+-103-90)$
510	1.8 (M^+-103)
598	0.5 (M^+-15)

[a] References 8, 16, and 19 give detailed
descriptions of fragmentation pathways
and identifications of ions.

TABLE IX

Major Fragment Ions Observed in
the Mass Spectrum of
2-Amino-2-deoxy-1,3,4,5,6-penta-O-
trimethylsilyl-D-glucitol
$(MW = 541)^a$

m/e	Relative intensity
73	83.2 $[(CH_3)_3Si]^+$
103	20.2
132	32.1
147	17.5
204	25.7
205	10.9
217	100.0
258	8.8
348	7.2 $(M^+-103-90)$
438	6.6 (M^+-103)
451	0.3 (M^+-90)
526	5.7 (M^+-15)

[a] References 8, 16, and 19 give detailed
descriptions of fragmentation pathways
and identifications of ions.

TABLE XI

Major Fragment Ions Observed in
the Mass Spectrum of
Methyl 2-Acetamido-2-deoxy-
3,4,6-tri-O-trimethylsilyl-D-
galactopyranoside $(MW = 451)^a$

m/e	Relative intensity
73	91.6 $[(CH_3)_3Si]^+$
75	14.4
131	26.5
147	20.9
173	100.0
204	10.9
217	7.6
218	10.9
226	0.8
247	11.2
259	3.9
314	2.8
330	0.5 $(M^+-31-90)$
346	0.4 $(M^+-15-90)$

[a] References 8, 16, 20, and 21 give
detailed descriptions of fragmentation
pathways and identifications of ions.

TABLE XII

Major Fragment Ions Observed in
the Mass Spectrum of
2-Acetamido-2-deoxy-1,3,4,6-
tetra-O-trimethylsilyl-D-
galactopyranoside $(MW = 509)^a$

m/e	Relative intensity
73	73.0 $[(CH_3)_3Si]^+$
103	5.7
117	5.8
131	21.7
147	13.8
173	100.0
204	10.4
217	9.5
233	3.9
305	1.5
314	3.2 (M^+-15-90-90)
404	0.8 (M^+-15-90)
494	1.6 (M^+-15)

^{a} References 8, 16, 20, and 21 give
detailed descriptions of fragmentation
pathways and identifications of ions.

Acknowledgments

We wish to thank Mr. Jack Harten and Dr. Frank Martin for their technical assistance in this work. This work was supported in part by grants from the National Institute of Arthritis, Metabolism and Digestive Diseases (AM 12434) and the Biotechnology Resources Branch (RR 00480) of the National Institutes of Health.

References

(1) J. D. Blake and G. N. Richards, Carbohydr. Res., 14, 375 (1970).
(2) C. T. Bishop, F. P. Cooper, and R. K. Murray, Can. J. Chem., 41, 2245 (1963).
(3) M. B. Perry, Can. J. Biochem., 42, 451 (1964).
(4) S. W. Gunner, J. K. N. Jones, and M. B. Perry, Can. J. Chem., 39, 1892 (1961).
(5) K. Stellner, H. Saito, and S.-I. Hakomori, Arch. Biochem. Biophys., 155, 464 (1973).
(6) C. C. Sweeley, R. Bentley, M. Makita, and W. W. Wells, J. Amer. Chem. Soc., 85, 2497 (1963).
(7) J. F. Klebe, H. Finkbeiner, and D. M. White, J. Amer. Chem. Soc., 88, 3390 (1966).
(8) J. Kärkkäinen and R. Vihko, Carbohydr. Res., 10, 113 (1969).
(9) B. Lindberg, Methods Enzymol., 28, 178 (1972).
(10) S.-I. Hakomori, J. Biochem. (Tokyo), 55, 205 (1964).
(11) T. Tai, K. Yamashita, and A. Kobata, J. Biochem. (Tokyo), 78, 679 (1975).

(*12*) W. Niedermeier and M. Tomana, *Anal. Biochem.*, **57**, 363 (1974).
(*13*) K. Heyns and H. Scharmann, *Justus Liebigs Ann. Chem.*, **667**, 183 (1963).
(*14*) R. C. Dougherty, D. Horton, K. D. Philips, and J. D. Wander, *Org. Mass Spectrom.*, **7**, 805 (1973).
(*15*) H. Björndal, C. G. Hellerqvist, B. Lindberg, and S. Svensson, *Angew. Chem., Internat. Ed.*, **9**, 610 (1970).
(*16*) J. Lönngren and S. Svensson, *Adv. Carbohydr. Chem. Biochem.*, **29**, 41 (1974).
(*17*) W. Stoffel and P. Hanfland, *Hoppe-Seyler's Z. Physiol. Chem.*, **354**, 21 (1973).
(*18*) G. O. H. Schwarzmann and R. W. Jeanloz, *Carbohydr. Res.*, **34**, 161 (1974).
(*19*) M. Dizdaroglu, D. Henneberg, and C. von Sonntag, *Org. Mass Spectrom.*, **8**, 335 (1974).
(*20*) D. C. DeJongh, T. Radford, J. D. Hribar, S. Hanessian, M. Bieber, G. Dawson, and C. C. Sweeley, *J. Amer. Chem. Soc.*, **91**, 1728 (1969).
(*21*) P. L. Coduti and C. A. Bush, *Anal. Biochem.*, **78**, 21 (1977).
(*22*) C. W. Gehrke and K. Leimer, *J. Chromatogr.*, **57**, 219 (1971).
(*23*) R. E. Hurst, *Carbohydr. Res.*, **30**, 143 (1973).

ANALYSIS OF POLYSACCHARIDES BY CHEMICAL, PHYSICAL, AND ENZYMIC METHODS

[7] Sequential Degradation of Methylated Polysaccharides

By Sigfrid Svensson

Department of Clinical Chemistry, University Hospital, Lund, Sweden

And

Lennart Kenne

Department of Organic Chemistry, Arrhenius Laboratory, University of Stockholm, Stockholm, Sweden

Introduction

Determination of the sequential arrangement of sugar residues in polysaccharides or glycoconjugates often requires specific degradation methods. In many cases, it is possible to prepare methylated polysaccharides with free hydroxyl groups at specified positions. Suitable methods to achieve this aim are acid-catalyzed hydrolysis of furanosidic linkages (*1*) or glycosidic linkages of 2-deoxysugars (*1*) and 3,6-dideoxysugars (*2*) in methylated polysaccharides. Other methods available are sulfone degradation (*3*), degradation of pyranosiduronate residues (Vol. VII [24]) or a modified Smith degradation (*4*). Pyruvate or other acetals can be specifically removed from methylated polysaccharides generating free hydroxyl groups (*5*). Terminal D-galactopyranose residues can be converted into D-galactopyranosiduronate residues, by oxidation with galactose oxidase and iodine, which then can be removed from the methylated polysaccharide (*6*).

Oxidation of free hydroxyl groups generate keto or aldehyde functions. Treatment of the oxidized, methylated sugar residue with base eliminates the methoxyl or glycosyl residue in the β-position to the keto or aldehyde group. The resulting α,β-unsaturated *keto*- or *aldehydo*-sugar formed is readily further degraded by mild acid-catalyzed hydrolysis, with release of the ring substituents. This oxidation—β-elimination—mild acid-catalyzed hydrolysis procedure has been extensively studied with low-molecular-weight model compounds; and it has been shown that the glycosidic linkage of methylated

67

pyranosidic sugar residues with a free hydroxyl group(s) on C-2, C-3, C-4, or C-6 or free hydroxyl groups on both C-4 and C-6 can be cleaved in almost quantitative yields at conditions which do not affect methylated glycosides (7–11). The degradations are depicted below.

Free Hydroxyl Group on C-2

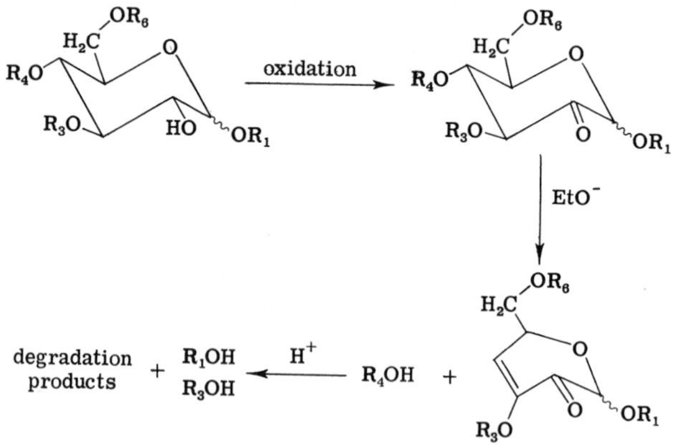

Free Hydroxyl Group on C-3

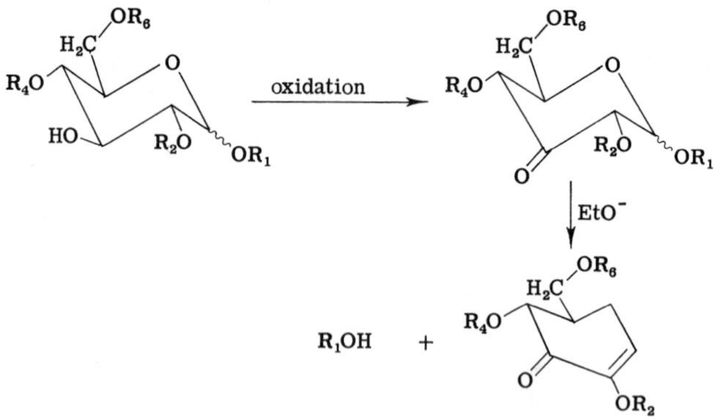

Free Hydroxyl Group on C-4

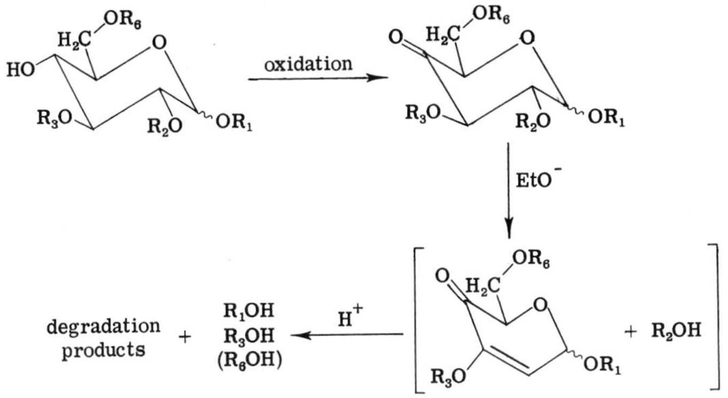

oxidation

EtO⁻

degradation products + R_1OH R_3OH (R_6OH) ←(H⁺)— + R_2OH

Free Hydroxyl Group on C-6

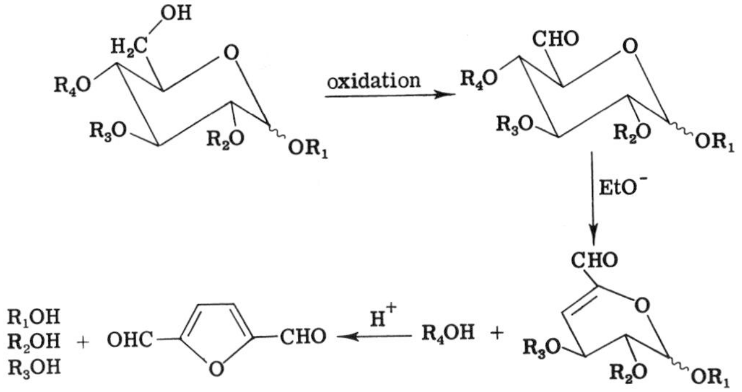

oxidation

EtO⁻

R_1OH R_2OH + OHC⟨⟩CHO ←(H⁺)— R_4OH + R_3OH

Free Hydroxyl Groups on C-4 and C-6

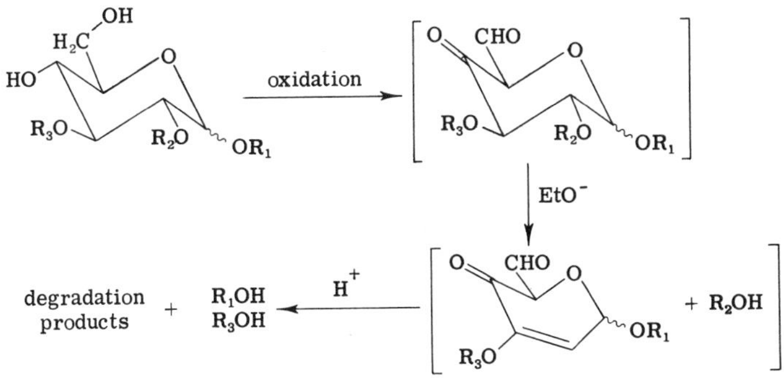

oxidation

EtO⁻

degradation products + R_1OH R_3OH ←(H⁺)— + R_2OH

When a sugar residue in a methylated polysaccharide is eliminated by the oxidation—β-elimination—mild acid-catalyzed hydrolysis procedure, a new free hydroxyl group is generated in the next sugar residue, and thus the procedure can be repeated. When a methylated sugar residue is eliminated from the oxidized sugar unit, it will undergo a second β-elimination (12).

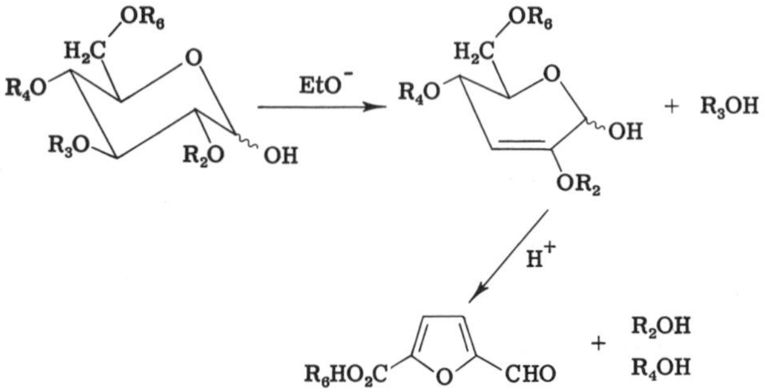

Each step in the degradation procedure is convienently followed by methylation analysis (13).

The oxidation—β-elimination—mild acid-catalyzed hydrolysis procedure has been used in sequence analysis of several bacterial polysaccharides (2, 14–16) and of hemicelluloses (17, 18).

Procedures

Oxidation (19)

An anhydrous 1 M solution of chlorine in dichloromethane (25 ml) is placed under nitrogen in a serum flask, sealed with a rubber cap, and cooled to $-45°$. Dry dimethyl sulfoxide (9 ml, Vol. VI [64], VII [26]; This Vol. [6]) is then added dropwise, with the aid of a syringe, under continuous stirring. A white precipitate is formed (chlorine–dimethyl sulfoxide complex). The methylated polysaccharide (100 mg) with free hydroxyl groups, at specific positions, in 10 ml of dichloromethane is then added dropwise. The reaction mixture is kept at $-45°$ under continuous stirring for 7 h. Triethylamine (10 ml) is then added, and the solution is allowed to attain room temperature (0.5 h). After concentration, the oxidized polysaccharide can be freed from reagents by dialysis (Vol. V [15]) or gel-permeation chromatography (Vol. V [7]) on Sephadex LH 20 (eluant: 1:2 v/v chloroform–acetone).

β-Elimination

The oxidized polysaccharide (100 mg) is dissolved in 10 ml of dichloromethane and 5 ml of 1 M sodium ethoxide in ethanol is added. The reaction mixture is kept at $\sim 20°$ for 1.5 h, neutralized with glacial acetic acid, and concentrated to dryness.

Mild Acid-Catalyzed Hydrolysis

The crude reaction product from the β-elimination step is dissolved in 10 ml of 50% aqueous acetic acid, and the solution is heated at 100° for 4 h. The reaction mixture is then cooled and concentrated to dryness. The degraded polysaccharide material can be recovered by dialysis or by gel-permeation chromatography on Sephadex LH 20 (eluant: 1:2 v/v chloroform–acetone).

References

(1) Unpublished results from the authors' laboratories.

(2) L. Kenne, J. Lönngren, and S. Svensson, *Acta Chem. Scand.*, 27, 3557 (1973).

(3) O. Larm, B. Lindberg, and S. Svensson, *Carbohydr. Res.*, 15, 339 (1971).

(4) B. Lindberg, J. Lönngren, U. Rudén, and W. Nimmich, *Acta Chem. Scand.*, 27, 3787 (1973).

(5) P.-E. Jansson, L. Kenne, B. Lindberg, H. Ljunggren, J. Lönngren, U. Rudén, and S. Svensson, *J. Amer. Chem. Soc.*, 99, 3812 (1977).

(6) B. Lindberg, J. Lönngren, and D. A. Powell, *Carbohydr. Res.*, 58, 177 (1977).

(7) L. Kenne, and S. Svensson, *Acta Chem. Scand.*, 26, 2144 (1972).

(8) L. Kenne, O. Larm, and S. Svensson, *Acta Chem. Scand.*, 26, 2473 (1972).

(9) L. Kenne, O. Larm, and S. Svensson, *Acta Chem. Scand.*, 27, 2797 (1973).

(10) P.-E. Jansson, L. Kenne, and S. Svensson, *Acta Chem. Scand.*, B30, 61 (1976).

(11) P.-E. Jansson, L. Kenne, B. Lindberg, and S. Svensson, *Acta Chem. Scand.*, B30, 631 (1976).

(12) E. F. L. J. Anet, *Chem. Ind.*, 1035, (1963).

(13) H. Björndal, C. G. Hellerqvist, B. Lindberg, and S. Svensson, *Angew. Chem., Int. Ed. Engl.*, 9, 610 (1970).

(14) M. Curwall, B. Lindberg, J. Lönngren, and W. Nimmich, *Carbohydr. Res.*, 42, 95 (1975).

(15) B. Lindberg, J. Lönngren, H. Rudén, and W. Nimmich, *Carbohydr. Res.*, 42, 83 (1975).

(16) P.-E. Jansson, L. Kenne, and B. Lindberg, *Carbohydr. Res.*, 45, 275 (1975).

(17) K.-G. Rosell and S. Svensson, *Carbohydr. Res.*, 42, 297 (1975).

(18) L. Kenne, K.-G. Rosell, and S. Svensson, *Carbohydr. Res.*, 44, 69 (1975).

(19) E. J. Corey and C. U. Kim, *Tetrahedr. Lett.*, 12, 919 (1973).

[8] Enzymic Method for Determination of Glycerol: Chain Lengths of Hexans

By R. J. Sturgeon

*Department of Brewing and Biological Sciences, Heriot-Watt University,
Edinburgh, Scotland*

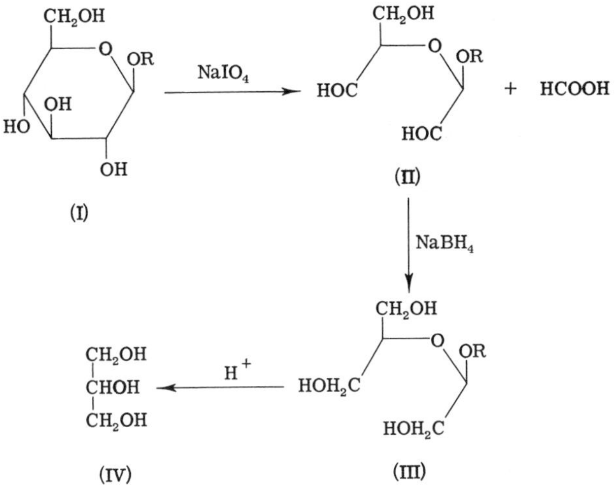

$$(1) \quad \text{Glycerol} + \text{ATP} \xrightarrow{\text{GK}} \begin{array}{c} \text{CH}_2\text{OH} \\ | \\ \text{HOCH} \\ | \\ \text{CH}_2\text{O}-\text{P} \end{array} + \text{ADP}$$

(V)

$$(2) \quad \begin{array}{c} \text{CH}_2\text{OH} \\ | \\ \text{HOCH} \\ | \\ \text{CH}_2\text{O}-\text{P} \end{array} + \text{NAD}^+ \underset{\text{GDH}}{\rightleftharpoons} \begin{array}{c} \text{CH}_2\text{OH} \\ | \\ \text{C}=\text{O} \\ | \\ \text{CH}_2\text{O}-\text{P} \end{array} + \text{NADH} + \text{H}^+$$

(VI)

73

Introduction

In the oxidation of oligo- and polysaccharides (I) containing only hexose residues, the terminal non-reducing units are oxidized to the dialdehyde (II), which in turn can be reduced with aqueous sodium borohydride to the corresponding alcohol (III). On total acid-catalyzed hydrolysis, one molecular proportion of glycerol (IV) is released from each non-reducing end group of these oxidized and reduced polymers (Vol. V [78]). Since formic acid is also released during the oxidation step, titrimetric (1, 2, Vol. V [75]), potentiometric (3, Vol. V [75]), manometric (4, Vol. V [75]), enzymic (5-7) and spectrophotometric (8, Vol. VI [13]) methods have been devised for measurement of this acid, and hence the chain lengths of the polymers.

The reactions employed in the determination of glycerol are based on two coupled enzyme reactions (9). Glycerol (IV) is phosphorylated by glycerokinase (GK) and adenosine triphosphate (ATP) to give L(−)-glycerol 3-phosphate (V) which in turn is oxidized with α-glycerophosphate dehydrogenase (GDH) and nicotinamide adenine dinucleotide (NAD^+). The amount of NADH formed is equivalent to the amount of glycerol present. The equilibrium of the indicator reaction (2), which lies far to the left, is displaced in the required direction by working at pH 9.8 and trapping the dihydroxyacetone phosphate (VI) with hydrazine.

Procedures

Reagents

The following reagents are required:

0.03 M Sodium metaperiodate .
0.08 M Sodium borohydride
Glycine buffer.—1.5 g of glycine, 20.8 g of hydrazine hydrate, and 0.02 g of magnesium chloride are dissolved in ∼50 ml of water. The solution is adjusted to pH 9.8 by addition of 2 N sodium hydroxide, and the final volume is adjusted to 100 ml with water. This solution, which contains 0.2 M glycine, 1 M hydrazine and 2 mM magnesium ions, is stored in a glass-stoppered brown bottle at 4°.
0.05 M Adenosine triphosphate (ATP)
0.02 M Nicotinamide adenine dinucleotide (NAD)
Glycerokinase (GK, about 85 units/ml[1]). Since the assay is started by the

---

[1] 1 unit is the amount of enzyme which catalyzes the formation of 1 μmole of product per minute.

addition of GK, the most concentrated solutions available should be used. Glycerol is occasionally used to stabilize some GK preparations. It is recommended that only GK which is free from glycerol be used.

Glycerol 3-phosphate dehydrogenase (GDH), also known as α-glycerophosphate dehydrogenase (40 units/ml[1]).

Procedure (7, Vol. V [76])

Oligosaccharides (1 mg) or polysaccharides (5 mg) are dissolved or suspended in 1 ml of water, and 1 ml of 0.03 M sodium metaperiodate solution is added. Oxidation is effected in the dark at ~20°. Aliquots (0.4 ml) are removed at intervals, and the excess of oxidant is destroyed by the addition of 0.2 ml of ethylene glycol. When this reaction is complete, in about 30 min, the polyaldehydes are reduced to the polyols by addition of 0.3 ml of 0.08 M sodium borohydride. The samples are allowed to stand for 18 h at ~20°, before being hydrolyzed by the addition of 1 ml of N sulfuric acid and heating at 100° for 1 h. After cooling the samples to room temperature, 1 N sodium hydroxide is added to pH 7.0; and the volumes are adjusted to 5 ml with water.

Determination of Glycerol

To 0.1-ml portions of the oxidized, reduced, and hydrolyzed samples, containing up to 5 μg of glycerol, the following additions are made: 2.7 ml of glycine–hydrazine buffer, 0.1 ml of 0.05 M ATP, 0.05 ml of 0.02 M NAD, and 0.02 ml of GDH solution. The initial absorbance is measured at 340 nm in an automatic ultraviolet spectrophotometer fitted with an external recorder capable of giving a full scale deflection of 0.2 absorbance units. The reaction is started by the addition of 0.02 ml of GK solution, and readings are taken until the extinction becomes constant, usually within 15–20 min. The total hexose content of the original oligo- or polysaccharide is obtained by use of one of the established methods such as the orcinol–sulfuric acid (10, This Vol. [11]) or the phenol–sulfuric acid (11, Vol. I [115], [116]) procedures.

Calculation

For each mole of glycerol, 1 mole of NADH is formed. The amount of glycerol (in μmoles) is given by the increase in absorbance multiplied by 0.482. There is a linear relation between the change in absorbance and the glycerol concentration up to 0.054 μmole.

The chain length = μmole of "anhydrohexose"/μmole of glycerol.

Discussion

The enzymic method for the estimation of glycerol is specific as α-glycero-phosphate dehydrogenase only reacts with L-glycerol 3-phosphate. Although glycerokinase will phosphorylate dihydroxyacetone to dihydroxyacetone phosphate in the presence of ATP (*12*), this reaction does not involve the production of NADH so there is no interference in the assay. Borate, iodate, and sulfate ions in the concentration used in the oxidation, reduction, and hydrolysis stages have no effect on the assay. Similarly, other fragments to be found after oxidation, reduction and hydrolysis of polysaccharides (for example erythritol, L-threitol, ethylene glycol, glycolaldehyde, D-glyceraldehyde) when present in 100-fold excess do not inhibit the enzymic reaction.

In polymers containing either $(1 \to 2)$- or $(1 \to 6)$-linkages, glycerol can be expected to be produced from these residues, as well as from the non-reducing end groups. The method can, however, be applied to measurements of the average chain lengths of hexose-containing polysaccharides with $(1 \to 3)$- and $(1 \to 4)$-glycosidic linkages.

References

(*1*) T. G. Halsall, E. L. Hirst, and J. K. N. Jones, *J. Chem. Soc.*, 1427 (1947).

(*2*) M. Abdel-Akher and F. Smith, *J. Amer. Chem. Soc.*, **73**, 994 (1951).

(*3*) D. M. W. Anderson, C. T. Greenwood, and E. L. Hirst, *J. Chem. Soc.*, **225** (1955).

(*4*) R. E. Asnis and M. C. Glick, *J. Biol. Chem.*, **220**, 691 (1956).

(*5*) D. H. Rammler and J. C. Rabinowitz, *Anal. Biochem.*, **4**, 116 (1962).

(*6*) E. Itagaki and S. Suzuki, *J. Biochem.* (Tokyo), **56**, 77 (1964).

(*7*) D. W. Noble and R. J. Sturgeon, *Carbohydr. Res.*, **12**, 448 (1970).

(*8*) S. A. Barker and P. J. Somers, *Carbohydr. Res.*, **3**, 220 (1966).

(*9*) O. Wieland, *Biochem. Z.*, **329**, 313 (1957).

(*10*) C. Francois, R. D. Marshall, and A. Neuberger, *Biochem. J.*, **83**, 335 (1962).

(*11*) M. Dubois, K. A. Gilles, J. K. Hamilton, P. A. Rebers, and F. Smith, *Anal. Chem.*, **28**, 350 (1956).

(*12*) O. Wieland and M. Suyter, *Biochem. Z.*, **329**, 320 (1957).

[9] Enzymic Method for Determination of the Degree of Polymerization of Glucans and Xylans

By R. J. Sturgeon

Department of Brewing and Biological Sciences,
Heriot-Watt University, Edinburgh, Scotland

Introduction

Analyses of the reducing end-group in oligo- and polysaccharides have proved to be of great value for the estimation of the number average degree of polymerization $(\overline{DP}_n)$ (Vol. V [64], [65]). For this purpose, the formaldehyde produced by the periodate oxidation of borohydride-reduced polysaccharides has been estimated (Vol. V [65], [77]). The terminal hexitol and pentitol residues give rise to one or two equivalents of formaldehyde, depending upon their linkage type. Thus, terminal hexitol residues linked at C-2, C-5 or C-6 give rise to one molecular proportion of formaldehyde, while residues linked at C-3 or C-4 yield two molecular proportions.

Alternatively, a comparison can be made of the reducing power of hydrolyzates of an oligosaccharide with that of the borohydride-reduced oligosaccharide (1). This method is very satisfactory for the estimation of $\overline{DP}_n$ of the smaller oligosaccharides, but the decrease in accuracy with increase in molecular weight prevents the method being applied to polysaccharides.

A number of chemical methods are available based on the relative proportion of terminal alditol residues in the reduced polysaccharides. Procedures involving the use of radioactive intermediates have been described. A polysaccharide is treated with [^{14}C] sodium cyanide and the net radioactivity of the polysaccharide is measured (Vol. V [64]). Reduction with sodium borotritide of a series of maltose oligosaccharides has yielded products whose residual radioactivity has a linear dependence on the DP_n (2; see also Vol. VII [15]).

The ratio of reducing sugars and alditols in hydrolyzates of borohydride-reduced polysaccharides has been achieved by means of gas-liquid chromatographic separations of the derived trimethylsilyl ethers (3, Vol. VI [1], [2]) and alditol acetates (4, Vol. VI [4]).

After reduction with borohydride, a glucan, on acid-catalyzed hydrolysis, produces D-glucitol (sorbitol) which can be determined by an enzyme-catalyzed oxidation reaction (5). Sorbitol dehydrogenase (SDH; EC 1.1.1.14,

77

also known as L-iditol:NAD oxidoreductase or L-iditol dehydrogenase) catalyzes the following reversible reaction:

$$\text{D-Glucitol} + \text{NAD}^+ \rightleftarrows \text{D-fructose} + \text{NADH} + \text{H}^+$$

The oxidation of D-glucitol is quantitative at pH 9–10 and in the presence of excess NAD$^+$. Since the enzyme can also oxidise xylitol to D-xylulose in a similar but not quantitative reaction, the following procedure can be used for the estimation of the DP of xylans (6).

Procedure

Using the highest grade chemicals, the following reagents are prepared:
0.5% w/v Aqueous sodium borohydride.
0.1 M, pH 9.5, Sodium pyrophosphate buffer.
0.02 M Nicotinamide adenine dinucleotide (NAD).
Sorbitol dehydrogenase (SDH, ~70 units/ml[1]) (C. F. Beohringer, Mannheim, Germany).

General Procedure for Reduction and Hydrolysis of Samples

The weights and volumes involved in the method should be adjusted to suit the quantities of D-glucitol and D-glucose being assayed. For an oligosaccharide, less than 2 mg is required; whereas for a large glucan, 10–20 mg may be necessary to give an adequate amount of D-glucitol.

To 2–10 mg of glucan is added 1 ml of water, followed by sufficient 0.5% w/v sodium borohydride solution to give a polysaccharide to sodium borohydride ratio of 5:1. The reduction is allowed to proceed at ~20° for 18–24 h. Then the polysaccharide is hydrolyzed for 2 h at 100° after the addition of 1 ml of 2 N hydrochloric acid. Borate ions are removed by the addition of five 2-ml portions of methanol and rotary evaporation. The product is dissolved in 5 ml of water and used for the estimation of D-glucose and D-glucitol.

Spectrophotometric Determination of D-Glucitol

The following solutions are added by pipet into a 1-cm silica cell: 2 ml of sodium pyrophosphate buffer, 2 ml of 0.02 M NAD, a sample from the reduced and hydrolyzed polysaccharide containing less than 20 μg of D-glucitol (0.05–0.5 ml), and water to give a total volume of 2.95 ml. The initial absorbance is measured at 340 nm in an automatic ultraviolet spectro-

---

[1] 1 unit is the amount of enzyme which catalyzes the formation of 1 μmole of product per minute.

photometer fitted with an external recorder capable of giving a full scale deflection of 0.20 absorbance units. The reaction is started by addition of 0.05 ml of sorbitol dehydrogenase, and readings are taken until the extinction becomes constant, usually within 10 min.

Calculation

The amount of D-glucitol (in μmoles) is given by the increase in absorbance multiplied by 0.482. There is a linear relation between the change in extinction and the D-glucitol concentration over the range 0.0275–0.110 μ moles. Aliquots of the hydrolyzate, after appropriate dilution, are used for determination of D-glucose by the D-glucose oxidase procedure (7) or by the hexokinase method (This Volume [17]).

The DP is calculated from the formula

$$DP = \frac{\mu\text{moles of D-glucose}}{\mu\text{moles of D-glucitol}} + 1$$

Estimation of the DP of Xylans (6)

The reduction with sodium borohydride and the initial acid-catalyzed hydrolysis conditions for the treatment of xylans are identical to those described for glucans. After removal of the borate ions, the product is dissolved in 5 ml of water (solution A); and 0.5-ml aliquots are reduced with 0.5 ml of 0.5% w/v sodium borohydride solution at $\sim 20°$ for 24 h. The solution is freed from borate as described above and is adjusted to a final volume of 5 ml (solution B). Xylitol is determined using the enzymic procedure for D-glucitol and the DP is calculated from the formula

$$DP = \frac{\mu\text{moles of NADH in solution B}}{\mu\text{moles of NADH in solution A}} + 1$$

Discussion

Although the sorbitol dehydrogenase assay is relatively specific, the fact that xylitol is also oxidized means that glucan preparations must be free from any contaminating xylan, and *vice versa*. It should be emphasized that the experimental conditions described here refer only to sheep liver SDH. It is known that SDH preparations from other sources, for example, from ram spermatozoa (8), guinea pig (9), and *Bacillus subtilis* (10), are not as specific as the sheep liver enzyme and will oxidize a variety of hexitols and pentitols.

The two procedures described above are suited for the study of neutral polysaccharides. Many polysaccharides contain uronic acid residues, such as D-glucuronic acid; and on acid-catalyzed hydrolysis, the xylose is not completely liberated. Some wood hemicelluloses yield D-xylose and aldobiouronic acids such as O-(D-glucopyranosyluronic acid)-D-xylose on acid-catalyzed hydrolysis; the aldobiouronic acid is not a substrate for sorbitol dehydrogenase. Occasionally, it is necessary to use formic acid for hydrolysis (6), and it is necessary to remove the inhibitory formate ions by evaporation before enzymic assay.

References

(1) S. Peat, W. J. Whelan, and J. G. Roberts, *J. Chem. Soc.*, **2**, 258 (1956).
(2) G. N. Richards and W. J. Whelan, *Carbohydr. Res.*, **27**, 185 (1973).
(3) G. G. S. Dutton, P. E. Reid, J. J. M. Rowe, and K. L. Rowe, *J. Chromatogr.*, **47**, 195 (1970).
(4) H. Yamaguchi, S. Inamura, and K. Makino, *J. Biochem.* (*Tokyo*), **79**, 299 (1976).
(5) D. J. Manners, A. J. Masson, and R. J. Sturgeon, *Carbohydr. Res.*, **17**, 109 (1971).
(6) R. J. Sturgeon, *Carbohydr. Res.*, **30**, 175 (1973).
(7) A. Dahlqvist, *Biochem. J.*, **80**, 547 (1961).
(8) T. E. King and T. Mann, *Proc. Roy. Soc. London B*, **151**, 226 (1959).
(9) H. G. Williams-Ashmann, J. Banks, and S. K. Wolfson, *Arch. Biochem. Biophys.*, **72**, 485 (1957).
(10) S. B. Horowitz and N. O. Kaplan, *J. Biol. Chem.*, **239**, 830 (1964).

[10] Enzymic Analysis of Acidic Glycosaminoglycans

BY KATSUMI MURATA

*Department of Medicine and Physical Therapy,
University of Tokyo School of Medicine, Tokyo, Japan*

Introduction

Enzymic assay of acidic glycosaminoglycans (AGAG)[1] can be used to determine the constituent units more accurately than other methods. Testicular hyaluronidase was first used in a turbidity reducing assay of hyaluronic acid by Meyer (*1*). Mathews and Inouye (*2*) developed an enzymic method differentiating chondroitin 6-sulfate from chondroitin 4-sulfate that involved digesting with testicular hyaluronidase, followed by determining the oligosaccharide produced with the Morgan–Elson reaction (*3*, This Vol. [11]). Enzymic assay of chondroitin sulfate using chondroitinases has been described (*4–13*).

Although enzyme-catalyzed hydrolysis of AGAG to oligosaccharides has been reported, no practical application to assay the AGAG has been established. The procedure described here is a reliable enzymic assay of AGAG, at the constitutional disaccharide level, combining with the differentiation of AGAG at the macromolecules. This enzymic procedure is well substantiated by the results of electrophoresis of the AGAG or of thin-layer chromatography of the hydrolyzates (*14–23*).

Procedures

Preparation of Enzymes

Chondroitinase-ABC (EC 4.2.2.4)[2] from *Proteus vulgaris* (*7*), chon-

---

[1] The following abbreviations are used in the text: acidic glycosaminoglycans = AGAG; 2-acetamindo-2-deoxy-3-*O*-(4-deoxy-α-L-*threo*-hex-4-enepyranosyluronic acid)-D-glucose = 4,5-unsaturated nonsulfated disaccharide; 2-acetamido-2-deoxy-3-*O*-(4-deoxy-α-L-*threo*-hex-4-enepyranosyluronic acid)-4-*O*-sulfo-D-galactose = 4,5-unsaturated 4-sulfated disaccharide; 2-acetamido-2-deoxy-3-*O*-(4-deoxy-α-L-*threo*-hex-4-enepyranosyluronic acid)-6-*O*-sulfo-D-galactose = 4,5-unsaturated 6-sulfated disaccharide; and 2-acetamido-2-deoxy-3-*O*-(4-deoxy-2- or 3-*O*-sulfo-α-L-*threo*-hex-4-enepyranosyluronic acid)-4- or -6-*O*-sulfo-D-galactose = 4,5-unsaturated disulfated disaccharide.

[2] The enzymes here mentioned are commercially available from Seikagaku Kogyo Co., 2–9, Nihonbashi Honcho, Chuo-ku, Tokyo.

81

droitinase-AC (EC 4.2.2.5) from *Flavobacterium heparinum* (*7*), and chon-droitinase-AC-II (EC 4.2.99.1) from *Arthrobacter aurescens* (*24*) are purified through DEAE-cellulose and phosphocellulose columns (*7*). The enzymic activity is examined using authentic chondroitin 4- and 6-sulfates as substrates when the enzyme is utilized for digesting the specimens. Chon-droitinase-AC catalyzes the cleavage of hexosamine to D-glucuronic acid residue linkages only in chondroitin sulfate chains and converts chondroitin sulfate to unsaturated disaccharide units. In addition, chondroitinase-ABC cleaves hexosaminide linkages of both D-glucuronic acid and L-iduronic acid of chondroitin sulfate and dermatan sulfate chains producing corresponding unsaturated disaccharide units. These chondroitinases, particularly the AC-II type (*8*), also degrade hyaluronic acid.

Chondrosulfatases are prepared from *Proteus vulgaris* and purified by phosphocellulose column chromatography (*7*). Chondro-4-sulfatase (EC 3.1.6.10) and chondro-6-sulfatase (EC 3.1.6.9) cleave the sulfate linkages at C-4 and C-6 positions, respectively, of the disaccharide units. These enzymes are useful for determination of the position of sulfate on the galactosamine moiety, after degrading chondroitin sulfate isomers to disaccharide units by chondroitinases.

Streptomyces hyaluronidase (EC 4.2.2.1), extracted from *Streptomyces hyalurolyticus nov.* (*25*), specifically attacks hyaluronic acid (*25*). The susceptibility of authentic hyaluronic acid to *Streptomyces* hyaluronidase is examined when the enzyme is used.

Preparation of Acidic Glycosaminoglycans (*23*, See also Vol. VII [20])

Proteoglycans (10 mg) are dissolved in 10 ml of 0.067 M phosphate buffer (pH 7.8) and digested with 0.5 g of pronase (70,000 PUK/g, Kaken Kagaku Co., Tokyo) for 12 h at 37°. The proteolytic digestion is stopped by placing the reaction mixture in a boiling water bath for 10 min, and the solution is cooled. The solution is made 0.3 M in NaOH and kept overnight at 5°. Cold trichloroacetic acid is then added at to final concentration of 7%, and the solution is kept at 5° overnight. After centrifugation, the supernatant is dialyzed against running and distilled water and concentrated by a micro-rotatory flash evaporator. AGAG are precipitated by adding 4 volumes of ethanol and dried. AGAG can also be prepared from various sources principally based on the following procedures: lipid extraction, protein digestion, β-elimination, and purification (*14–23*).

AGAG specimens can be subjected to enzymic assay after initial preparation by several fractionation procedures: ion-resin exchange column chromatography, gel filtration, precipitating with quaternary ammonium salt solution, and fractional precipitation with ethanol (*14–23*).

The AGAG standards (chondroitin 4- and 6-sulfate, hyaluronic acid, heparan sulfates, keratan sulfate, and dermatan sulfate) can be obtained from Dr. M. B. Mathews and Dr. J. A. Cifonelli, University of Chicago, Illinois, or from the Tokyo Institute of Seikagaku Kogyo Co., Tokyo, Japan.

Enzymic Analyses (*14–23*)

Enzymic Digestion

The enzymic procedure with chondroitinases is done as follows: The AGAG (approximately 500 μg as uronic acid in 50 μl water) are exhaustively digested with 1–1.5 units of chondroitinase-ABC in 20 μl of Tris-HCl[3] buffer (pH 8.0) at 37° for 120 min. They are similarly digested with 1.5–2.0 units of chondroitinase-AC in 20 μl of Tris-HCl buffer (pH 7.0),[4] and stopped by heating in a boiling water bath for 10 min. The resulting unsaturated disaccharides and undigested AGAG are used for successive analyses.

Paper Chromatographic Separation

The unsaturated disaccharides derived from chondroitin sulfate chains by digestion with the enzymes are spotted by a micropipette as bands (2.5 cm/500 μg) to Whatman No. 1 filter paper (25 cm × 55 cm). Descending paper chromatography (Vol. I [6]) is done at room temperature, as follows. Paper chromatography for desalting is initially done in 13:8:4 v/v 1-butanol–ethanol–water for 24 h. Succeeding paper chromatographic separation of the individual disaccharides is carried out in 5:3 v/v *n*-butyric acid–0.5 *M* ammonia for 72 h or in 2:3:1 v/v 1-butanol–acetic acid–1 *M* ammonia for 24 h.

Localization of spots of the separated unsaturated disaccharides is detected and photographed (if necessary) by exposuring to ultraviolet radiation at 232 nm (Fig. 1). The order of migration (in decreasing order) is unsaturated nonsulfated, 4-sulfated, 6-sulfated, and disulfated disaccharides.

The visualized spot and the area of origin are marked by a pencil, cut out, and eluted with small amounts (1.0–1.5 ml) of distilled water in test tubes. Aliquots of the eluted individual disaccharides are prepared at several different concentrations. The content of unsaturated disaccharides is then

---

[3] Tris-HCl buffer (pH 8.0) is prepared by mixing 3 g of tris(hydroxymethyl)aminomethane, 2.4 g of sodium acetate, 1.46 g of sodium chloride, and 50 mg of crystalline bovine serum albumin (Sigma) in 100 ml of 0.13 *N* hydrochloric acid.

[4] Recommended to be used at pH 6.0 for chondroitinase-AC II.

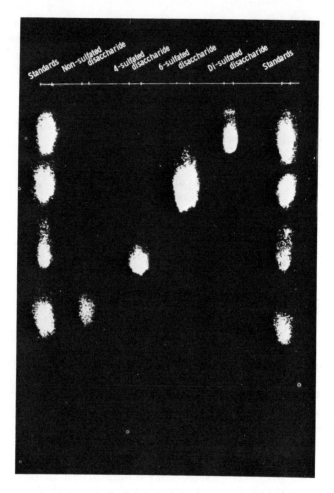

FIG. 1.—Paper chromatographic separation of unsaturated nonsulfated, 4-sulfated, 6-sulfated, and disulfated disaccharides prepared from human urine after digesting with chondroitinase-ABC (*10*). The chromatogram has been photographed by exposure to ultraviolet radiation at 232 nm.

determined by the carbazole reaction (*26*, This Volume [11]) so as to eliminate a nucleate reaction (*10*) or directly at 232 nm when they are purified.

Chondroitin and chondroitin 6-sulfate can be estimated from the amount of unsaturated nonsulfated and 6-sulfated disaccharides respectively produced by digestion with chondroitinase-ABC and/or chondroitinase-AC. Chondroitin 4-sulfate can be measured from the amount of unsaturated 4-sulfated disaccharide produced by digestion with chondroitinase-AC. Dermatan sulfate can be calculated from subtracting the amount of un-

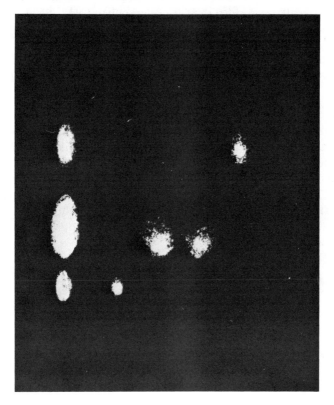

Fig. 2.—Paper electrophoretic identification of the unsaturated disaccharides identical with those in Figure 1 (10).

saturated 4-sulfated disaccharide produced by digestion with chondroitinase-AC from that with chondroitinase-ABC. Oversulfated chondroitin sulfate can be estimated from the values of unsaturated disulfated disaccharide produced by digestion with each enzyme.

For further identification of the disaccharide units, high-voltage electrophoresis and an assay system using chondrosulfatases can be used.

High-Voltage Electrophoresis (See also Vol. VII [20])

Unsaturated disaccharides can also be separated into disulfated, monosulfated (4-sulfated and 6-sulfated) and nonsulfated disaccharide fractions (in decreasing order of migration) by high-voltage paper chromatography (8, 10) (Fig. 2).

After production of unsaturated disaccharides with chondroitinases in Tris-HCl, the resulting disaccharides (10–20 μg each as uronic acid) are applied as spots to Whatman No. 1 filter paper. Electrophoresis is done in 3:1:16 v/v acetic acid–formic acid–water buffer (pH 2.0) at a potential gradient of 45 V/cm for 90 min. The paper is dried immediately under a stream of cool air. The spots of the separated disaccharides are visualized by exposure to a ultraviolet radiation at 232 nm. Electrophoresis can also be done in 0.05 M sodium citrate–citric acid buffer (pH 5.0) (8, 10).

Assay System Using Chondrosulfatases

After degradation of chondroitin sulfates with chondroitinases, the resulting unsaturated disaccharides (10–20 μg each as uronic acid) are digested with either 0.20 unit of chondro-4-sulfatase or 0.15 unit of chondro-6-sulfatase or both in 10 μl Tris-HCl buffer (pH 8.0) and in 10 μl of distilled water for 120 min (8, 10). After complete digestion, the resulting unsaturated disaccharides are applied as spots to Whatman No. 1 filter paper. They can be identified by separation either by paper chromatography or by high-voltage paper chromatography.

Enzymic Assay with Other Analyses

Differentiation of AGAG from the Enzymically Degraded Disaccharides

By the enzymic assay mentioned above, the AGAG are differentiated into two categories: (a) AGAG susceptible to the enzymes and (b) AGAG undigested by the enzymes. Amounts of the susceptible polysaccharides (chondroitin sulfates, dermatan sulfate, and hyaluronic acid) can be determined from yields of the disaccharide units obtained by the specific enzyme or by subtraction of the yields obtained by two enzymes. In order to characterize undigested AGAG, electrophoresis and thin-layer chromatography are used.

Electrophoretic Characterization (See also Vol. VII [20])

AGAG (~5 μg as uronic acid) can be characterized by electrophoresis prior to and after enzymic digestion with chondroitinase-ABC or -AC and *Streptomyces* hyaluronidase. Electrophoresis of AGAG is performed on cellulose acetate membranes, Separax (6 cm × 10 cm, Jōkō Sangyo Co., Tokyo) in 0.1 M formic acid–pyridine buffer (pH 3.1) at 0.5 mA/cm for 60 min, in 0.1 M calcium acetate at 0.5 mA/cm for 180 min (27), and in 0.1 M barium acetate (pH 7.4) at 5 V/cm for 180 min (28). For staining the membranes are immersed in 0.05% toluidine blue and/or 0.05% alcian blue in 0.5% acetic acid solution for 1–3 h, followed by washing with 0.5% acetic acid for 10 sec.

Combining these electrophoretic characterizations, the individual AGAG may be estimated semiquantitatively. This procedure is also useful for examining the susceptibility of each enzyme.

Thin-Layer Chromatography of Hexosamines

For hexosamine determination, a specimen (20–50 μg as uronic acid) is treated with 100 μl of 4 N hydrochloric acid in a sealed tube at 100° for 6 h. The hydrolyzates (2–5 μg) are spotted on an Eastman Kodak 511V sheet (Eastman Kodak Co., Rochester, New York) and ascending thin-layer chromatography is done in 170:1:29 v/v ethanol-25% ammonia–water for 50 min at ∼20° by the method described by Moczar *et al.* (*29*). After drying, the sheet is dipped in 1:1 v/v 1% ethanolic triphenyltetrazolium chloride–1% methanolic NaOH and heated at 95° for 15 min to develop the reddish color.

Galactosamine (5 μg) can be separated from glucosamine by thin-layer chromatography using Eastman Kodak 511V sheets (*29*). Galactosaminoglycans and glucosaminoglycans are separately estimated by this method. The separation can be used not only for macromolecular AGAG but also for degraded disaccharides.

Other Chemical Analyses

To confirm the composition of the disaccharides and the undigested AGAG, the extracted specimens may be analyzed by the methods given in this Volume [11].

The enzymic assay is based on the release of disaccharide units using easily available enzymes. Heterogeneity of the AGAG preparation can, therefore, be a problem. Therefore, the enzymic analyses are recommended for use with these other analytical methods.

References

(*1*) K. Meyer, Hyaluronidase *in* "The Enzymes," P. D. Boyer, ed., Vol. 5, Academic Press, New York, N.Y., 3rd Ed., 1971, pp. 307–320.

(*2*) M. B. Mathews and M. Inouye, *Biochim. Biophys. Acta*, **53**, 509 (1961).

(*3*) J. L. Reissing, J. L. Strominger, and L. F. Leloir, *J. Biol. Chem.*, **217**, 959 (1955).

(*4*) J. S. Mayes and R. G. Hansen, *Anal. Biochem.*, **10**, 15 (1965).

(*5*) H. Saito, T. Yamagata, and S. Suzuki, *J. Biol. Chem.*, **243**, 1936 (1968).

(*6*) S. Suzuki, H. Saito, T. Yamagata, K. Anno, N. Seno, Y. Kawai, and T. Furuhashi, *J. Biol. Chem.*, **243**, 1543 (1968).

(*7*) K. Murata, T. Harada, and K. Okubo, *J. Atheroscler. Res.*, **8**, 951 (1968).

(8)   T. Harada, K. Murata, T. Fujiwara, and T. Furuhashi, *Biochim. Biophys. Acta*, **177**, 676 (1969).

(9)   D. P. Varadi and C. Griffiths, *Biochim. Biophys. Acta*, **230**, 248 (1971).

(10) K. Murata, T. Harada, T. Fujiwara, and T. Furuhashi, *Biochim. Biophys. Acta*, **230**, 583 (1971).

(11) K. Anno, N. Seno, M. B. Mathews, T. Yamagata, and S. Suzuki, *Biochim. Biophys. Acta*, **237**, 173 (1971).

(12) N. Seno, K. Anno, Y. Yaegashi, and T. Okuyama, *Connect. Tiss. Res.*, **3**, 87 (1975).

(13) N. Taniguchi, T. Okamoto, and N. Okuda, *Biochem. Med.*, **15**, 169 (1976).

(14) K. Murata, T. Ishikawa, and Y. Oshima, *Clin. Chim. Acta*, **28**, 213 (1970).

(15) K. Murata, *Anal. Lett.*, **5**, 93 (1972).

(16) K. Murata, T. Ogura, and T. Okuyama, *Biochem. Med.*, **6**, 223 (1972).

(17) K. Murata, T. Ishikawa, and H. Ninomiya, *Biochem. Med.*, **8**, 472 (1973).

(18) K. Murata, Y. Yukiyama, and Y. Horiuchi, *Clin. Chim. Acta*, **49**, 129 (1973).

(19) K. Murata, T. Ogura, and T. Okuyama, *Connect. Tiss. Res.*, **2**, 101 (1974).

(20) K. Murata, *Clin. Chim. Acta*, **57**, 115 (1974).

(21) K. Murata, *Clin. Chim. Acta*, **63**, 157 (1975).

(22) K. Murata, *Connect. Tiss. Res.*, **4**, 131 (1976).

(23) K. Murata and A. O. Bjelle, *J. Biochem.*, **80**, 203 (1976).

(24) K. Hiyama and S. Okada, *J. Biol. Chem.*, **250**, 1824 (1975).

(25) T. Ohya and Y. Kaneko, *Biochim. Biophys. Acta*, **198**, 607 (1970).

(26) T. Bitter and H. M. Muir, *Anal. Biochem.*, **4**, 330 (1962).

(27) N. Seno, K. Anno, K. Kondo, S. Nagase, and S. Saito, *Anal. Biochem.*, **37**, 197 (1970).

(28) E. Wessler, *Anal. Biochem.*, **26**, 439 (1968).

(29) E. Moczar, M. Moczar, G. Schillinger, and L. Robert, *J. Chromatogr.*, **31**, 561 (1967).

[11] Constituent Analysis of Glycosaminoglycans

E. V. Chandrasekaran

Department of Biological Chemistry, The Milton S. Hershey Medical Center,
The Pennsylvania State University, Hershey, Pennsylvania

AND

James N. BeMiller

Department of Chemistry and Biochemistry,
Southern Illinois University at Carbondale,
Carbondale, Illinois

Introduction

The fine structural details of glycosaminoglycans (GAG) remain to be elucidated. Glycosaminoglycans are, in essence, composed of specific disaccharide repeating units containing hexosamine and hexuronic acid (hexose). The unique occurrence of uronic acid in GAG imparts to them an anionic charge, facilitates their separation from proteins and glycoproteins, and aids in their identification. In some GAG, the anionic charge is further elevated by the presence of sulfate groups. Based on the anionic constituents of their repeating units, GAG fall into four categories: (a) hyaluronic acid and chondroitin (anionic charge due to hexuronic acid residues only); (b) chondroitin 4-sulfate, chondroitin 6-sulfate, and dermatan sulfate (anionic charge contributed by both uronic acid residues and O-sulfate half-ester groups); (c) heparin and heparan sulfate (anionic charge conferred by uronic acid residues, O-sulfate groups, and N-sulfate groups); and (d) keratan sulfate (charge solely due to O-sulfate groups).

Glycosaminoglycans can be separated from each other using anion exchangers. Dowex-1(Cl⁻) (1), DEAE-Sephadex A-25 (2–6), and ECTEOLA-cellulose (7) columns have been used for this purpose. Depending on the tissue of origin, there is always microheterogeneity in one or more GAG, resulting from (a) the degree and difference in sulfation of the repeating unit (8), (b) copolymeric structure (9) caused by the occurrence of both D-glucuronic acid and L-iduronic acid residues in the same chain, and (c) chains of different length. So any method of fractionation of GAG from a tissue always results in the isolation of two or more GAG as a mixture. Hence, a quantitative picture of GAG in a tissue can be obtained only by subjecting the isolated fractions to constituent analysis. A number of methods

89

METHODS IN CARBOHYDRATE
CHEMISTRY, VOL. VIII

can be applied to analyze each constituent. It is essential that the method used give a high extinction coefficient (absorptivity) and be insensitive to interference by salts. The methods in use in our laboratories and in general use elsewhere are described here; they give quite consistent results. They can be used for preliminary characterization. Enzymic and electrophoretic methods that give absolute identification will be detailed in later volumes in this series.

Procedures

Uronic Acid

Outline of the Method

The glycosaminoglycan sample is heated in concentrated sulfuric acid containing borate and then reacted with carbazole (*10*).

Reagents

(1) Borate–sulfuric acid: 3.82 g of sodium borate ($Na_2B_4O_7 \cdot 10\ H_2O$) is dissolved in 10 ml of hot water, and 390 ml of well-cooled conc. sulfuric acid is added by stirring with a glass rod. This reagent is stored in a refrigerator in a glass bottle.
(2) 0.2% Carbazole: 100 mg of carbazole is dissolved in 50 ml of absolute ethanol and stored in a refrigerator in a brown glass bottle.
(3) Standard glucuronolactone: 17.6 mg of D-glucuronolactone is dissolved in 100 ml of deionized water (1 μmole/ml) just before use.

Procedure

Aliquots of the standard solution of glucuronolactone (0 to 0.2 ml by 0.02-ml increments) are transferred by pipet to 1.75 × 15-cm test tubes. After bringing the volume in each tube to 0.5 ml with deionized water, 3 ml of cold borate–sulfuric acid reagent is added with immediate mixing. The tubes are heated in boiling water for 20 min. After cooling to 0°, 0.1 ml of 0.2% carbazole is added. The solution is shaken well and heated again in the boiling water bath for 10 min. After cooling at room temperature for 15 min, the color is read at 530 nm against the blank (Tube 1). Three different aliquots of the GAG sample are subjected to this reaction to determine the amount of uronic acid present.

Hexosamines

A GAG sample containing 0.3–0.5 μmole of uronic acid in 0.2 ml of water is placed in a small ampule (2 ml capacity), mixed with 0.2 ml of conc. hydrochloric acid, sealed, and heated in a thermoblock at 100° for 4 h. After hydrolysis, the ampule is opened and placed in a vacuum desiccator containing sodium hydroxide pellets in a beaker. After complete drying, the residue in the ampule is dissolved in 0.3 ml of water. An aliquot is used for identification of hexosamines by paper chromatography and the rest for estimation of hexosamines.

Outline of the method

Hexosamines are reacted with an alkaline solution of 2,4-pentanedione (acetylacetone) to form chromogens, which yield color with N,N-dimethyl-p-aminobenzaldehyde in acid. Both glucosamine and galactosamine form chromogens when this reaction is done at 100°; whereas only galactosamine forms chromogens when the reaction is done at 0° (*11*). Hence, the method can be used to give the glucosamine:galactosamine ratio. Molar ratios of glucosamine:galactosamine can also be obtained after separating the two by ion-exchange chromatography (Vol. VII [42]).

Reagents

(1) Alkaline 2,4-pentanedione: 0.15 ml of 2,4-pentanedione is diluted to 5 ml with 0.7 M sodium carbonate. This reagent is prepared just before use.
(2) Ehrlich reagent: 100 mg of N,N-dimethyl-p-aminobenzaldehyde is added to 3.2 ml of 60% perchloric acid and then made up to 10 ml with 95% ethanol. This reagent must be freshly prepared before use.
(3) 90% ethanol
(4) Hexosamine hydrochloride solution, 2 μmole/ml.

Total hexosamine

Aliquots of the standard solution of hexosamine hydrochloride (0 to 0.1 ml by 0.01-ml increments) are transferred by pipet to screw-cap tubes. After making up the solutions to 0.1 ml with water, 0.025 ml of 1.5 M hydrochloric acid is added with thorough mixing. Then 0.25 ml of pentanedione solution is added with thorough mixing. The tubes are capped and heated in boiling water bath for 20 min, then cooled to $\sim 20°$ by immersion in cold water. Then 2 ml of 90% ethanol and 0.5 ml of Ehrlich reagent are added with thorough mixing, and the tubes are allowed to stand at 20°–25° for 1 h. The color is read at 535 nm against the blank.

Galactosamine

Standards of galactosamine are prepared as described for total hexosamine; a few blanks containing 0.2 μmole or more of standard glucosamine are also prepared. The tubes are placed in ice. To the tubes at 0° are added 0.025 ml of 1.5 M hydrochloric acid (0°) and then 0.25 ml of pentanedione (0°) reagent. The tubes are capped; the contents are mixed thoroughly, and the tubes are kept for 18 h in ice. After incubation, 2 ml of 90% ethanol and 0.5 ml of Ehrlich reagent are added; the contents are mixed, and the tubes are heated in a water bath at 70° for 1 h. The tubes are then kept at 20°–25° for 1 h and read at 525 nm against a blank.

O-Sulfate

Outline of the method

Sulfate is estimated turbidimetrically as barium sulfate. Gelatin is used as a cloud-stabilizer and absorbance is measured at 360 nm (*12*).

The GAG sample (containing 0.2–1.0 μmole of uronic acid) in 0.3 ml of water is mixed with 0.3 ml of 0.5 N hydrochloric acid in an ampule. The ampule is sealed and heated in a boiling water bath for 4 h. After hydrolysis, 0.5 ml of the hydrolyzate is taken for determination of sulfate.

Reagents

(1) Gelatin: 200 mg of gelatin is dissolved in 40 ml of 60°–70° water; the solution is placed in a refrigerator overnight.
(2) BaCl$_2$–Gelatin: 100 mg of barium chloride is dissolved in 20 ml of the gelatin solution; the resulting solution is kept in a refrigerator for 3 h before using.
(3) 10% w/v Trichloroacetic acid
(4) Standard sulfate solution: Na$_2$SO$_4$ or K$_2$SO$_4$ (0.2 μmole/ml of water).

Procedure

To avoid any sulfate contamination, all glassware must be washed with hot nitric or hydrochloric acid, rinsed with deionized water, and dried in an oven.

Aliquots of the standard sulfate solution (0 to 1.0 ml by 0.1-ml increments) are transferred by pipet into test tubes, and the volume is made up to 1.5 ml with deionized water. Trichloroacetic acid solution (0.9 ml) is added, and the contents are mixed thoroughly. Then, 0.6 ml of the barium chloride–gelatin solution is added and the contents are mixed. The tubes are allowed to stand for 20 min at 20°–25°; then the absorbance is measured at 360 nm against the blank.

Correction is made for any other ultraviolet absorbing materials as follows: The hydrolyzed GAG sample is mixed with the gelatin rather than the barium chloride–gelatin solution, and the absorbance is measured at 360 nm against a blank prepared with gelatin solution instead of with the barium chloride–gelatin solution. This value is subtracted from the reading obtained when the barium chloride–gelatin solution is used.

N-Sulfate

Outline of the method

Hexosamine *N*-sulfate is desulfated and deaminated by nitrous acid and the resulting anhydrohexose (Vol. VII [14], [23]) is estimated by indole–hydrochloric acid reagent (*13*, Vol. VII [23]).

Reagents

(1) 5% Sodium nitrite (NaNO$_2$, freshly prepared)
(2) 33% Acetic acid: 1 volume of glacial acetic acid is diluted with 2 volumes of water.
(3) 12.5% Ammonium sulfamate
(4) 5% Hydrochloric acid: 1 volume of conc. hydrochloric acid is mixed with 7 volumes of water.
(5) 1% Indole in ethanol (freshly prepared)
(6) Standard glucosamine solution: 8.95 mg of D-glucosamine or 10.77 mg of D-glucosamine hydrochloride is dissolved in 100 ml of water (0.5 μmole/ml) just before use.

Procedure

Aliquots of standard glucosamine solution (0 to 0.25 ml by 0.025-ml increments) are transferred by pipet into screw-cap tubes. The volume is made up to 0.25 ml with water. Sodium nitrite solution (0.25 ml) is added, and the contents are mixed. Then 0.25 ml of the acetic acid solution is added. After thorough mixing, the tubes are kept at 20°–25° for 90 min with shaking at 15-min intervals. Ammonium sulfamate solution (0.25 ml) is added, and the contents are well shaken and left at 20°–25° for 30 min. Then 2 ml of hydrochloric acid and 0.1 ml of indole solution are added, and the contents are thoroughly mixed. The tubes are capped and heated in boiling water bath for *exactly* 5 min. Ethanol (1 ml) is added with immediate mixing. The absorbance is measured at 492 nm against the blank.

It is good practice to check this reaction with hyaluronic acid and chondroitin sulfate standards each time it is done. These GAG will not give any color unless a mistake in the determination has been made.

Hexose

Outline of the method

Other than keratan sulfate, GAG do not contain hexose, except in the linkage region to serine. The sample is heated with orcinol–sulfuric acid reagent (*14*).

Reagents

(1) 60% Sulfuric acid: 300 ml of conc. sulfuric acid is added slowly with stirring to 200 ml of water cooled in ice.

(2) 1.6% Solution of orcinol in water

(3) Orcinol–sulfuric reagent: 7.5 volumes of A is mixed with 1 volume of B, just before use.

(4) D-Galactose standard: 36 mg of D-galactose is dissolved in 100 ml of water (2 μmole/ml) just before use.

Procedure

Aliquots of the standard solution of galactose (0 to 0.40 ml by 0.04-ml increments) are transferred by pipet into test tubes. The contents are made up to 0.40 ml with water. Then 3.4 ml of the orcinol–sulfuric acid reagent is added, and the tubes are heated in a water bath at 80° for 15 min. After allowing the tubes to cool for 15 min, absorbance is measured at 505 nm against the blank. If galactose has already been identified in a GAG, the hexose value obtained by the orcinol method can be checked by hydrolyzing a sample of the GAG in 1 *M* hydrochloric acid for 3 h in sealed ampules and then determining D-galactose with galactose oxidase (*15*), which is commercially available in assay kits or as described in this Volume [16].

Pentose, 6-Deoxyhexose, and Sialic Acid

As fucose (*16*) and sialic acid (*16, 17*) have been reported to be present in keratan sulfate preparations and arabinose (*18, 19*) in hyaluronic acid, it is obvious to look for these sugars in a GAG fraction.

Pentose

Outline of the method

The GAG sample is heated with orcinol in hydrochloric acid containing ferric chloride (*16*). The use of anthrone in sulfuric acid provides an alternative method (*17*).

Reagents

(1) Arabinose or xylose as standard: 2 mg of D-xylose or L-arabinose is dissolved in 100 ml of water (20 μg/ml) just before use.
(2) $FeCl_3$–HCl–orcinol: 80 ml of conc. hydrochloric acid are mixed with 16 ml of water. To this solution are added 200 mg of orcinol and 5 ml of 1.5% ferric chloride solution with thorough mixing. This reagent is prepared immediately before use.

Procedure

Aliquots of standard pentose solution (0 to 1.0 ml by 0.1-ml increments) is transferred by pipet to test tubes. Reagent B is added, and the tubes are heated in boiling water bath for 20 min. After allowing the tubes to cool at room temperature for 15 min, the absorbance is measured at 670 nm against the blank.

6-Deoxyhexose

Outline of the method

The GAG sample is heated with sulfuric acid and then reacted with thioglycolic acid (*18*).

Reagents

(1) Sulfuric acid: conc. sulfuric acid (90 ml) is added to 15 ml of distilled water cooled in ice.
(2) Thioglycolic acid: 0.1 ml of thioglycolic acid is diluted to 3 ml with water just before use.
(3) 6-Deoxyhexose standard: 4 mg of L-fucose or L-rhamnose is dissolved in 100 ml of water (40 μg/ml) just before use.

Procedure

Aliquots of the standard 6-deoxyhexose solution (0 to 1.0 ml by 0.1-ml increments) are transferred by pipet to test tubes, and the volume is brought to 1 ml with water. Sulfuric acid solution (4.5 ml) is added. The solution is then heated in boiling water for 10 min. The tubes are cooled in cold water. Thioglycolic acid solution (0.1 ml) is added, and the contents are mixed well. The tubes are kept in the dark for 3 h. The absorbance is measured at both

400 nm and 430 nm against the blank. Because 6-deoxyhexose has a very low absorbance as compared to hexose when read at 430 nm, a correction can be made for the presence of other sugars by reading at both wavelengths.

Sialic acid

See Volume VII [40].

References

(1) S. Schiller, G. A. Slover, and A. Dorfman, *J. Biol. Chem.*, **236**, 983 (1961).

(2) M. Schmidt, *Biochim. Biophys. Acta*, **63**, 346 (1962).

(3) M. Singh and B. K. Bachhawat, *J. Neurochem.*, **15**, 249 (1968).

(4) E. V. Chandrasekaran and B. K. Bachhawat, *Biochim. Biophys. Acta*, **177**, 265 (1969).

(5) M. Singh, E. V. Chandrasekaran, R. Cherian, and B. K. Bachhawat, *J. Neurochem.*, **16**, 1157 (1969).

(6) R. Cherian, E. V. Chandrasekaran, and B. K. Bachhawat, *Indian J. Biochem.*, **7**, 174 (1970).

(7) N. R. Ringertz and P. Reichard, *Acta Chem. Scand.*, **14**, 303 (1960).

(8) C. A. Antonopoulos, B. Engfeldt, S. Gerdell, S.-O. Hjertquist, and K. Solheim, *Biochem. Biophys. Acta*, **101**, 140 (1965).

(9) L.-A. Fransson, *J. Biol. Chem.*, **243**, 1504 (1968).

(10) T. Bitter and H. M. Muir, *Anal. Biochem.*, **4**, 330 (1962).

(11) J. Ludoweig and J. D. Benmanan, *Carbohydr. Res.*, **8**, 185 (1968).

(12) K. S. Dodgson and R. G. Price, *Biochem. J.*, **84**, 106 (1962).

(13) D. Lagunoff and G. Warren, *Arch. Biochem. Biophys.*, **99**, 396 (1962).

(14) R. J. Winzler, *Methods Biochem. Anal.*, **2**, 290 (1955).

(15) D. Amaral, F. Kelly-Falcoz, and B. L. Horecker, *Methods Enzymol.*, **9**, 87 (1966).

(16) A. H. Brown, *Arch. Biochem.*, **11**, 269 (1946).

(17) C. P. Tsiganos and H. Muir, *Anal. Biochem.*, **17**, 495 (1966).

(18) M. N. Gibbons, *Analyst*, **80**, 268 (1955).

[12] Determination of Polysaccharide Structures with ^{13}C NMR

By Harold J. Jennings and Ian C. P. Smith

Division of Biological Sciences, National Research Council of Canada,
Ottawa, Ontario, Canada

Introduction

Carbon-13 nuclear magnetic resonance has proven to be a powerful tool in studies on biopolymers (1–5) in general, and recent work has established it as a valuable technique in the structural determination of polysaccharides (6–18). Using the Fourier transform method, it allows spectra of the polysaccharides to be obtained using only their natural abundance carbon-13 atoms; it complements [^{1}H] nmr spectroscopy in that it gives better signal separation due to the wider range of chemical shifts involved (19, Vol. VII [19]). The technique is rapid, non-destructive and can be used on relatively small amounts of material. In this regard, it has great potential in the study of polysaccharides of biological origin (12–18) and has already proven of particular value in bacteriological research (12, 13, 15, 16, 18), especially in monitoring minor structural changes initiated by changing physiological conditions (15). Although information can be obtained directly from proton-coupled natural abundance spectra of polysaccharides, the method is time consuming and is limited by the complexity of the spectra obtained. Therefore most of the studies to date have been concerned with proton-decoupled spectra. Although ^{13}C–^{31}P couplings (12, 16) can still be employed for the detailed interpretation of the proton-decoupled spectra, the most generally informative method of analysis is based on the correlation of the chemical shifts of the carbon atoms of the polysaccharides with those of their previously assigned monosaccharide and oligosaccharide units. Related monosaccharides, substituted monosaccharides, and polysaccharides can also be employed as model compounds. Experience indicates that the chemical shifts of the monosaccharides are similar to those of the monosaccharide units within the polysaccharide except for substituent effects. These effects produced by the attachment of any substituent to a sugar moiety cause an increase in chemical shift of the carbon directly involved in the linkage; this is usually accompanied by a decrease of smaller magnitude (sometimes an increase) in the chemical shifts of the neighboring β-carbon atoms. Thus, these patterns of chemical shift differences between monosaccharide and

97

METHODS IN CARBOHYDRATE
CHEMISTRY, VOL. VIII

polysaccharide can be used to determine the position of linkages; also similarities in chemical shifts, especially on selective carbon atoms known to be sensitive to change in anomeric configuration, can be employed to determine the configuration of linkages. This general approach is illustrated in the following complete structural elucidations of the serogroup A (*12*), B, and C (*13*) polysaccharide antigens of *Neisseria meningitidis*. Some polysaccharides are structurally more complex than the examples cited above and defy complete analysis by this technique. Even in these cases carbon-13 nmr spectroscopy provides a rich additional source of information to add to that obtained by other means of analysis.

The spectrum of the group A polysaccharide is shown in Figure 1, and although complex due to the presence of *O*-acetyl groups, it is considerably

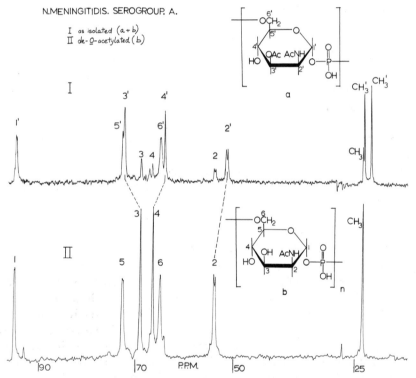

FIG. 1.—Carbon-13 nmr spectrum of the native polysaccharide antigen from *Neisseria meningitidis* serogroup A (upper) and its de-*O*-acetylated form (lower). The spectra were obtained on a Varian XL-100 Fourier transform spectrometer operating at 25.2 MHz at 31° with complete proton decoupling, using 71,361 accumulations of free induction decay for the upper spectrum and 70,528 for the lower. Both spectra were obtained using a 12-mm tube (o.d.) containing 100 mg of each polysaccharide dissolved in D$_2$O at pH 7.0. From reference (*12*).

TABLE I

^{13}C Chemical Shifts of the De-O-acetylated A Polysaccharide and those of
2-Acetamido-2-deoxy-α- and β-D-mannopyranose (ManNAc)

Polysaccharide and monomers	C-1	C-2	C-3	C-4	C-5	C-6	CH_3 $(NHCOCH_3)$	$C{=}O$ $(NHCOCH_3)$
De-O-acetylated A	96.4	54.3	69.7	67.1	73.5	65.6	23.3	175.8
α-ManNAc	94.3	54.4	70.1	68.0	73.2	61.7	23.2	175.9
β-ManNAc	94.3	55.3	73.2	67.8	77.5	61.7	23.2	176.8

simplified to an eight-resonance spectrum (carbonyl signal at 175.8 ppm not shown) on removal of these O-acetyl groups. This simplicity indicates that the polysaccharide is linear and free of other substituents; the number of signals is consistent with those of the individual carbon atoms of a 2-acetamido-2-deoxymannopyranosyl phosphate repeating unit. Thus, a comparison of the chemical shifts of the de-O-acetylated A polysaccharide with those of 2-acetamido-2-deoxy-α- and β-D-mannopyranose (Table I) shows greater similarity with C-2, C-3 and C-5 of the α-anomer than with the β-anomer and that C-6 of the polysaccharide has experienced a large increase of 3.9 ppm in chemical shift in comparison with both anomers. Thus the A polysaccharide is α-D-(1 → 6) linked. A similar analysis using 2-acetamido-2-deoxy-D-glucopyranose as the model indicates that the X polysaccharide is α-D-(1 → 4) linked (12).

Because the A and X polysaccharides contain unique phosphodiester bonds, it was also possible to confirm these structural assignments by an analysis of the ^{31}P–^{13}C couplings. These couplings are manifest as small splittings (in some cases barely resolved, or suggested by increased resonance width with a concomitant decrease in vertical intensity) in the spectrum of the de-O-acetylated A polysaccharide (Fig. 1) at C-1, C-2, C-5, and C-6. The pattern of coupling can be used to determine the position of linkage, and the large value of the coupling to C-2 found in both the spectra of the A and X polysaccharides (8.6 and 8.0 Hz respectively) is indicative of an extended conformation for these polysaccharides (12). This technique has also been applied recently to the structural determination of a teichoic acid (16).

In solving the structure of polysaccharides containing sialic acid residues, [^{13}C] nmr has proved to be particularly valuable as the assignment of anomeric configuration was not possible using [^1H] nmr due to the lack of protons on the anomeric center (13, 18). In addition, other methods of analysis did not permit an unambiguous linkage analysis. The B and de-O-acetylated C polysaccharides produced simple eleven-resonance spectra indicating that they were linear homopolymers of sialic acid with complete anomeric

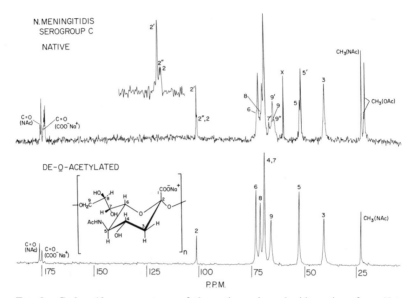

FIG. 2.—Carbon-13 nmr spectrum of the native polysaccharide antigen from *Neisseria meningitidis* serogroup C (upper) and its de-*O*-acetylated form (lower). All other operating conditions were the same as for Fig. 1 except that 54,000 accumulations were used for the upper spectrum and 257,000 for the lower. From reference *(13)*.

homogeneity *(13)*; the spectra of the native and de-*O*-acetylated C polysaccharide are shown in Figure 2. In this case, the methyl α- and β-glycosides of sialic acid were used as models, and the differences in chemical shift between some of their relevant carbon atoms and those of the polysaccharides are shown in Table 2. The chemical shift differences independent of anomeric configuration indicate the carbon atoms involved in the linkages; thus, the

TABLE II

^{13}C Chemical Shift Differences (ppm)[a] between the B and De-O-acetylated C Polysaccharides and Methyl α- and β-D-N-Acetylneuraminic Acid

Polysaccharides–methyl ketosides	C-1	C-4	C-5	C-6	C-7	C-8	C-9
B–α anomer	+2.1	+0.9	+0.7	+0.3	+0.9	+6.8	−1.9
B–β anomer	+0.3	+1.6	+0.5	+3.1	+1.0	+7.0	−2.3
C^b–α anomer	+2.6	+1.0	+0.1	+0.4	0.0	−0.6	+2.0
C^b–β anomer	+0.8	+1.7	+0.1	+2.4	+0.1	−0.4	+1.6

[a] A positive difference indicates an increase in the chemical shift of the polysaccharides in relation to the monomers.

[b] The de-*O*-acetylated C polysaccharide.

large differences for C-8 of B and C-9 of C indicate their respective $(2 \rightarrow 8)$ and $(2 \rightarrow 9)$ linkages, confirmed by the presence of smaller shifts in the neighboring β-carbon atoms. In addition, both polysaccharides can be assigned the α-D configuration as C-4 and C-6 have much smaller differences in chemical shifts with respect to the corresponding resonances of the α-D-ketoside than with respect to those of the β-D-ketoside. It is interesting to note that the carboxyl carbon atom (C-1) appears to be anomolous in this analysis; this was thought to be due to differing aglycones (13). However, recently, it has been found that it is due to comparing the sodium salt of the polysaccharides with the free acid form of the monomers. When the sodium salt forms of the monomers were employed as models, unambiguous assignments of anomeric configuration could be made based on the carboxylate carbon atom (20).

The A and C polysaccharides are O-acetylated, and the degree of O-acetylation in each can be determined by a comparison of the intensities of the characteristic methyl signals of the O-acetyl and N-acetyl groups. In the case of the A polysaccharide (Fig. 1), this ratio is 0.7 to 1.0, respectively, indicating that approximately 70% of the mannosamine residues are O-acetylated. That this is so is confirmed by the fact that the C-2 resonance of the native polysaccharide is split; 70% of the signal (C-2′) undergoing a decrease in chemical shift of 2.6 ppm. Thus, employing the previously defined substituent rules, C-2′ must be adjacent to the position of O-acetylation. This was confirmed by finding a decrease in chemical shift (2.6 ppm) of the signal assigned to the other neighboring carbon atom (C-4′) and an increase (3.0 ppm) in the signal assigned to the position of O-acetylation (C-3′). The location of the O-acetyl groups in the native C polysaccharide was more difficult. However, it was ascertained that the C-4 position of the sialic acid residues was free of O-acetyl groups, as the adjacent carbon atom (C-3) had an identical chemical shift in both the native and de-O-acetylated C polysaccharides and that the O-acetyl substituents are restricted to C-7 and C-8 of the sialic acid residues (13).

The examples described above involve polysaccharides having very simple sequences of single sugar-residues. However, by utilizing the number and intensity of the anomeric signals (other signals can be employed for confirmation), more complex sequences can be determined (7, 8, 11, 18). Due to the empirical nature of this analysis, the assignments made on the polysaccharides are dependent on the accuracy of the assignments made on the resonances of the monosaccharides. Early assignments were based on the large amount of data already accumulated on a variety of model compounds and on specifically derivatized sugars (for leading references, see reference 21). Although this method has proved to be effective, it has been justifiably criticized because some of the more recent work using ^{2}H- (22–24) and

[13]C-labeled hexoses (21, this Volume [19]) has necessitated the reassignment of some of the carbon atoms of the hexoses (21), albeit those involving carbon atoms having very similar chemical shifts. The isotopic labeling of complex sugars (e.g., sialic acid) presents extensive synthetic problems; in many cases the ambiguities they resolve are not critical for the purposes of polysaccharide structural determinations. However, to avoid errors, great care must be exercised in employing tentative assignments; use must be made of as much confirmatory data as can be obtained from the spectrum of the compound, including the [1]H-coupled spectrum (11, 13), from the spectra of related polysaccharides and monomers, and, most importantly, from other conventional analytical techniques.

Another limiting factor is the signal-to-noise ratio obtained in the spectra. Under normal operating conditions (overnight run), it would be difficult to detect fine structural detail represented by individual sugar units or carbon-containing substituents which are present in a polysaccharide to the extent of less than 5% (11).

Procedures

Instrumentation

Carbon-13 spectra of polysaccharides can be obtained satisfactorily from instruments operating in the frequency range 15–25 MHz (corresponding to proton frequencies of 60–100 MHz). Greater sensitivity and separation of resonances can be attained with higher-frequency spectrometers, but problems are encountered with temperature control due to intense heating of the aqueous sample by the high power necessary to decouple hydrogen atoms. The low natural abundance (1.1%) and intrinsic sensitivity (1.6% of that of ^{1}H) of ^{13}C necessitates considerable time averaging of spectra to achieve acceptable signal-to-noise ratios. This is best done in the Fourier transform mode where a repetitive series of high-power radiofrequency pulses are applied to the sample and the free induction decay after each pulse is time-averaged. When sufficient intensity in the averaged free induction decay has been realized, a Fourier transform is performed on the decay to produce the usual intensity versus frequency spectrum. This procedure results in an improvement in signal-to-noise for a given time spent of approximately 80 with respect to the continuous sweep time-averaging procedure. The spectra are usually obtained with complete proton decoupling for simplification and to gain intensity via multiplet collapse and the nuclear Overhauser effect. Proton-coupled spectra can be useful for assignment of resonances and to derive conformational information.

To stabilize the magnetic field of the spectrometer, one normally uses the

^{2}H resonance of a deuterated solvent, usually D_2O for polysaccharides. Although an internal reference can be used to calibrate chemical shifts, it is usually not necessary as the magnetic field and the radiofrequency are locked together via the deuterium resonance of the solvent. One merely needs to calibrate the radiofrequency occasionally with an external reference. (Over six years, such a calibration on our spectrometers has changed less than 1 Hz, or 0.04 p.p.m.). This procedure avoids contamination of the sample by an internal reference and loss of sensitivity by displacing the sample with a coaxial tube containing an external reference. For work with aqueous solutions, our spectrometers are calibrated with respect to neat tetramethylsilane (TMS) in a coaxial 5-mm tube; this calibration must be established for all temperatures used, as the susceptibility correction for external references is slightly temperature-dependent. Although early work used carbon disulfide as a reference, TMS is now accepted as the universal standard for ^{13}C nmr. The use of water-soluble compounds as secondary references to TMS should be avoided, as the correction factor for comparison with data obtained in other solvents varies greatly with the other solvent.

Solutions

Spectra of polysaccharides are routinely obtained from solutions of 50–100 mg/ml in D_2O (volumes of 0.8–1.0 ml) in tubes of outside diameter 10–12 mm. For compounds of low solubility, 18-mm tubes and 3-ml volumes are used. Teflon vortex plugs must be used with these minimal volumes. With 1-mm tubes and sample volumes of 25 μl, as little as 1 mg of polysaccharide can be studied.

For solubilization of the polysaccharide, addition of the compound (freeze-dried) in small quantities to the full sample volume of D_2O with continuous stirring is preferred. If problems occur with thixotropy or high viscosity, mechanical agitation or mild sonification is useful. Even with solutions of high viscosity, acceptable spectra are obtained providing a rigid tertiary structure is not formed. In extreme cases of low solubility, D_2O at pD 14 has been found to be extremely useful; in these cases, the addition of a trace of sodium borohydride assists in minimizing degradation. Under these conditions, resonances are usually narrower, and better resolution of sequence-specific chemical shift differences can be ascertained (11). In contrast to ^{1}H nmr, particles or undissolved material make very little difference to the quality of the spectra.

Integration of Spectra

For rapid quantitation of relative amounts of components, the resonances heights printed out by the computer can be used. Improved accuracy is achieved by means of digital integration of the peaks in the computer. In both cases, care must be taken to ensure that any effects of different nuclear Overhauser enhancements for the resonances of interest have been suppressed. This is conveniently accomplished by using a gated decoupling technique where the decoupling is on during data accumulation but off during a delay which is at least seven times as long as the longest spin-lattice relaxation time (T_1) of the carbon atoms of interest. Acquiring data too rapidly can also cause intensity aberrations, as can the use of insufficient computer data points to characterize a resonance. These techniques are described in a brief review by Gray (25) and in the recent text by Wehrli and Wirthlin (26).

References

(1)   T. L. James, "Nuclear Magnetic Resonance in Biochemistry," Academic Press, New York, 1975.
(2)   G. Gray, Crit. Rev. Biochem., 1, 247 (1977).
(3)   A. S. Perlin, "MTP International Review of Science, Organic Chemistry, Series Two," Vol. 7, 1977, p. 1.
(4)   "Topics in Carbon-13 NMR Spectroscopy," G. C. Levy, ed., Wiley Interscience, New York, Vol. 1, 1974; Vol. 2, 1975.
(5)   I. C. P. Smith, H. J. Jennings, and R. Deslauriers, Acc. Chem. Res., 8, 306 (1975).
(6)   D. E. Dorman and J. D. Roberts, J. Amer. Chem. Soc., 92, 1355 (1970).
(7)   A. S. Perlin, N. M. K. Ng Ying Kin, S. S. Bhattacharjee, and L. F. Johnson, Can. J. Chem., 50, 2437 (1972).
(8)   H. J. Jennings and I. C. P. Smith, J. Amer. Chem. Soc., 95, 606 (1973).
(9)   T. Usui, N. Yamaoka, K. Matusda, K. Tuzimura, H. Singiyama, and S. Seto, J. Chem. Soc., Perkins Trans., 1, 2425 (1973).
(10) P. A. J. Gorin, Can. J. Chem., 51, 2375 (1973).
(11) P. Colson, H. J. Jennings, and I. C. P. Smith, J. Amer. Chem. Soc., 96, 8081 (1974).
(12) D. R. Bundle, I. C. P. Smith, and H. J. Jennings, J. Biol. Chem., 249, 2275 (1974).
(13) A. K. Bhattacharjee, H. J. Jennings, C. P. Kenny, A. Martin, and I. C. P. Smith, J. Biol. Chem., 250, 1926 (1975).
(14) P. A. J. Gorin, Carbohydr. Res., 39, 3 (1975).
(15) H. J. Jennings, A. K. Bhattacharjee, D. R. Bundle, C. P. Kenny, A. Martin, and I. C. P. Smith, J. Infect. Dis., 136, Suppl. 78 (1977).
(16) W. R. De Boer, F. J. Kruyssen, J. T. M. Wouters, and C. Kruk, Eur. J. Biochem., 62, 1 (1976).
(17) G. K. Hamer and A. S. Perlin, Carbohydr. Res., 49, 37 (1976).
(18) A. K. Bhattacharjee, H. J. Jennings, C. P. Kenny, A. Martin, and I. C. P. Smith, Can. J. Biochem., 54, 1 (1976).
(19) J. B. Stothers, "Carbon-13 NMR Spectroscopy," Academic Press, New York, 1972.
(20) H. J. Jennings and A. K. Bhattacharjee, Carbohydr. Res., 55, 105 (1977).
(21) T. E. Walker, R. E. London, T. W. Whaley, R. Barker, and N. A. Matwiyoff, J. Amer. Chem. Soc., 98, 5807 (1976).
(22) H. J. Koch and A. S. Perlin, Carbohydr. Res., 15, 403 (1970).

(*23*) D. R. Bundle, H. J. Jennings, and I. C. P. Smith, *Can. J. Chem.*, **51**, 3812 (1973).

(*24*) P. A. J. Gorin and M. Mazurek, *Can. J. Chem.*, **53**, 1212 (1975) and references therein.

(*25*) G. A. Gray, *Anal. Chem.*, **47**, 546 (1975).

(*26*) F. W. Wehrli and T. Wirthlin, "Interpretation of Carbon-13 NMR Spectra," Heydon and Sons, London, 1976.

[13] Determination of the Sugar Sequence and Glycosidic Bond Arrangement of Heteroglycans by an Integrated Analytical Scheme

By John H. Pazur and L. Scott Forsberg

Department of Biochemistry and Biophysics,
The Pennsylvania State University, University Park, Pennsylvania

Introduction

In studies on immunogenic streptococcal heteroglycans (*1, 2*), it was necessary to determine the sugar sequence of the glycans in order to allow for correlations of chemical structure with antibody specificity (*3*). An integrated analytical scheme (*4*) based on methylation analysis, periodate oxidation, enzymic hydrolysis, and chemical degradation has been developed for determining the complete sugar sequence and glycosidic bond arrangement of heteroglycans. The microtechniques of combined gas-liquid chromatography and mass spectrometry (*5*) have been employed for the identification and the characterization of the methyl derivatives from the native, the enzymicly modified, and the chemically degraded glycan. Application of the method is described for the determination of the sugar sequences and glycosidic bond arrangements of two type-specific cell wall carbohydrates, a diheteroglycan of D-glucose and D-galactose from *Streptococcus faecalis*, strain N (*6*) and a tetraheteroglycan of 6-deoxy-L-talose, L-rhamnose, D-galactose, and D-glucuronic acid from *Streptococcus bovis*, strain C3 (*2*).

Procedure

Heteroglycans

The diheteroglycan of D-glucose and D-galactose was extracted from the cell wall of *Streptococcus faecalis*, strain N, with 10% trichloroacetic acid and purified by alcohol fractionation as described in This Volume [29]. The tetraheteroglycan of 6-deoxy-L-talose, L-rhamnose, D-galactose, and D-glucuronic acid was extracted from the wall of *Streptococcus bovis*, strain C3 with 0.05 *M* potassium chloride–0.01 *M* hydrochloric acid of pH 2.0 by heating in a boiling water bath for short periods. Several antigenic glycans

METHODS IN CARBOHYDRATE
CHEMISTRY, VOL. VIII

are obtained in the initial extract of the cells. Isolation and purification of the tetraheteroglycan was achieved by adsorption on DEAE-cellulose and elution with a sodium chloride gradient in 0.001 M potassium phosphate buffer of pH 7.5 as described in This Volume [29].

Methylation Analysis

Methylation of the antigenic glycans was achieved by the Hakomori method (7) with the modifications described in recent publications (1, 2). In a typical methylation, a sample of 5 mg of the glycan is dissolved in 1 ml of dry dimethyl sulfoxide (Vol. VI [64], Vol. VII [26], This Volume [6]); 1 ml of dimethylsulfinyl carbanion (Vol. VI [64]) is added, and the reaction mixture is maintained for 6 h at 20°–25°. At this point, 0.5 ml of dry redistilled methyl iodide is added slowly, and the reaction is allowed to proceed at 20°–25° for 2 h. Excess dimethylsulfinyl carbanion is then neutralized by addition of water. The methylated glycan is purified and recovered by filtration through Sephadex LH20 in a 2:1 v/v chloroform–acetone solution. A preliminary hydrolysis of the methylated glycan is effected in 90% formic acid for 60 min; and after removal of the formic acid by evaporation, the hydrolysis is completed in 0.1 M sulfuric acid (15 h at 100°). The reaction mixture is neutralized with barium hydroxide; the precipitate is removed by filtration, and the supernatant is evaporated to dryness under reduced pressure. The residue is dissolved in 2 ml of water, and the solution is stirred with sodium borohydride (10 mg) at 20°–25° for 12 h. Borate ion is removed by evaporation several times following dissolution in methanol, and the dried residue is acetylated with 1 ml of 1:1 v/v acetic anhydride–dry pyridine (Vol. II [53]). Excess reagents are removed by evaporation several times after dissolution in ethyl ether, and the acetylated products are dissolved in chloroform at concentrations suitable for gas-liquid chromatography (glc).

The glc analysis may be performed in a metal column (1/8″ × 6′) packed with 3% OV 225 on 80/100 Supelcoport (Supelco, Belletonte, PA, 16823). (See also Vol. V [5].) The resolution of the di-O-methylrhamnose derivatives was superior on a column of OS 138 on 100/120 Chromosorb WHP (Supelco). The column temperature for glc analysis was 170°. Quantitative values for the methylated products can be obtained by use of a integrator. A mass marker was used to obtain m/e values for the various fragments. The glc patterns for the analysis of the native diheteroglycan and tetraheteroglycan are shown in Figure 1.

Table I contains the data on retention times relative to that for 1,5-di-O-acetyl-2,3,4,6-tetra-O-methyl-D-glucitol and the m/e fragments characteristic of the various derivatives. The values represent relative abundance with the most abundant fragment being assigned a value of 100.

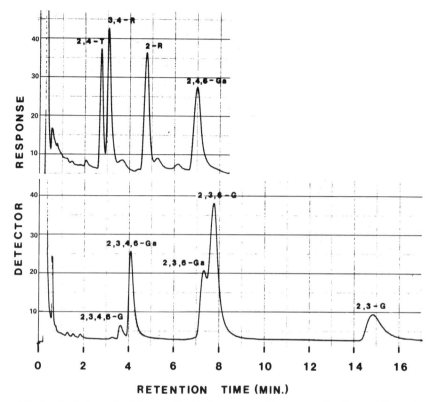

RETENTION TIME (MIN.)

FIG. 1.—A photograph of the gas–liquid chromatographic patterns for the partially methylated alditol acetates from the native tetraheteroglycan (top frame) and the native diheteroglycan (bottom frame). The abbreviations are recorded in Table I (4).

Enzymic Hydrolysis

A sample of 60 mg of the diheteroglycan was dissolved in 1.5 ml of water and mixed with 0.75 ml of 4% almond emulsion (Nutritional Biochemical Co., Cleveland, OH) in 0.1 M phosphate buffer of pH 6.8. The digest was incubated at 20°–25° for 44 h and analyzed periodically for the liberation of reducing sugars by paper chromatography. The qualitative paper chromatograms of the digest revealed that D-galactose was liberted from the glycan by the enzyme treatment. At this point, the enzyme was inactivated and precipitated with an equal volume of 10% trichloroacetic acid. The precipitate was removed by centrifugation; the supernatant was dialyzed for 48 h against distilled water and lyophilized to dryness to recover the enzyme-modified glycan. Qualitative precipitin tests showed that the enzyme modified glycan still yielded a positive precipitin test with homologous antisera.

TABLE I

The Characteristic Fragments of the Mass Spectra and the Retention Times on OV-225 at 170° Relative to 1,5-Di-O-acetyl-2,3,4,6-tetra-O-methyl-D-glucitol of Some Partially Methylated Alditol Acetates

Compound	Abbreviation	Retention time	m/e								
			45	117	131	161	175	189	205	233	261
1,5-Di-O-acetyl-6-deoxy-2,3,4-tri-O-methyl-L-talitol	2,3,4-T	0.40	—	100[a]	40	10	10	—	—	—	—
1,3,5-Tri-O-acetyl-6-deoxy-2,4-di-O-methyl-L-talitol	2,4-T	0.74	—	100	60	10	—	—	—	20	—
1,2,5-Tri-O-acetyl-3,4-di-O-methyl-L-rhamnitol	3,4-R	0.84	—	—	100	—	—	20	—	—	—
1,3,5-Tri-O-acetyl-2,4-di-O-methyl-L-rhamnitol	2,4-R	0.89	—	100	70	—	—	—	—	20	—
1,5-Di-O-acetyl-2,3,4,6-tetra-O-methyl-D-glucitol	2,3,4,6-G	1.00	90	100	—	70	—	—	30	—	—
1,5-Di-O-acetyl-2,3,4,6-tetra-O-methyl-D-galactitol	2,3,4,6-Ga	1.18	80	100	—	60	—	—	30	—	—
1,3,4,5-Tetra-O-acetyl-2-O-methyl-L-rhamnitol	2-R	1.30	—	100	—	—	—	—	—	—	—
1,3,5-Tri-O-acetyl-2,4,6-tri-O-methyl-D-galactitol	2,4,6-Ga	1.95	50	100	—	40	—	—	—	10	—
1,4,5-Tri-O-acetyl-2,3,6-tri-O-methyl-D-galactitol	2,3,6-Ga	2.15	50	100	—	—	—	—	—	30	—
1,4,5-Tri-O-acetyl-2,3,6-tri-O-methyl-D-glucitol	2,3,6-G	2.29	40	100	—	—	—	—	—	20	—
1,4,5,6-Tetra-O-acetyl-2,3-di-O-methyl-D-glucitol	2,3-G	4.42	—	100	—	—	—	—	—	—	10

[a] The figures represent relative abundance with the most abundant fragment assigned an arbitrary value of 100.

TABLE II

Moles of Methyl Monosaccharides per Mole of Glycan in Acid
Hdrolyzates of the Native and the Enzymicly Modified Diheteroglycan

Compound	Native	Modified	Difference
2,3,4,6-Tetra-O-methyl-D-glucose	4.0	15.9	+ 11.9
2,3,4,6-Tetra-O-methyl-D-galactose	17.5	5.0	− 12.5
2,3,6-Tri-O-methyl-D-galactose	17.6	17.8	+ 0.2
2,3,6-Tri-O-methyl-D-glucose	35.0	21.9	− 13.1
2,3-Di-O-methyl-D-glucose	17.1	16.8	− 0.3

The enzyme treatment was repeated two more times as described above.
The yield of enzyme-modified glycan after the third treatment was 30 mg.
This preparation gave only a small amount of precipitin complex with the
homologous antiserum. A sample of 2 mg of enzyme modified glycan was
subjected to methylation analysis, and the products were identified by glc
and mass spectrometry. The quantitative values for the products were obtained by integration of the areas on the glc chart corresponding to the
various components. The values for the methylated sugars from the native
and enzymicly modified glycan are recorded in Table II.

Periodate Oxidation

A sample of 40 mg of the diheteroglycan was dissolved in 50 ml of 0.02 M
sodium periodate of pH 4.5 and maintained in the dark at 4° for 18 h. At
this point, excess periodate was decomposed by addition of ethylene glycol;
low-molecular-weight materials were removed from the reaction mixture by
dialysis against distilled water for 48 h, and the oxidized glycan was recovered by lyophilization. A sample of 20 mg of the glycan was reduced
with 5 mg of sodium borohydride at 20°–25° for 24 h. The reduced product
was purified by dialysis and lyophilization. A 10-mg sample of the oxidized
and reduced glycan was dissolved in 1 ml of 0.02 N hydrochloric acid and
heated in a boiling water bath for 20 min. A second sample of 2 mg of the
modified glycan was dissolved in 0.1 ml of 0.1 N hydrochloric acid and heated
for 3 h in a boiling water bath. Analysis of the hydrolyzate for carbohydrates
was performed by paper chromatography, two ascents in the solvent system
of 6:4:3 v/v 1-butanol–pyridine–water. In the hydrolyzate of the native
glycan, the following products (R_f values) were present: glucose (0.55),
galactose (0.50), disaccharides (0.36), trisaccharides (0.25), tetrasaccharides
(0.19), and pentasaccharides (0.11). In the hydrolyzate of the oxidized and
reduced glycan the following products were present: glycerol (0.76), threitol

(0.67), glucose (0.55), disaccharides (0.35), and trisaccharides (0.25). Galactose was not present in the latter hydrolyzate.

The chromatogram of the 0.02 N acid hydrolyzate of the oxidized and reduced glycan showed only trace amounts of D-glucose and D-galactose but large amounts of a hydrolytic fragment which moved on paper in the region of trisaccharides. The trisaccharide was isolated by preparative paper chromatography (6, Vol. I [6]) and subjected to methylation analysis as described in the preceding section. The methyl alditol acetates obtained from this compound were 1,5-di-O-acetyl-2,3,4,6-tetra-O-methylglucitol and 1,4,5,6-tetra-O-acetyl-2,3,di-O-methylglucitol in a 2:1 molar ratio.

A sample of 40 mg of the tetraheteroglycan was subjected to periodate oxidation in 20 ml of 0.1 M sodium periodate of pH 4.5 for 48 h at 4° in the dark. The oxidized glycan was recovered by dialysis and lyophilization and subsequently reduced with 50 mg of sodium borohydride. Because analysis of the product by the carbazole method (8, This Vol. [11]) revealed that some of the D-glucuronic acid was still unoxidized, the modified glycan was subjected to a second oxidation by the above procedure. Analysis of the resulting product by the carbazole method (8, This Vol. [11]) and the cysteine–sulfuric acid method (9, Vol. I [138]) showed that over 95% of the D-glucuronic acid and 35% of the 6-deoxyhexose had been destroyed by the oxidation. A sample of 10 mg of the oxidized and reduced glycan was dissolved in 1 ml of 0.02 N hydrochloric acid and heated at room temperature for 15 min. Qualitative paper chromatograms showed that, under the above conditions, the acetal linkages were cleaved, yielding oligosaccharides, but the glycosidic linkages were not hydrolyzed. Two oligosaccharide fragments were detected in the hydrolyzate of the oxidized and reduced glycan, and these were isolated by preparative paper chromatography in the above solvent system. The individual oligosaccharides were subjected to methylation analysis. The acetyl and methyl products identified in the methylation analysis were 1,5-di-O-acetyl-6-deoxy-2,3,4-tri-O-methyl talitol, 1,3,5-tri-O-acetyl-2,4,di-O-methyl-rhamnitol, and 1,3,5-tri-O-acetyl-2,4,6-tri-O-methylgalactitol in equal molar ratios from both oligosaccharides. The oligosaccharide with the higher R_f value is probably a glycoside of the other compound with the R group being a three carbon oxidation and reduction fragment.

Acetolysis

A sample of 10 mg of the diheteroglycan was dissolved in 1 ml of acetolysis mixture (10 parts of acetic anhydride, 10 parts of glacial acetic acid, and 1 part of conc. sulfuric acid) and heated in a stoppered reaction vessel at 40° for 3 h. The deacetylated products from the acetolysis mixture were obtained following a published procedure (10). The deacetylated products were separated and identified by paper chromatography. Small amounts of D-glucose and D-galactose were detectable on the chromatogram, but the major reducing products were a disaccharide which moved on paper at the same R_f value as lactose and a fragment which did not move from the origin. The latter products were isolated by preparative paper chromatography (6, Vol. I [6]) and subjected to methylation analysis. The products identified from the oligosaccharide were 1,5-di-O-acetyl-2,3,4,6-tetra-O-methyl-D-galactitol and 1,4,5-tri-O-acetyl-2,3,6-tri-O-methyl-D-glucitol in a 1:1 molar ratio. The deacetylated product in the acetolysis mixture isolated from the origin yielded 1,4,5-tri-O-acetyl-2,3,6-tri-O-methyl-D-galactitol and 1,4,5-tri-O-acetyl-2,3,6-tri-O-methyl-D-glucitol. The above compounds were present in a ratio of 1:1 and accounted for approximately 80% of the methylated products. The remaining 20% was composed of 1,5-di-O-acetyl-2,3,4,6-tetra-O-methyl-D-glucitol,1,4,5,6-tetra-O-acetyl-2,3-di-O-methyl-D-glucitol, and of 1,5-di-O-acetyl-2,3,4,6-tetra-O-methyl-D-galactitol. These products probably arise from an acetolysis fragment produced by a random cleavage of glycosidic linkages other than the α-D-(1 → 6) linkages.

β-Elimination

A sample of 12 mg of the tetraheteroglycan was subjected to β-elimination reactions (11, This Volume [7]). The native glycan was first methylated by the modified Hakomori procedure described above. Following purification of the methylated product on a Sephadex LH20, the eluate containing the methylated glycan was transferred to a small flask and evaporated to dryness under nitrogen in a water bath at 35°. Subsequently, the flask was fitted with a serum cap and flushed with nitrogen. The following reagents were then added by syringe through the serum cap: (A) 2 ml of a solution of dry dimethyl sulfoxide and 2,2-dimethoxypropane (19:1 by volume) containing a trace amount of p-toluenesulfonic acid, followed by (B) 0.5 ml of freshly prepared methylsulfinyl carbanion. The mixture was subjected to sonication for 30 min and maintained at 20°–25° overnight. The mixture was then placed in an ice bath, and the reaction was stopped by addition of excess 50% acetic acid. Purification of the glycan was effected by pouring the entire reaction mixture into water and extracting the methylated glycan with three 10-ml portions of chloroform. The combined chloroform extracts were washed three times with water and evaporated to dryness. The resulting residue was hydrolyzed in 2 ml of 10% acetic acid in a sealed ampule under nitrogen for 1 h at 100°. The reaction mixture was taken to dryness by lyophilization, and the methylated glycan was purified by gel filtration on Sephadex LH20. The product was dried and remethylated by the modified Hakomori method. The remethylated glycan was hydrolyzed, and the products were converted to their alditol acetates as described in an earlier section. The products identified in the mixture were 1,3,5-tri-O-acetyl-6-deoxy-2,4-di-O-methyl talitol, 1,2,5-tri-O-acetyl-3,4,di-O-methylrhamnitol, 1,3,5-tri-O-acetyl-2,4-di-O-methylrhamnitol, and 1,3,5-tri-O-acetyl-2,4,6-tri-O-methylgalactitol.

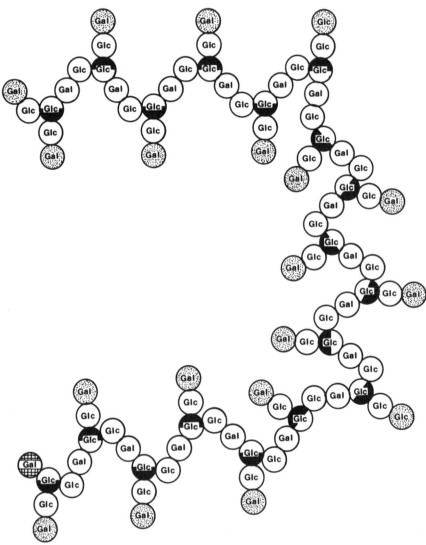

Fig. 2.—Diagrammatic representation of the structure of a typical molecule of the diheteroglycan. Dotted circles indicate unsubstituted residues; open circles indicate residues substituted at C-4; open and dark half circles indicate residues substituted at C-4 (open half) and C-6 (dark half), and the hatched circle indicates the residue with the reducing group and substituted at C-4 (4).

Conclusion

Data on the structure of the diheteroglycan from *Streptococcus faecalis* and the tetraheteroglycan from *Streptococcus bovis* have been obtained by a combined analytical scheme based on methylation analysis, periodate oxidation, enzymic hydrolysis, and chemical degradation. The molecular structures consistent with the data obtained to date for typical molecules of these glycans are shown diagrammatically in Figures 2 and 3. The diheteroglycan consists of 91 monosaccharide residues with the calculated molecular weight of 14,760 in agreement with the experimentally determined value of 15,000 (*1*), while the tetraheteroglycan consists of 40 monosaccharide residues with the calculated molecular weight of 6,226 in agreement with the experimentally determined value of 6,000 (*2*). The combined analytical scheme should be of general applicability for the structural analysis of heteroglycans for which selective degradation procedures can be devised.

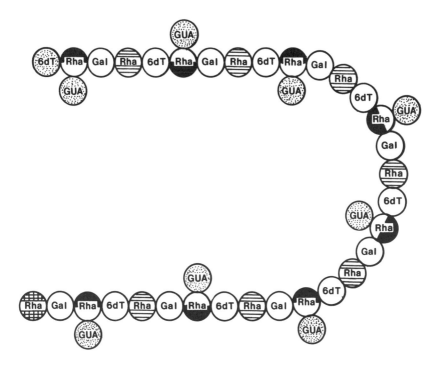

FIG. 3.—Diagrammatic representation of the structure of a typical molecule of the tetraheteroglycan. Dotted circles indicate unsubstituted residues; open circles indicate residues substituted at C-3; open and dark half circles indicate residues substituted at C-3 (dark half) and C-4 (open half); lined circles indicate residues substituted at C-2, and the hatched circle indicates the residue with the reducing group and substituted at C-2.

References

(*1*) J. H. Pazur, A. Cepure, J. A. Kane, and C. G. Hellerquist, *J. Biol. Chem.*, **248,** 279 (1973).
(*2*) J. H. Pazur, D. J. Dropkin, K. L. Dreher, L. S. Forsberg, and C. S. Lowman, *Arch. Biochem. Biophys.*, **176,** 257 (1976).
(*3*) J. H. Pazur, K. B. Miller, K. L. Dreher, and L. S. Forsberg, *Biochem. Biophys. Res. Commun.*, **70,** 545 (1976).
(*4*) J. H. Pazur and L. S. Forsberg, *Carbohydr. Res.*, **60,** 167 (1978).
(*5*) B. Lindberg, *Methods Enzymol.*, **28,** 178 (1972).
(*6*) J. H. Pazur, J. S. Anderson, and W. W. Karakawa, *J. Biol. Chem.*, **246,** 1793 (1971).
(*7*) S.-I. Hakomori, *J. Biochem.* (Tokyo), **55,** 205 (1964).
(*8*) Z. Dische, *J. Biol. Chem.*, **167,** 189 (1947).
(*9*) Z. Dische and L. B. Shettles, *J. Biol. Chem.*, **175,** 595 (1948).
(*10*) T. S. Stewart, P. B. Menderhausen, and C. E. Ballou, *Biochemistry*, **7,** 1843 (1968).
(*11*) L. Kenne, B. Lindberg, and S. Svensson, *Carbohydr. Res.*, **40,** 69 (1975).

[14] Oxidation of Acetylated Carbohydrates with Chromium Trioxide in Acetic Acid

By James Hoffman and Bengt Lindberg

Department of Organic Chemistry, Arrhenius Laboratory,
University of Stockholm, Stockholm, Sweden

Introduction

Angyal and James (*1*) have shown that fully acetylated aldopyranosides with equatorially attached aglycons in the most stable chair form (generally the β-anomers, for example, I) are oxidized by chromium trioxide in acetic acid to 5-aldulosonates (for example, II). The anomers with axially attached aglycons (generally the α-anomers, for example, III) are oxidized only slowly under the same conditions. Acetylated furanosides are oxidized irrespective of their anomeric configurations.

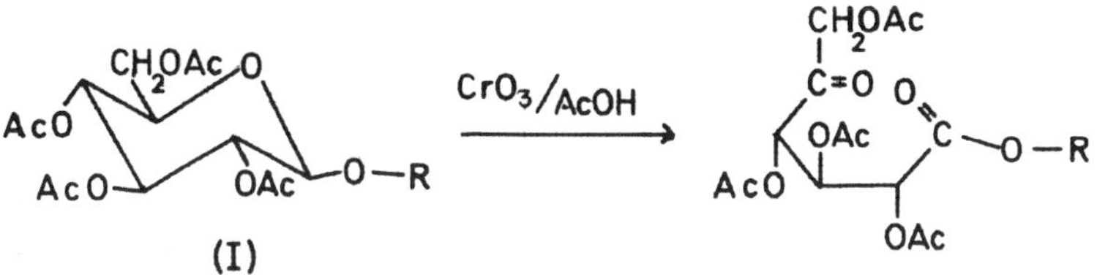

(I)   Cro₃/AcOH →   (II)

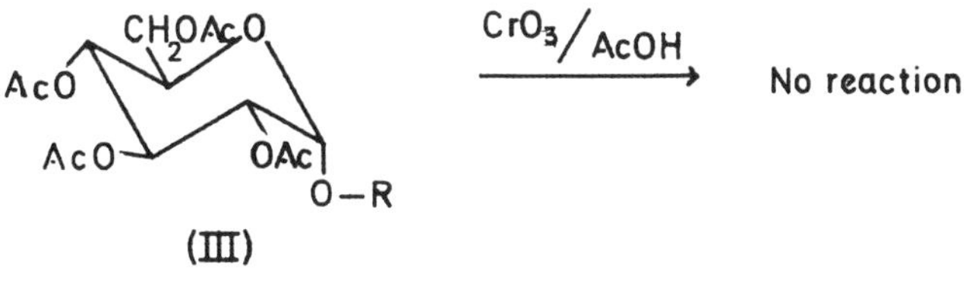

(III)   Cro₃/AcOH →   No reaction

The chromium trioxide oxidation has been used to determine the anomeric natures of sugar residues in oligo- and polysaccharides (*2*). The materials are fully acetylated, oligosaccharides having first been reduced to their

117

alditols, and the products are then treated with chromium trioxide in acetic acid in the presence of an internal standard (for example, *myo*-inositol hexaacetate). A comparison of sugar analyses, performed before and after oxidation, reveals which sugar residues have been oxidized.

A prerequisite for obtaining reliable results is, of course, that the carbohydrate does not contain free hydroxyl groups but is fully acetylated. It is also important that the conformational stability of the chain forms having axially attached aglycons is large enough to ensure that the proportion of the alternate form is negligible. This supposition is valid for the α-pyranosides of the common hexoses, 6-deoxyhexoses, 2-acetamido-2-deoxyhexoses, and xylose. It does not seem impossible, however, that substitution may increase the percentage of the alternate forms of α-fucopyranosides and α-rhamnopyranosides in complex carbohydrates, thus rendering them susceptible to oxidation.

The first example given below illustrates the application of this technique to bacterial lipopolysaccharides from *Salmonella typhi* and *Salmonella strasbourg* (*3*). In both polymers, the O-specific side chains are composed of oligosaccharide repeating units containing 3,6-dideoxy-D-*arabino*-hexose (tyvelose), L-rhamnose, D-mannose, D-galactose, and D-glucose residues. The anomeric configurations of the D-mannopyranosyl residues in both polymers were unknown. Oxidation of the fully acetylated polymers and sugar analyses (Table I) revealed that the D-mannopyranosyl residues in the *S. typhi* lipopolysaccharide are α-linked and that those in the *S. strasbourg* lipopolysaccharide are β-linked. The results have been confirmed by enzymic analysis (*4*).

The 3,6-dideoxy-D-*arabino*-hexoses (not included in Table I) in the lipopolysaccharides, which are pyranosidic and α-linked, are also oxidized to a considerable extent. This is not unexpected as there is likely to be only a small difference in energy between the two chain forms of this sugar (IVA and IVB).

TABLE I

Sugar Compositions of Original and Oxidized Polysaccharides

Strain	Oxidation time (hour)	Relative proportions			
		D-*Rha*	D-*Man*	D-*Gal*	D-*Glu*
S. typhi	0	20	21	23	17
S. typhi	1	19	21	20	16
S. strasbourg	0	18	19	27	9
S. strasbourg	1	18	3	18	3

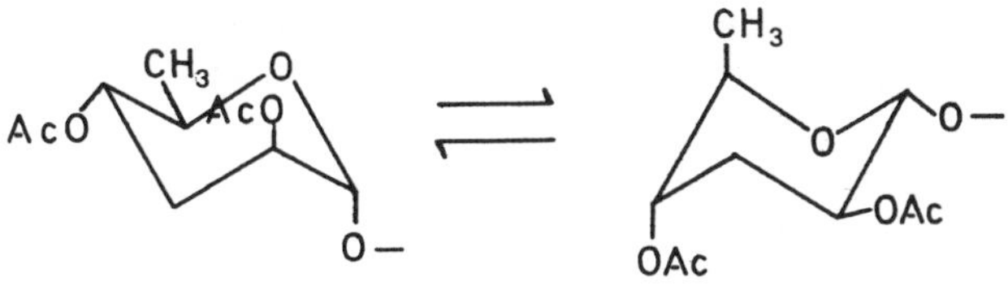

(IVa) (IVb)

The method may also, in favorable circumstances, be used for sequence analysis. Thus, in the hexasaccharide (V) repeating unit (5, 6) of the *Pneumococcus* type 2 capsular polysaccharide, all sugar residues, apart from one L-rhamnopyranosyl residue, are α-linked. On treatment of the carboxyl

$$\rightarrow 3)\text{-}\alpha\text{-}L\text{-Rha}p\text{-}(1 \rightarrow 3)\text{-}\alpha\text{-}L\text{-Rha}p\text{-}(1 \rightarrow 3)\text{-}\beta\text{-}L\text{-Rha}p\text{-}(1 \rightarrow 4)\text{-}\alpha\text{-}D\text{-Glc}p\text{-}(1 \rightarrow$$

2

↑

1

α-D-Glcp

6

↑

1

α-D-GlcAp

(V)

reduced (performed using deuterated reagent) and acetylated polysaccharide (partial structure VIB) with chromium trioxide in acetic acid, this residue was oxidized to a 5-hexulosonate residue (as in partial structure VII). On methylation, using sodium methylsulfinylmethanide and methyl iodide (Vol. VI [64]), all ester linkages were cleaved and the resulting hydroxyl groups methylated. Partial structure VII, therefore, yielded a fully methylated

TABLE II

Partially Methylated Sugars from the Hydrolyzate of Carboxyl-Reduced, Methylated S 2 (A) and of Carboxyl-Reduced, Acetylated, Oxidized, and Methylated S 2 (B)

Sugars	A	B
2,3,4,6-Tetra-*O*-Me-D-Glc[a]	—	0.8
2,3,4,6-Tetra-*O*-Me-D-Glc[6-^{2}H$_2$][a]	1.0	1.0
2,3,4-Tri-*O*-Me-D-Glc	1.0	1.0
2,3,6-Tri-*O*-Me-D-Glc	1.0	0.2
2,4-Di-*O*-Me-L-Rha	2.0	1.1
4-Mono-*O*-Me-L-Rha	1.0	1.0

[a] The amounts of these sugars were determined by comparing the relative intensities of appropriate ions in the ms of the mixture.

α-D-glucopyranosyl group VIII. Methylation analysis, before and after oxidation (Table II), also showed that a 3-linked L-rhamnopyranosyl residue (oxidized) and a 4-linked D-glucopyranosyl residue disappeared after oxidation and that a terminal D-glucopyranosyl group had been formed. The sequence VIA in the polysaccharide was thereby established.

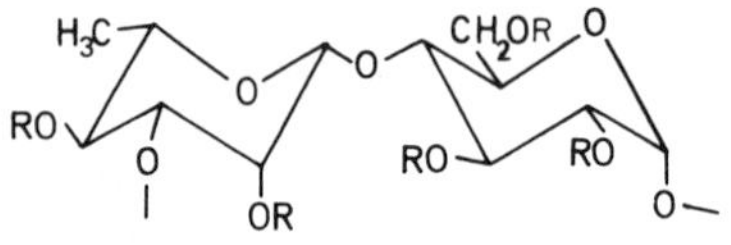

(VII)

(VIa, R=H)
(VIb, R=Ac)

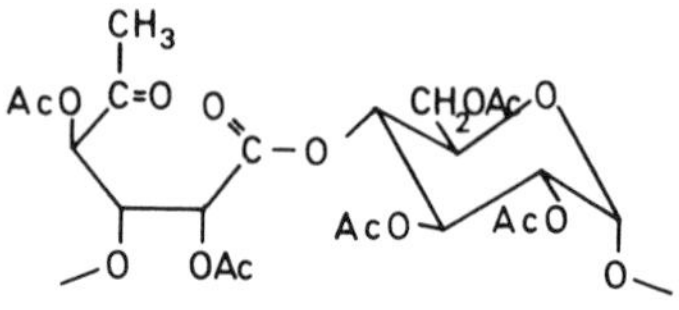

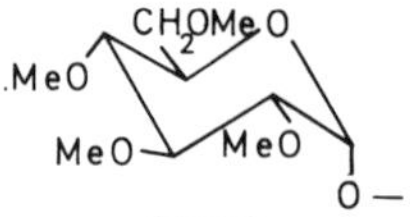

(VIII)

Oxidation with chromium trioxide in acetic acid has also been used in structural studies of oligosaccharides (7) and glycolipids (8).

Procedure

Acetylation and Oxidation of the *S. typhimurium* and *S. strasbourg* O-Specific Polysaccharides

The lipid-free polysaccharide (50 mg), prepared from the lipopolysaccharide by mild acid hydrolysis (Vol. IX), is dissolved in 10 ml of formamide; and 5 ml of pyridine and 5 ml of acetic anhydride are added. The mixture is kept at room temperature for 20 h and then applied to the top of a 60 × 4-cm column of Sephadex LH 20 which is irrigated with acetone. The acetylated polysaccharide, detected by its optical rotation, is eluted with the void volume free from the acetylating reagents; yield 60 mg.

The acetylated polysaccharide (10 mg), together with 5 mg of *myo*-inositol hexaacetate as an internal standard, is dissolved in 1 ml of chloroform in a small test tube, and 0.1 ml of this solution is withdrawn for sugar analysis. The main part of the solution is evaporated; the residue is dissolved in 0.27 ml of glacial acetic acid, and 27 mg of powdered chromium trioxide is added. The test tube is closed, and the suspension is agitated in an ultrasonic bath at 50° for 1 h, after which time 2 ml of water is added and the material is recovered by extraction with five 1-ml portions of chloroform. The oxidation product is evaporated twice with 1-ml portions ethanol to remove remaining acetic acid and treated with 0.5 M sulfuric acid at 100° overnight to effect hydrolysis. The sugars are analyzed as their alditol acetates by glc using an OV-225 column at 200° (Vol. VI [4]). The results for the two polysaccharides are given in Table I.

The procedures for oligosaccharide alditols and glycosides are analogous, except that the acetylations are performed in the usual manner, using 1:1 v/v acetic anhydride–pyridine at 100°.

Oxidation and Methylation Analysis of the Carboxyl-reduced *Pneumococcus* Type 2 Capsular Polysaccharide (5)

Carboxyl-reduced polysaccharide (10 mg), prepared using LiAlD$_4$, is acetylated and purified as described above. Powdered chromium trioxide (30 mg) is added to a solution of 10 mg of the acetylated polysaccharide in 0.3 ml of glacial acetic acid, and the oxidation and product isolation are performed as described above.

The oxidized material (10 mg), in a 5-ml serum bottle sealed with a rubber cap, is dissolved in 1 ml of dry dimethyl sulfoxide (Vol. VI [64]). The bottle is flushed with nitrogen, and 2 M sodium methylsulfinylmethanide (methylsulfinyl anion) in 1 ml of dimethyl sulfoxide is added using a syringe (Vol. VI [64]). The gelatinous solution is agitated in an ultrasonic bath for 30 min and kept at 20°–25° overnight. Methyl iodide (1 ml) is added dropwise with external cooling, and the resulting turbid solution is agitated for 30 min in the ultrasonic bath. The methylated material is isolated from the reaction mixture by partition between water and chloroform. A solution of the methylated polysaccharide in 2 ml of 90% formic acid is kept for 2 h at 100° and concentrated to dryness, and the residue dissolved in 0.5 M sulfuric acid and kept overnight at 100°. The hydrolyzate is neutralized with barium carbonate and the sugars are converted into alditol acetates and analyzed by glc-ms (*10*, Vol. VI [4], [94]).

The result of this analysis and of the analysis of carboxyl-reduced methlated S 2 are given in Table II.

References

(*1*)   S. J. Angyal and K. James, *Aust. J. Chem.*, **23**, 1209 (1970).

(*2*)   J. Hoffman, B. Lindberg, and S. Svensson, *Acta Chem. Scand.*, **26**, 661 (1972).

(*3*)   C. G. Hellerqvist, J. Hoffman, B. Lindberg, Å. Pilotti, and A. A. Lindberg, *Acta Chem. Scand.*, **25**, 1512 (1971).

(*4*)   M. Fukuda, F. Egami, G. Hammerling, O. Lüderitz, G. Bagdian, and A.-M. Staub, *Eur. J. Biochem.*, **20**, 438 (1971).

(*5*)   O. Larm, B. Lindberg, and S. Svensson, *Carbohydr. Res.*, **31**, 120 (1973).

(*6*)   L. Kenne, B. Lindberg, and S. Svensson, *Carbohydr. Res.*, **40**, 69 (1975).

(*7*)   S.-I. Hakomori, *J. Biochem.* (*Tokyo*), **55**, 205 (1964).

(*8*)   N. E. Nordén, A. Lundblad, S. Svensson, P.-A. Öckerman, and S. Autio, *J. Biol. Chem.*, **248**, 6210 (1973).

(*9*)   R. A. Laine and O. Renkonen, *J. Lipid Res.*, **16**, 102 (1975).

(*10*) H. Björndal, C. G. Hellerqvist, B. Lindberg, and S. Svensson, *Angew. Chem., Int. Ed. Engl.*, **9**, 610 (1970).

[15] Specific Cleavage of Polysaccharides Containing Uronic Acid Residues by the Hoffman–Veerman Reaction

By N. K. Kochetkov, O. S. Chizhov, and A. F. Sviridov

N. D. Zelinsky Institute of Organic Chemistry
of Academy of Science of U.S.S.R., Moscow

Introduction

Specific cleavage of polysaccharides is becoming a method of choice for their structural examination. Several excellent reviews have appeared recently (*1–3*). The deoxyaminosugar or uronic acid residues are usually the specific points of cleavage of polysaccharide chains.

Polysaccharides can be classified into two groups according to the positions of uronic acid residues in the chain. The polysaccharides that have uronic acid moieties in side chains belong to the first group; glucuronoxylan is an example of this group (*4*). The second group includes polysaccharides with uronic acid residues in the main chain; units of uronic acids can either alternate with other monosaccharide residues, as in polysaccharides of *Lipomyces* (*5*) or form long sequences as in pectic and alginic acids (*6*).

Application of the Hoffman-Veerman procedure to these two groups of polysaccharides to effect cleavage at the site of uronic acid residues leads to different results. Specific removal of side chains and formation of a modified polysaccharide is achieved with the first group, whereas selective degradation of polysaccharides into oligosaccharides occurs with the second group. Preliminary partial reduction of carboxyl groups is necessary with polysaccharides of the pectic acid type to escape total destruction (*7*).

The following sequence of reactions is used to attain selective cleavage: (a) all the uronic acid units are converted into amides by any suitable method; (b) the resulting amides are transformed by the Hoffman–Veerman reaction into 5-aminopentopyranoses; (c) the latter compounds are hydrolyzed with dilute acids.

123

In this way, selective cleavage of glycuronic linkages is achieved, whereas, all other types of glycosidic bonds survive. The method has been successfully used both for the polysaccharides of glucuronoxylan type (*8, 9*) and for those containing uronic acid units in the main chain (*10, 11*).

The Lossen reaction can be applied to the selective splitting of uronic acid residues in polysaccharide chains, but this method has not been used until recently. A modification of this reaction including treatment of the polysaccharides with hydroxylamine and a water-soluble carbodiimide is suitable for these purposes. Successful employment of this procedure has been used for the removal of the D-glyceric acid unit at the terminal position of the polysaccharide produced by *Mycobacterium phlei* (*12, 13*).

An approach, including β-elimination of a substituent from C-4 of the uronic acid residue, simulates the above discussed procedures, but can be applied only to O-substituted derivatives of polysaccharides (*1, 14*, Vol. VI [24]).

Procedure

Specific Cleavage of White Birch Xylan (*8*)

To 0.5 g of the amide of white birch xylan (*4*) in 15 ml of water, 15 ml of a 0.5 N solution of sodium hypochlorite (pH 13.5) is added with stirring at 0° under nitrogen. After 1 h stirring, the mixture is heated on a boiling water bath for 5 min, cooled, acidified with acetic acid to pH 5, and held at 4° for 24 h. The reaction mixture is dialyzed against tap water and then poured into 11 ml of 1:4 v/v methanol–butanol; polysaccharide precipitating during 24 h is collected by filtration, washed with absolute ethanol, and dried in a vacuum desiccator; yield, 0.43 g (86%); $[\alpha]^{18}_D$ −114° (c 1.4, 2 N KOH). The relative decrease in uronic acid content is 83% as determined by periodate oxidation of both the original and modified polysaccharide. Repetition of the amidation and the hypochlorite treatment decreases the relative uronic acid content to 94%.

Specific Cleavage of the Amide of Extracellular Glucuronomannan from *Lipomyces lipofer* st. 133 (*10*)

Polysaccharide permethylated by the Hakomori procedure (*15*) (0.5 g) is dried over phosphorus pentaoxide under reduced pressure for 10 h at 60° is dissolved in 100 ml of neat methanol, saturated by anhydrous ammonia at 0°, and kept 24 h at 20°. After concentration under reduced pressure, the residue is dissolved in 5 ml of methanol; 50 ml of 2 N sodium hypochlorite is added, and the mixture is stirred for 4 h. Then 0.2 g of sodium borohydride is added to destroy the excess of hypochlorite; the solution is acidified with acetic acid to pH 5, and the mixture is stored overnight. It is then extracted with chloroform, and the chloroform solution is dried over sodium sulfate. After evaporation under reduced pressure, the residue is chromatographed on silica gel; yield of disaccharide I, 0.16 g, characterized as its deutero (OCD_3) derivative by mass spectrometry.

(I)

The same disaccharide can be isolated by degrading the permethylated amide of the extracellular polysaccharide from *Lipomyces tetrasporus* st. 70 (*11*).

References

(*1*) B. Lindberg, J. Lonngren, and S. Svensson, *Advan. Carbohydr. Chem. Biochem.*, **31**, 185 (1975).
(*2*) J. M. Williams, *Advan. Carbohydr. Chem. Biochem.*, **31**, 9 (1975).
(*3*) A. F. Sviridov and O. S. Chizhov, *Bioorganicheskaja chimija* (*USSR*), **2**, 315 (1976).
(*4*) T. E. Timell, *Advan. Carbohydr. Chem.*, **19**, 282 (1964).
(*5*) A. F. Sviridov, O. D. Gikia, S. E. Gorin, O. S. Chizhov, I. P. Bab'eva, and N. K. Kochetkov, *Bioorganicheskaja chimija* (*USSR*), **3**, 232 (1977).
(*6*) A. Haug, B. Larsen, and O. Smidsrod, *Carbohydr. Res.*, **32**, 214 (1974).
(*7*) R. L. Whistler and G. A. Towle, *Arch. Biochem. Biophys.*, **138**, 39 (1970).
(*8*) N. K. Kochetkov, O. S. Chizhov, and A. F. Sviridov, *Izvest. Akad. Nauk SSSR, Ser. Chim.*, 2089 (1968).
(*9*) A. Main and E. Percival, *Carbohydr. Res.*, **26**, 147 (1973).
(*10*) N. K. Kochetkov, S. E. Gorin, A. F. Sviridov, O. S. Chizhov, V. I. Golubev, I. P. Bab'eva, and A. Ja. Podel'ko, *Izvest. Akad. Nauk SSSR, Ser. Chim.*, 2304 (1973).
(*11*) N. K. Kochetkov, O. S. Chizhov, A. F. Sviridov, S. E. Gorin, and I. P. Bab'eva, *Izvest. Akad. Nauk SSSR, Ser. Chim.*, 2774 (1975).
(*12*) M. H. Seier and C. E. Ballou, *J. Biol. Chem.*, **243**, 992 (1968).
(*13*) C. E. Ballou, *Acc. Chem. Res.*, **1**, 3666 (1968).
(*14*) B. Lindberg, J. Lönngren, and J. L. Thompson, *Carbohydr. Res.*, **28**, 351 (1973).
(*15*) S.-I. Hakomori, *J. Biochem.*, **55**, 205 (1964).

[16] Determination of Carboxymethyl Ether Groups In Sodium O-Carboxymethylcellulose (CMC)

By Donald F. Durso

Johnson and Johnson, New Brunswick, New Jersey

Introduction

The most efficient and precise method for carboxymethyl group determination is a modification of the original procedure of Francis (*1*). The method is simple, eliminating tedious filtration problems.

Procedure

Uranyl nitrate reagent is made by dissolving 40 g of reagent-grade uranyl nitrate hexahydrate in 800 ml of distilled water and diluting to 1 liter with neat methanol or 95% ethanol.

A weighted sample (0.250–0.500 g) of sodium O-carboxymethylcellulose (NaCMC) is transferred into a 600-ml beaker provided with a stirring bar and moistened with 10–20 ml of alcohol. Distilled water (95 ml) is added very slowly, and the mixture is stirred gently at 60° until no further dissolution occurs. Then 200 ml of water is added, and the stirring bar is removed with thorough rinsing. The beaker is stored in a refrigerator overnight to allow insoluble matter to settle. The mixture is then filtered through a Corning C-grade, 30-ml, glass, filtering crucible; and the crucible washed with 100 ml of distilled water. The filtrate is placed in a 600-ml beaker with a stirring bar and heated to 60° with gentle stirring. Uranyl nitrate reagent (25 ml) is added slowly by pipet, holding the tip under the surface of the stirred solution. The heat source is then removed while gentle stirring is continued for 5–10 min. The combination of temperature, rate of addition, and rate of stirring produces a granular precipitate of uranyl CMC; this granular form is vital to permit simple filtration and to prevent loss of semi-colloidal uranyl CMC. The mixture must stand for at least 2–4 h at this point for complete settling.

The supernatant liquid is next filtered through a tared glass crucible retaining the major precipitate in the beaker. This precipitate is rinsed with three 200-ml portions of distilled water, followed by two 100-ml portions of alcohol with all washings decanted through the filtering crucible. Lastly, the remaining precipitate of uranyl CMC is washed into the crucible with alcohol. Residual liquid is pulled through the filtering crucible by use of

127

reduced pressure. The crucible and contents are dried at 105° overnight, or to constant weight. The crucible is cooled in a desiccator and weighed. Record as "UCMC found."

The dry precipitate of uranyl CMC is transferred as quantitatively as possible to a tared porcelain crucible. This crucible, with its cover, is dried at 105° for 2 h, cooled, and weighed. The weight of the contents is designated "UCMC used." The crucible is placed in a muffle furnace at 300°–400°; the temperature is then raised to 750° and held for 20–30 min. This heating converts the crucible contents to uranium oxide, U_3O_8. After turning off the furnace and after the temperature has dropped to 300°–400°, the crucible is removed and placed in a desiccator. It is weighed after room temperature is attained. The crucible content weight is recorded as U_3O_8.

Calculations

$$\text{Uranyl fraction (UF)} = \frac{\text{wt. of } U_3O_8 \times 0.961}{\text{UCMC used}}$$

$$\text{Degree of substitution (DS)} = \frac{162 \times \text{UF}}{135 - 192(\text{UF})}$$

$$\% \text{ Active agent (NaCMC)} = \frac{\text{UCMC found} \times \dfrac{\text{NaCMC}}{\text{UO}_2(\text{CMC})_2} \times 100}{\text{wt. of sample}}$$

where $\text{NaCMC}/\text{UO}_2(\text{CMC})_2 = 162 + 80(\text{DS})/162 + 192(\text{DS})$. To calculate results to a dry basis, dry the sample in an oven for 2 h at 105° before weighing or calculate the sample dry weight from the moisture content.

Discussion

The sample weight and dilutions indicated are optimum to produce a light, easily filtered precipitate. When the uranyl reagent is added, turbidity develops prior to precipitation; if large particles or "fibers" appear at this point, further dilution is indicated. For extremely viscous samples, greater dilution is required; for material of very low viscosity, less dilution and a shorter filtering time can be used. At the time of filtering the UCMC, any turbidity present in the supernatant liquor indicates the need for additional stirring at 50°–70° (or perhaps at higher temperatures) to agglomerate the UCMC. For best results in filtering, the supernatant liquor should be clear, with no more than a few small particles in suspension.

If the final alcohol wash is omitted, the uranyl carboxymethylcellulose remains slightly sticky and is more difficult to transfer from the beaker.

On drying, such sticky material forms a hard cake which may be difficult to remove from the glass filtering crucible. These difficulties are eliminated by washing with the specified amount of alcohol. Most errors in the method result from improper washing of the precipitate.

By working deliberately over a 3-day period, excellent results are obtained.

Reference

(*1*)   C. V. Francis, *Anal. Chem.*, **25,** 941 (1953).

OTHER CHEMICAL, PHYSICAL, AND ENZYMIC METHODS

[17] Enzymic Determination of D-Galactose

BY R. J. STURGEON

Department of Brewing and Biological Sciences,
Heriot-Watt University, Edinburgh, Scotland

Introduction

The methods commonly used for the quantitative analysis of mixtures of neutral sugars frequently involve the resolution of the mixture of the individual sugars by techniques such as paper (Vol. I [6], [116]) or thin-layer (Vol. VI [6], [7]) chromatography, ion-exchange chromatography (Vol. VI [9]), or gas-liquid chromatography (Vol. VI [1], [2], [4]). Many non-specific colorimetric methods can be applied to mixtures of sugars without prior purification (Vol. I [131]–[140], This Vol. [11]). Enzymic methods, on the other hand, have a number of theoretical and practical advantages, and with the commercial availability of a wide range of highly purified enzymes, a variety of methods are currently used by carbohydrate chemists. For example, enzymic methods are used for the assay of D-glucose and D-galactose by D-glucose oxidase (1, 2, Vol. I [117]) and D-galactose oxidase (3, 4), respectively, in which determinations the product is measured using spectrophotometric, polarimetric or fluorimetric techniques (2).

A wider range of sugar analyses may now be made using spectrophotometric techniques, with the use of nicotinamide adenine dinucleotide (NAD)- and nicotinamide adenine dinucleotide phosphate (NADP)-dependent dehydrogenases (5). Assuming the molar extinction coefficient for NADH and NADPH (6) at 340 nm is 6.22×10^3, it is possible to calculate the extent of conversion of substrate to product at completion of the reaction. To illustrate these principles, specific methods for the analysis of three commonly occurring hexoses are given in this article and in This Volume [15], and methods for estimating the average chain length and degree of polymerization of polysaccharides are given in This Volume [8] and [9].

D-Galactose can be determined with either D-galactose oxidase (D-galactose: oxygen 6-oxidoreductase, E.C.1.1.3.9) from *Polyporus circinatus* (4, 7) or with D-galactose dehydrogenase (8) from *Pseudomonas saccharophila* or

131

Pseudomonas fluorescens. In the presence of oxygen, galactose oxidase oxidizes D-galactose at the C-6 position to form D-*galacto*-hexdialdo-1,5-pyranose and hydrogen peroxide, the latter being determined colorimetrically with peroxidase in the presence of hydrogen donors such as *o*-dianisidine and *o*-toluidine. This enzyme is of limited use in the quantitative determination of D-galactose because other galactose-containing compounds are oxidized, often at higher rates (*4, 9*). The enzyme, which was originally demonstrated to oxidize D-galactopyranoside residues in various polysaccharides (*10*), glycoproteins (*11*) and glycolipids (*12*), will also oxidize certain D-galactofuranose residues at the 6-6 position (*13*).

The D-galactose dehydrogenase (D-galactose:NAD 1-oxidoreductase, EC 1.1.1.48) from *Ps. fluorescens* has a broad pH optimum, and shows no reaction with galactosides, catalyzing the oxidation of β-D-galactose, α-L-arabinose and β-D-fucose, and to a lesser extent 2-amino-2-deoxy-D-galactose, only.

Procedure (6)

$$\beta\text{-D-Galactose} + \text{NAD}^+ \xrightarrow[\text{dehydrogenase}]{\text{galactose}} \text{D-galactono-1,5-lactone} + \text{NADH} + \text{H}^+$$

The assay is done at a relatively high pH, so that the galactonolactone is spontaneously hydrolyzed; the reaction thus becomes irreversible and proceeds to completion. The formation of NADH is measured by the change in absorbance at 340 nm.

The following reagents are required: (A) 0.1 M, pH 8.6 Tris buffer; (B) 0.017 M nicotinamide adenine dinucleotide (NAD); (C) D-galactose dehydrogenase (approximately 5 units/mg). Into a 1-cm silica cell 2.68 ml of reagent A, 0.1 ml of reagent B, and 0.2 ml of a sample containing 0–0.2 μmole of D-galactose are transferred by pipet. The initial absorbance of this solution is measured at 340 nm in an automatic ultraviolet spectrophotometer. Then, 0.02 ml of reagent C is added to start the reaction; the contents of the cell are mixed, and the change in absorbance is followed until it is constant.

The amount of D-galactose in μmoles is given by the change in absorbance multiplied by $3/6.22 = 0.482$.

References

(*1*) A. Dahlqvist, *Biochem. J.*, **80,** 547 (1961).
(*2*) J. Okuda and I. Miwa, *Methods Biochem. Anal.*, **21,** 156 (1973).
(*3*) G. Avigad, C. Asensio, D. Amaral, and B. L. Horecker, *J. Biol. Chem.*, **237,** 2736 (1962).

(*4*)   D. Amaral, F. Kelly-Falcoz, and B. L. Horecker, *Methods Enzymol.*, **9,** 87 (1966).
(*5*)   H. Schachter, *Methods Enzymol.*, **41,** 3 (1975).
(*6*)   B. L. Horecker and A. Kornberg, *J. Biol. Chem.*, **175,** 385 (1948).
(*7*)   G. Avigad, D. Amaral, C. Asensio, and B. L. Horecker, *J. Biol. Chem.*, **237,** 2736 (1962).
(*8*)   K. Wallenfels and G. Kurz, *Biochem. Z.*, **335,** 559 (1962).
(*9*)   G. Avigad, C. Asensio, D. Amaral, and B. L. Horecker, *Biochem. Biophys. Res. Commun.*,
        4, 474 (1961).
(*10*) J. K. Rogers and N. S. Thompson, *Carbohydr. Res.*, **7,** 66 (1968).
(*11*) C. G. Gahmberg and S.-I. Hakomori, *J. Biol. Chem.*, **248,** 4311 (1973).
(*12*) A. K. Hajra, D. M. Brown, Y. Kishimoto, and N. S. Radin, *J. Lipid Res.*, **7,** 379 (1966).
(*13*) W. Jack and R. J. Sturgeon, *Carbohydr. Res.*, **49,** 335 (1976).

[18] Enzymic Determination of D-Glucose, D-Fructose, and D-Mannose

By R. J. Sturgeon

Department of Brewing and Biological Sciences,
Heriot-Watt University, Edinburgh, Scotland

Introduction

The majority of chemical methods for the determination of D-glucose, D-fructose and D-mannose, in a mixture, depend on their initial separation, one from another, before measurement of the reducing action of the sugar. Enzymic methods are rapid and specific. Many methods have been devised for the estimation of D-glucose with D-glucose oxidase using an oxygen electrode (*1*), colorimetry (*1, 2*), fluorimetry (*1, 3*), electrochemical methods (*1*), and immobilized enzyme methods (*4–6*).

Methods for the enzymic determination of D-mannose, D-fructose and D-glucose in the same assay mixture are based on the phosphorylation of these sugars by hexokinase to the corresponding hexose 6-phosphates which are then converted to D-glucose 6-phosphate. This intermediate is assayed with glucose-6-phosphate dehydrogenase, G-6-PDH (D-glucose 6-phosphate: NADP-1-oxidoreductase, EC 1.1.1.49) as described by Warburg *et al.* (*7*).

$$\text{D-Mannose} + \text{ATP} \xrightarrow[\text{ADP}]{\text{hexokinase}} \text{D-mannose 6-phosphate} \qquad (1)$$

$$\Big\Updownarrow \text{PMI (4)}$$

$$\text{D-Fructose} + \text{ATP} \xrightarrow[\text{ADP}]{\text{hexokinase}} \text{D-fructose 6-phosphate} \qquad (2)$$

$$\Big\Updownarrow \text{PGI (5)}$$

$$\text{D-Glucose} + \text{ATP} \xrightarrow[\text{ADP}]{\text{hexokinase}} \text{D-glucose 6-phosphate} \qquad (3)$$

$$\text{D-Glucose 6-phosphate} + \text{NADP}^+ \xrightarrow{\text{G-6-PDH}} \text{6-phospho-D-gluconate} + \text{NADPH} + \text{H}^+ \qquad (6)$$

135

METHODS IN CARBOHYDRATE
CHEMISTRY, VOL. VIII

Addition of hexokinase, HK (ATP:D-hexose-6-phosphotransferase, EC 2.7.1.1) to solutions of D-glucose, D-fructose, and D-mannose containing adenosine triphosphate (ATP) results in the production of the corresponding hexose 6-phosphate and adenosine diphosphate (ADP) as illustrated in reactions 1, 2, and 3. The increase in extinction occurring after the addition of G-6-PDH is due to glucose (reaction 6). Addition of phosphoglucose isomerase, PGI (D-glucose-6-phosphate ketol-isomerase, EC 5.3.1.9) results in the conversion of the D-fructose 6-phosphate to D-glucose 6-phosphate (reaction 5) with a further change in absorbance due to the reaction of this latter product with G-6-PDH. The D-mannose 6-phosphate is then converted to D-fructose-6-phosphate with phosphomannose isomerase, PMI (D-mannose-6-phosphate ketol-isomerase, EC 5.3.1.8) (reaction 4), with the resulting change in absorbance due to mannose. NADPH is produced in proportion to the quantity of hexoses and is measured by the changes in absorbance at 340 nm. The reactions proceed optimally at pH 7.5, with the equilibria of reactions 1, 2, 3, and 6 lying far to the right.

Procedure

The following reagents are prepared from the highest grade chemicals:

(1) 0.05 M Triethanolamine, pH 7.6, containing 0.07 M magnesium chloride and 0.001 M mercaptoethanol.
(2) 0.017 M Adenosine triphosphate (ATP).
(3) 0.011 M Nicotinamide adenine dinucleotide phosphate (NADP).
(4) Hexokinase, HK, ≥ 140 units/mg,[1] 10 mg protein/ml.
(5) Glucose-6-phosphate dehydrogenase, G-6-PDH, ≥ 140 units/mg,[1] 1 mg protein/ml.
(6) Phosphomannose isomerase, PMI, ≥ 60 units/mg,[1] 10 mg protein/ml.
(7) Phosphoglucose isomerase, PGI, ≥ 350 units/mg,[1] 1 mg protein/ml.

Into a 1-cm silica cell, 2.7 ml of reagent 1, 0.1 ml of reagent 2, 0.1 ml of reagent 3, 0.01 ml of reagent 4, and 0.1 ml of a solution of D-glucose, D-fructose, and D-mannose containing up to a total of 0.3 μmoles of sugar are transferred successively by pipet. The solutions are mixed and the initial absorbance (A_1) at 340 nm is measured. Then 0.02 ml of reagent 5 is added and the absorbance is followed until it is constant (usually 5–10 min). The final absorbance is A_2. The absorbance change, $A_2 - A_1 = \Delta A$, is used in the calculation of the amount of D-glucose in the sample. Reagent 7 (0.02 ml) is added, and the absorbance is followed until it is constant (usually

---

[1] 1 unit is the amount of enzyme which catalyzes the formation of 1 μmole of product per minute.

about 10 min). The final absorbance is A_3. The absorbance change $A_3 - A_2 = \Delta A'$ is used in the calculation of the amount of D-fructose in the cell.

Reagent 6, 0.02 ml is added and the absorbance is followed until it is constant (usually about 30 min). The final absorbance is A_4. The absorbance change $A_4 - A_3 = \Delta A''$ is used in the calculation of the amount of D-mannose in the cell.

Highly purified samples of HK and G-6-PDH may contain traces of phosphoglucose isomerase. Thus, if samples contain D-fructose, these will react slowly, with conversion of D-fructose 6-phosphate into D-glucose 6-phosphate before the addition of PGI to the assay mixture. For this reason, A_2 must be determined exactly by extrapolation.

The amount of sugar, in μmoles, in the cell is given by $(A \times v)/(\varepsilon \times d)$, where v is the volume of liquid in the cell, $\varepsilon = 6.22$ cm^2/μmole, and $d =$ light path in cm.

Discussion

Special attention must be paid to the purity of the enzyme preparations used for these estimations. HK, G-6-PDH, PGI, and PMI must not contain more than 0.01% 6-phosphogluconate dehydrogenase. HK and G-6-PDH must each contain not more than 0.05% PGI and 0.05% PMI, and PGI not more than 0.05% PMI. These percentages are relative to the activity of the corresponding enzymes. In addition, the enzymes must be free from NADPH oxidase and glycosidases which would liberate any of the three hexoses from glycosides. The substrate specificity of hexokinase is rather low; it acts on D-glucose, D-mannose, D-fructose, 2-deoxy-D-*arabino*-hexose (2-deoxy-D-glucose), and 2-amino-2-deoxy-D-glucose. However, G-6-PDH is specific for G-6-P; other hexose or pentose phosphates do not react. D-Glucose 6-phosphate, D-fructose 6-phosphate, and D-mannose 6-phosphate can also be estimated by measurement of individual reaction steps after separate addition of the appropriate enzymes in the absence of HK.

References

(1)   J. Okuda and I. Miwa, *Methods Biochem. Anal.*, **21**, 155 (1973).
(2)   J. Schreiber and R. Lachenicht, *Z. Klin. Chem. Klin. Biochem.*, **11**, 31 (1973).
(3)   A. J. Tomisek and S. Natelson, *Microchem. J.*, **19**, 54 (1974).
(4)   G. Nagy, L. H. Von Storp, and G. G. Guilbault, *Anal. Chim. Acta*, **66**, 443 (1973).
(5)   S. W. Kiang, J. W. Kuan, S. S. Kuan, and G. G. Guilbault, *Clin. Chem.*, **22**, 1378 (1976).
(6)   M. K. Weibel, W. Dritschilo, H. J. Bright, and A. E. Humphrey, *Anal. Biochem.*, **52**, 402 (1973).
(7)   O. Warburg, W. Christian, and A. Giese, *Biochem. Z.*, **282**, 157 (1935).

[19] Microenzymic Procedures for the Identification of Carbohydrates

By John H. Pazur and Kevin L. Dreher

Department of Biochemistry and Biophysics,
The Pennsylvania State University,
University Park, Pennsylvania

Introduction

Microenzymic procedures based on the use of enzyme sprays and paper chromatography have proven useful for the identification of monosaccharide constituents of glycans, glycolipids, and glycoproteins. Such techniques augment the classic enzymic methods (Vol. VII [38]) for the identification of carbohydrates and the determination of the molecular structure. French and Wild (*1*) introduced the enzyme spray method in studies of the oligosaccharide primer specificity of potato phosphorylase. In these experiments phosphorylase and α-D-glucose 1-phosphate were sprayed directly on the paper chromatograms on which the oligosaccharides under study were separated. The enzymic reaction occurred in the time interval that the chromatogram dried. After the chromatogram was sprayed with a dilute iodine solution, the areas of the chromatogram at which the synthesis of α-D-(1→4)-glucans had occurred appeared as blue spots. Results showed that malto-oligosaccharides containing four or more glucose residues functioned as good primers for the phosphorylase reaction. An adaptation of this method was used by Pazur and Kleppe (*2*) for identification of derivatives and stereoisomers of D-glucose which were oxidized by glucose oxidase. Since glucose oxidase is very sensitive, it has been especially useful for detecting small amounts of D-glucose that may be produced from starch and glycogen by various amylases (*3*).

Recently an enzyme spray technique similar to that utilized with glucose oxidase has been developed with galactose oxidase (*4*). Since this enzyme oxidizes D-galactose and *N*-acetyl-D-galactosamine at C-6, it can oxidize such residues which are terminal units of oligosaccharides and polysaccharides (*5*). By utilizing the spray method, the substrate specificity of galactose oxidase has been expanded to include glycerol, D-glyceraldehyde, D-threitol, uridine diphosphate galactose, thymidine diphosphate galactose and uridine diphosphate-*N*-acetyl-D-galactosamine (*4*). Accordingly, in the quantitative

139

determination of D-galactose by the use of galactose oxidase (6), corrections will need to be applied if the latter compounds are present.

Enzyme test strips available under the trade names Clinistix and Galactostix have proven to be very useful for the detection of microquantities of D-glucose, D-galactose, and their derivatives (7). Details of this method are described in the section on Procedure. The method when coupled with specific hydrolases (8) often yields additional information on the structure of D-glucose and D-galactose containing compounds. Both the test-strip method and enzyme spray method are suitable for detecting the presence of free D-glucose, D-galactose, and N-acetyl-D-galactosamine in acid hydrolyzates of naturally occurring carbohydrate polymers. In addition, the galactose oxidase methods can be used to detect glycans, glycoproteins, and sugar nucleotides which contain terminal D-galactose or N-acetyl-D-galactosamine residues (4).

The use of two-dimensional paper chromatography interspersed with an enzyme spray step is another important development in this area. French et al. (9) used this method to determine the nature of branched oligosaccharides isolated from enzymic hydrolysates of starch and glycogen. Pazur and Okada (10) used the technique to identify and classify the amylases which act on malto-oligosaccharides. Briefly the procedure involves separating the compounds under study in one direction on paper chromatograms, spraying the area of the dried chromatogram with a suitable enzyme, and identifying the products of enzyme action by chromatography in the second direction. Up to the present, this method has been used primarily with carbohydrates and hydrolytic carbohydrases. However, by selecting the proper solvent for paper chromatography and the appropriate enzymes, such procedures should be applicable for the identification of the constituent units of other classes of compounds including peptides, oligonucleotides, sugar nucleotides, and complex lipids.

Procedure

Identification of Monosaccharides and Monosaccharide Derivatives by Use of Glucose Oxidase (3)

Samples of 2–5 mg of the materials under study are dissolved in 0.2 ml of 0.1 M hydrochloric acid and heated in a boiling water bath for 2 h. Samples of 5 μL of the hydrolyzates and reference compounds are placed on duplicate paper chromatograms (27 cm × 25 cm, Whatman No. 1) 6:4:3 v/v and the chromatograms are developed in a suitable solvent system such as 1-butanol–pyridine–water. One chromatogram is stained by the silver nitrate procedure (11, Vol. I [6]), and the other is sprayed with a glucose oxidase–peroxidase solution followed by an alcoholic solution of *o*-tolidine (4,4'-diamino-3,3'-dimethylbiphenyl). The enzyme solution is prepared in 0.1 *M* sodium acetate buffer of pH 5.0 and contains 1% of crude glucose oxidase (Miles Laboratories Inc., Elkhart, IN) and 0.1% of purified horse radish peroxidase (Sigma Chemical Co., St. Louis, MO). The *o*-tolidine is dissolved in ethanol in final concentration of 0.5%. These reagents were devised and first used by White and Secor (*12*) for the detection of D-glucose on paper chromatograms. Generally, 10 ml of the enzyme solution and 20 ml of the *o*-tolidine solution is sufficient for spraying one chromatogram. The reducing sugars, on staining with silver nitrate, appear as brown spots. Those compounds which react with glucose oxidase appear as blue spots. Since the spots and the paper become discolored on standing, the chromatogram should be photographed shortly after development of the color for a permanent record. The length of time for appearance of a blue color is characteristic of the individual compounds being oxidized by the enzyme. On the basis of the R_f values of the compounds and the time of appearance of the color, the identification of the compounds can be established. The relative rates of oxidation by glucose oxidase for a number of monosaccharides and derivatives are recorded in Table I.

TABLE I

Relative Rates of Oxidation of Hexoses, Alditols, and their Derivatives by Glucose Oxidase and by Galactose Oxidase

Glucose oxidase		Galactose oxidase	
Compound	Relative rate[a]	Compound	Relative rate[a]
D-Glucose	100	D-Galactose	100
L-Glucose	0	L-Galactose	0
2-Deoxy-D-glucose	20	D-Glucose	0
(2-deoxy-D-*arabino*-hexose)		D-Talose	50
D-Mannose	1	2-Deoxy-D-galactose	70
D-Allose	0.02	(2-deoxy-D-*lyxo*-hexose)	
3-*O*-Methyl-D-glucose	0.02	*N*-Acetyl-D-galactosamine	100
4-Deoxy-D-glucose	2	Methyl α-D-galactoside	150
D-Galactose	0.5	D-Glyceraldehyde	60
4-*O*-Methyl-D-glucose	15	Glycerol	20
5-Deoxy-D-glucose	0.05	D-Threitol	50
6-Deoxy-D-glucose	10	D-Threose	0
6-*O*-Methyl-D-glucose	1	Galactitol	0
1,5-Anhydro-D-glucitol	0	1,5-Anhydro-D-galactitol	90

[a] Values calculated from the times required for the development of a color comparable to a standard color chart with the values for D-glucose and glucose oxidase and for D-galactose and galactose oxidase assigned arbitrary values of 100.

Identification of Aldoses, Alditols, and Sugar Nucleotides
by Use of Galactose Oxidase (4)

These tests are performed in a manner comparable to the glucose oxidase tests. The aldoses under test are separated on paper chromatograms using 6:4:3 v/v 1-butanol–pyridine–water, the sugar alcohols using 5:1:4 v/v 1-butanol–acetic acid–water (top layer), and the sugar nucleotides using 7:3 v/v ethanol–0.1 M ammonium acetate, pH 7.5. The enzyme solution is prepared in 0.1 M phosphate buffer of pH 7.0 and contains 0.01% of purified galactose oxidase (Sigma Chemical Co., St. Louis, MO) and 0.1% of horseradish peroxidase. The chromatogram is first sprayed with the enzyme solution and then with 0.5% o-tolidine in ethanol. Color development may take up to half an hour for the slow reacting compounds. For permanent records, the chromatograms should be photographed. It may be desirable to spray the chromatogram with a second portion of enzyme solution if color spots have not appeared in a few minutes with known standard compounds. The aldoses are located on the paper with the silver nitrate reagent (11, Vol. I [6]), the sugar alcohols by the periodate–permanganate reagent (13), and the sugar nucleotides by uv absorption (14). Hydrolysis of glycans and other compounds can be effected by the method described in the previous section; the analysis of the hydrolyzates can be performed as described here. The relative rates of oxidation of compounds by galactose oxidase aid in the identification of the substance. Relative rates for a number of compounds are recorded in Table I.

Use of Enzyme Test Strips for Detection of D-Glucose
and D-Galactose (7)

In the test-strip method, samples of 5 μl of the sugar solutions or neutralized hydrolyzates of carbohydrate polymers are placed on enzyme test strips (Clinistix or Galactostix, Miles Laboratories Inc., Elkhart, Indiana), and the times for appearance of a blue color comparable to that on the reagent bottle are recorded. Relative velocities of enzyme action on the various substrates can be calculated from these times and the known concentrations of the substrates. These values are used in the identification of unknowns. An adaptation of this technique for identifying glucosides, glucosyl oligosaccharides, and glucans involves the hydrolysis of these compounds with an appropriate hydrolase directly on the test strip. As the hydrolysis occurs, the liberated D-glucose is oxidized by the glucose oxidase and the color change on the test strip is obtained. Since galactose oxidase oxidizes D-galactose and N-acetyl-D-galactosamine residues that are terminal units of complex compounds, solutions of such compounds can be placed

TABLE II

Reactivity of Some D-Galactose-Containing Compounds with Galactose Oxidase

Compound	Galactostix test	Partially methylated galactose	Reference
Diheteroglycan of *S. faecalis*	Positive	2,3,4,6- 2,3,6-	(*16*)
Tetraheteroglycan of *S. faecalis*	Negative	2,3,6-	*a*
Tetraheteroglycan of *S. bovis*	Negative	2,4,6-	(*17*)
Glucoamylase of *A. niger*	Positive	2,3,4,6-	*a*
Carcinoembryonic antigen	Positive	2,3,4,6- 3,4,6- 2,4,6- 2,3,4-	(*18*)
Lactosyl ceramide	Positive	2,3,4,6-	(*19*)
Guaran	Positive	2,3,4,6-	(*20*)
Galactocarolose	Negative	2,3,5,6- 2,3,6-	(*21*)
Uridine diphosphate galactose	Positive	*b*	(*4*)
Thymidine diphosphate galactose	Positive	*b*	*a*
Uridine diphosphate *N*-acetyl-D-galactosamine	Positive	*b*	*a*

a Unpublished data of J. H. Pazur and L. S. Forsberg.
b Methylation data not available.

directly on the Galactostix. Results of such tests with a few D-galactose con-taining compounds are recorded in Table II. Also recorded in the Table are the types of partially methylated derivatives of D-galactose which are ob-tained from a methylation analysis on the compounds.

Identification of Glucosyltransferase Products from Maltotetraose and Maltoheptaose by Enzyme Spray and Two Dimensional Paper Chromatography (*15*)

Samples of 5 μl of the digest of maltotetraose or maltoheptaose with glucosyl transferase (*15*) are placed on chromatograms (27 cm × 25 cm) and developed in one direction using 6:4:3 v/v 1-butanol–pyridine–water by five ascents of the solvent (Vol. I [6]). The areas of the chromatogram con-taining the compounds are sprayed with 5 ml of a 0.1% solution of beta-amylase (Wallerstein Laboratories, New York, NY) in 0.05 M acetate buffer of pH 5.0. Another chromatogram is sprayed with 5 ml of a 0.01% solution of purified glucoamylase (Miles Laboratories Inc., Elkhart, Indiana) in 0.05 M actate buffer of pH 5.0. A reference mixture of malto-oligosaccharides is placed on the chromatogram in line with the products in the maltotetraose or maltoheptaose digests. The chromatogram is then developed in the second direction by three ascents of the solvent system. The finished chromatograms are stained with silver nitrate reagent (*11*, Vol. I [6]) to locate the reducing products. Of the series of compounds in the maltotetraose digest, only maltotetraose is hydrolyzed by β-amylase to yield maltose. Of the series of compounds in the maltoheptaose digest, maltotetraose and maltohexaose are hydrolyzed to maltose while maltopentaose and maltoheptaose are hydrolyzed to maltose and maltotriose by the β-amylase. In both digests, all the compounds are hydrolyzed to D-glucose by the glucoamylase. However, since the α-D-$(1 \rightarrow 6)$ glucosidic linkage is hydrolyzed at a slower rate than the α-D-$(1 \rightarrow 4)$ linkage by the glucoamylase, oligosaccharides with α-D-$(1 \rightarrow 6)$ linkages are not completely hydrolyzed to glucose. Thus, oligosaccharides terminated at the non-reducing end by α-D-$(1 \rightarrow 6)$-linked glucose residues are produced from both maltotetraose and maltoheptaose by the glucosyl transferase. That the terminal residues of these oligosaccharides are linked by α-D-$(1 \rightarrow 6)$ linkages to the remainder of the compound has been verified by methylation analysis (*15*).

References

(*1*)   D. French and G. M. Wild, *J. Amer. Chem. Soc.*, **75**, 4490 (1953).

(*2*)   J. H. Pazur and K. Kleppe, *Biochemistry*, **3**, 578 (1964).

(*3*)   J. J. Marshall and W. J. Whelan, *Arch. Biochem. Biophys.*, **161**, 234 (1974).

(*4*)   J. H. Pazur, H. R. Knull, and G. E. Chevalier, *J. Carbohydr. Nucleos. Nucleot.*, **4**, 129 (1977).

(*5*)   G. Avigad, D. Amaral, C. Asensio, and B. L. Horecker, *J. Biol. Chem.*, **237**, 2736 (1962).

(*6*)   H. Roth, S. Segal, and D. Bertoli, *Anal. Biochem.*, **10**, 32 (1965).

(*7*)   J. H. Pazur, M. Shadaksharaswamy, and G. E. Meidell, *Arch. Biochem. Biophys.*, **99**, 78 (1962).

(*8*)   J. H. Pazur and A. Cepure, *Carbohydr. Res.*, **5**, 359 (1967).

(*9*)   D. French, A. P. Pulley, M. Adbullah, and J. C. Linden, *J. Chromatogr.*, **24**, 271 (1966).

(*10*) J. H. Pazur and S. Okada, *J. Biol. Chem.*, **241**, 4146 (1966).

(*11*) F. C. Mayer and J. Larner, *J. Amer. Chem. Soc.*, **81**, 188 (1959).

(*12*) L. M. White and G. E. Secor, *Science*, **125**, 495 (1957).

(*13*) R. U. Lemieux and H. F. Bauer, *Anal. Chem.*, **26**, 920 (1954).

(*14*) J. H. Pazur and E. W. Shuey, *J. Biol. Chem.*, **236**, 1780 (1961).

(*15*) J. H. Pazur, A. Cepure, S. Okada, and L. S. Forsberg, *Carbohydr. Res.*, **58**, 193 (1977).

(*16*) J. H. Pazur, A. Cepure, J. A. Kane, and C. G. Hellerquist, *J. Biol. Chem.*, **248**, 279 (1973).

(*17*) J. H. Pazur, D. J. Dropkin, K. L. Dreher, L. S. Forsberg, and C. S. Lowman, *Arch. Biochem. Biophys.*, **176**, 257 (1976).

(*18*) S. Hammarstrom, E. Engvall, B. G. Johansson, S. Svensson, G. Sandland, and I. J. Goldstein, *Proc. Natl. Acad. Sci. U.S.*, **72**, 1528 (1975).

(*19*) R. V. P. Tao, C. C. Sweeley, and G. A. Jamieson, *J. Lipid Res.*, **14**, 16 (1973).

(*20*) Z. F. Ahmed and R. L. Whistler, *J. Amer. Chem. Soc.*, **72**, 2524 (1950).

(*21*) J. F. Preston and J. E. Gander, *Arch. Biochem. Biophys.*, **124**, 504 (1968).

[20] Method for Measuring the Affinity between Carbohydrate Ligands and Certain Proteins

By Cornelis P. J. Glaudemans

National Institutes of Health, Bethesda, Maryland

AND

Michael E. Jolley

Abbott Laboratories, Chicago, Illinois

Introduction

When protein solutions are irradiated at 280 nm, they usually emit (fluorescent) radiation with a wavelength varying from 330 to 350 nm. This fluorescence (F) is mostly due to tryptophanyl and tyrosyl residues.

In 1973, we observed (*1, 2*) a change in the fluorescence of many antibodies on binding simple carbohydrates (themselves transparent to ultraviolet radiation). This is unusual. Energy emitted by aromatic amino acids in proteins can be transferred readily to bound ligands if such ligands contain aromatic (or nitro) groups themselves (*3*). In this case of carbohydrate ligand-induced changes in the fluorescence of proteins (immunoglobulins), the ligands do not contain these groups which would facilitate the radiationless transfer of energy from the proteins. These changes, whether they be in the intensity or the fluorescence maximum, may be induced by changes in (*a*) conformation of the protein, (*b*) solvation (hydration), (*c*) temperature, (*d*) pH, and (*e*) changes in dipole moments in the aromatic nuclei of the protein. Some of these conditions can in turn be induced by the binding of small molecules to the protein. Another example is the changed fluorescence exhibited by yeast enolase on binding magnesium (*4*).

We have found that it is the tryptophanyl residues in the immunoglobulins examined which, most probably, have their fluorescence altered by ligand binding (*1*). In addition to our findings, it was independently shown by others that phosphorylcholine can induce fluorescence changes in immunoglobulins with antiphosphorylcholine specificity (*5, 6*).

Not all proteins can be expected to show carbohydrate ligand-induced changes in fluorescence. In our laboratory, we have dealt with approximately fifteen immunoglobulins capable of binding carbohydrates. About half of these show the above phenomenon. Some of them showed a ligand-induced

145

METHODS IN CARBOHYDRATE
CHEMISTRY, VOL. VIII

ISBN 0-12 746208-2

increase in fluorescence while the others showed a ligand-induced decrease in fluorescence. If the proteins in question do show carbohydrate induced changes in fluorescence, the affinity constant of the equilibrium

$$\text{Ligand} + \text{protein} \rightleftharpoons \text{Protein–ligand}$$

$$K_a = \frac{[\text{Protein–ligand}]}{[\text{Protein}][\text{Ligand}]}$$

may be determined by the procedure reported here. The method is well suited to carbohydrate-binding proteins, since carbohydrates do not themselves absorb in the ultraviolet or visible spectrum and, hence, cannot interfere with, or contribute to, observed fluorescence.

Figure 1 shows a typical plot for the data obtained by adding increasing amounts of ligand to a protein and monitoring the intensity of the fluorescent radiation. The maximum change in fluorescence, ΔF_{max} can be found by either of two means, by the addition of solid ligand (large excess) or by plotting the changed fluorescence over the free ligand concentration at that value versus ΔF (Fig. 2). At any point b, the fraction, $\bar{v}$, of available sites occupied by ligand equals $a/\Delta F_{max}$ (Fig. 1). The concentration of free ligand at any time is $C_{\text{free ligand}} = [\text{ligand added} - \bar{v}[\text{total sites}]$.[1] Thus, a Scatchard plot ($\bar{v}/C$ versus $\bar{v}$) of the data, using the formula $\bar{v}/C = K_a - \bar{v}K_a$, will yield a line whose slope equals $-K_a$.

The method has been extensively used (1, 2, 5–10) and has proved extremely reliable and sensitive. This method has many advantages over that of equilibrium dialysis, the foremost alternative for the determination of affinity constants. The method (a) is simple and quick, (b) requires very small amounts of proteins and ligands, (c) is extremely accurate and reproducible, (d) does not require the preparation of radioactive derivatives of ligands, (e) does not require that the protein be nondialyzable, and (f) has no restriction on the value of the K_a measured; we have measured values as low as 150 (11).

In Table I is given one example of a determination of the affinity constant between methyl β-D-galactopyranoside and an immunoglobulin capable of binding β-(1 → 6)-linked D-galactans.

---

[1] Usually the concentration of protein is kept low to avoid protein–protein interaction (0.05 A_{280}). The second term reduces to zero and $C_{\text{free ligand}} = [\text{ligand added}]$, if the affinity constant is $2 \times 10^5 \, M^{-1}$ or lower. However, when $K_a > 2 \times 10^5 \, M^{-1}$, the second term becomes significant.

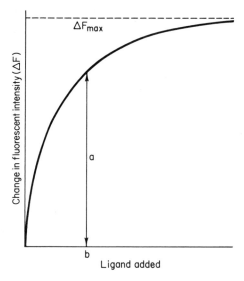

FIG. 1.—Typical change in fluorescent intensity when ligand solution is added to a myeloma immunoglobulin.

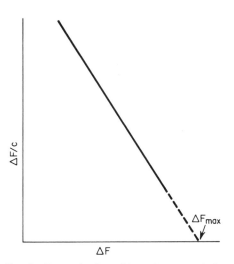

FIG. 2.—Determination of ΔF_{max} by extrapolation.

TABLE I

Titration of J539 Fab' With Methyl β-D-Galactopyranoside[a]

Ligand added, μl	Final volume, μl	Correction factor	F_{obs}	F_{obs}^{corr}	ΔF	$\bar{v}$	$C(\times 10^3)$	$\bar{v}/C$
3	1503	1.0020	81.0	81.16	3.36	.184	.21903	843.4
3	1506	1.0040	83.1	83.43	5.63	.309	.43723	707.8
3	1509	1.0060	84.4	84.91	7.11	.390	.65459	596.5
3	1512	1.0080	85.4	86.08	8.28	.455	.87110	522.5
3	1515	1.0100	86.3	87.16	9.36	.514	1.08676	473.4
5	1520	1.0133	87.1	88.26	10.46	.574	1.44433	397.9
10	1530	1.0200	88.3	90.07	12.27	.673	2.15246	313.1
15	1545	1.0300	89.0	91.67	13.87	.762	3.1975	238.3
30	1575	1.0500	88.6	93.03	15.23	.836	5.2279	160.1
45	1620	1.0800	87.3	94.28	16.48	.905	8.13268	111.4

[a] ΔF_{max} = [Last corr. factor $\times$ (F_0 + $\Delta F_{max\ obs}$)] $-$ F_0. C_{prot} = 0.05 A J539–Fab' in 0.05 M TRIS buffer, pH 7.4. C_{ligand} = 0.1098 M methyl β-D-galactopyranoside. Initial F_0 = F_{obs} $-$ F_{buffer} = 77.8.

Procedure

Interaction of Mouse Myeloma Immunoglobulin J539 (Fab') with Methyl β-D-Galactopyranoside

The fluorescence titrations are performed on a fluorescence spectrophotometer (equivalent to a Perkin-Elmer MPF-3L) equipped with a thermostated, sample cell assembly, through which water at 25° is circulated. Protein solution (1.5 ml, absorbance $\leq$ 0.05 at 280 nm) is added to each of two cells (10 $\times$ 10 $\times$ 45-mm), one to be used for ligand addition (test cell) and the other as the reference cell. A magnetic stirring-bar (6 $\times$ 1 $\times$ 1-mm) is placed in the test cell. After thermal equilibration, the difference in intensities of fluorescence of the test and reference solution are measured at 340 nm (excitation at 295 nm). The fluorescence of a buffer blank is also determined. Ligand solution is then added in small aliquots to the test solution; the solution is briefly stirred, and the increased fluorescence is determined against the reference cell. (It is best to use a concentration for the ligand solution such that half the protein sites are saturated after the addition of approximately 20 μl). The additions are continued until 100–150 μl is added. Small quantities of solid ligand are then added to the test cell until the fluorescence no longer increases. This value (when corrected for dilution, see Table I) is taken as the maximal fluorescence (ΔF_{max}) corresponding to protein site-saturation. (If not enough ligand is available for the addition

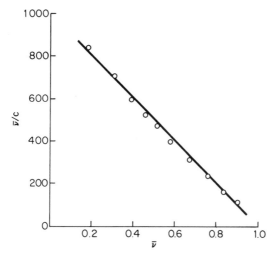

FIG. 3.—Scatchard plot of $\bar{v}/C$ versus $\bar{v}$ for immunoglobulin J539–Fab' and methyl β-D-galactopyranoside.

of solid, ΔF_{max} can be obtained by extrapolation of a $\Delta F/C$ versus ΔF plot. See introduction.)

The increase in fluorescence intensity due to each addition of ligand is corrected for dilution of the sample (Table I). It may be derived that the concentration of free ligand C is essentially the same as that of the ligand added in this case, since the association constant measured here is fairly low.

The values for $\bar{v}/C$ are plotted versus $\bar{v}$ (Fig. 3) and the association constant is determined from the slope of the Scatchard plot and found to be $1.00 \times 10^3\ M^{-1}$.

References

(1)   M. E. Jolley, S. Rudikoff, M. Potter, and C. P. J. Glaudemans, *Biochemistry*, **12**, 3039 (1973).

(2)   M. E. Jolley and C. P. J. Glaudemans, *Carbohydr. Res.*, **33**, 377 (1974).

(3)   J. E. McGuigan, *Methods Immunol. Immunochem.*, **3**, 384 (1971).

(4)   J. M. Brewer and B. Weber, *J. Biol. Chem.*, **241**, 2550 (1966).

(5)   R. Pollet and H. Edelhoch, *J. Biol. Chem.*, **248**, 5443 (1973).

(6)   R. Pollet, H. Edelhoch, S. Rudikoff, and M. Potter, *J. Biol. Chem.*, **249**, 5188 (1974).

(7)   M. E. Jolley, C. P. J. Glaudemans, and S. Rudikoff, *Biochemistry*, **13**, 3179 (1974).

(8)   C. P. J. Glaudemans, E. Zissis, and M. E. Jolley, *Carbohydr. Res.*, **40**, 129 (1975).

(9)   B. N. Manjula, C. P. J. Glaudemans, E. Mushinski, and M. Potter, *Carbohydr. Res.*, **40**, 137 (1975).

(10)   B. N. Manjula, C. P. J. Glaudemans, E. B. Mushinski, and M. Potter, *Proc. Natl. Acad. Sci. U.S.*, **73**, 932 (1976).

(11)   D. G. Streefkerk and C. P. J. Glaudemans, *Biochemistry* **16**, 3760 (1977).

[21] ^{13}C NMR Spectroscopy of Isotopically Enriched Carbohydrates

Department of Biochemistry, Michigan State University,
East Lansing, Michigan

AND

T. E. Walker

Los Alamos Scientific Laboratory, University of California,
Los Alamos, New Mexico

Introduction

The use of ^{13}C nmr spectroscopy in the elucidation of carbohydrate structures and conformations has increased rapidly in the past few years, and has been the subject of previous articles in this series (Vol. VI [93]; Vol. VII [17], [19]; This Vol. [12]). Several other valuable reviews have also appeared (*1–3*). This article deals principally with two aspects of the technique that are greatly facilitated by the use of isotopically enriched (^{13}C and ^{2}H) compounds: (a) assignment of chemical shifts and (b) evaluation of the factors which influence the magnitude of coupling constants in monosaccharides and some of their derivatives. These matters are important in the characterization of carbohydrate derivatives and the elucidation of carbohydrate conformations.

^{13}C-Chemical Shifts of the Hexopyranoses

The chemical shifts of a number of monosaccharides are listed in Table I. The resonances of C-1 and C-6 are readily identified as the farthest downfield and farthest upfield peaks, respectively, by comparison to the chemical shifts of simple hemiacetals and primary alcohols. The resonances due to carbon atoms 2–5 fall in the chemical shift range of 68–78 ppm and positive identification is much more difficult. Early assignments were based on substituent effects (for example, effects of *O*-methylation and anomeric dependent shifts), single frequency proton decoupling experiments, and comparisons of observed chemical shifts with expected electron densities. Unequivocal assignments, however, require isotopic labeling, particularly for closely

CHEMISTRY, VOL. VIII

TABLE I

^{13}C *Nmr Resonance Assignments*

	C-1	C-2	C-3	C-4	C-5	C-6	OMe
Glucopyranoses[a]							
α-D-Glucose	92.7	72.14[b]	73.4	70.4	72.10	61.3	
β-D-Glucose	96.5	74.8	76.4	70.3	76.6	61.5	
Methyl α-D-glucopyranoside	100.3	72.5[b]	74.2[b]	70.6	72.7	61.7	56.2
Methyl β-D-glucopyranoside	104.3	74.2	76.9	70.8	76.9	61.9	58.3
6-Deoxy-α-D-glucose	93.1	72.9	73.6	76.4	68.6	18.0	
6-Deoxy-β-D-glucose	96.8	75.6	76.1	76.1	73.0	18.0	
Methyl 6-deoxy-α-D-glucopyranoside	100.3	72.6	73.9	76.2	68.7	17.6	56.2
Methyl 6-deoxy-β-D-glucopyranoside	104.3	74.5	76.7	76.2	73.0	17.8	58.3
Methyl α-D-glucopyranosiduronic acid	100.7	71.9	73.8	72.5	71.9	—[d]	56.7
Methyl β-D-glucopyranosiduronic acid	104.3	73.8	76.5	72.3	75.6	—[d]	58.5
Methyl (methyl α-D-glucopyranosid)uronate	100.8	71.9[d]	73.7	72.4	71.9	—[d]	56.8
Methyl (methyl β-D-glucopyranosid)uronate	104.6	73.7	76.3	72.4	75.7	—[d]	58.7
Mannopyranoses[a]							
α-D-Mannose	95.0	71.7	71.3	68.0	73.4	62.1	
β-D-Mannose	94.6	72.3	74.1	67.8	77.2	62.1	
Methyl α-D-mannopyranoside	101.9	71.2[c]	71.8[c]	68.0	73.7	62.1	55.9
Methyl β-D-mannopyranoside[e]	102.0	71.4	74.2	68.1	77.3	62.1	—[d]
α-L-Rhamnose	95.0	71.9	71.1	73.3	69.4	18.0	
β-L-Rhamnose	94.6	72.4	73.8	72.9	73.1	18.0	
Methyl α-L-rhamnopyranoside	101.9	71.0	71.3	73.1	69.4	17.7	55.8
Galactopyranoses[a]							
α-D-Galactose	93.6	69.8[b,c]	70.56[b]	70.63[c]	71.7	62.5	
β-D-Galactose	97.7	73.3	74.2	70.1	76.3	62.3	
Methyl α-D-galactopyranoside	100.5	69.4[c]	70.6[c]	70.4	71.8	62.3	56.3
Methyl β-D-galactopyranoside	104.9	71.8	73.9	69.8	76.2	62.1	58.3
α-D-Fucose	93.3	69.2[b]	70.4[b]	73.0	67.4	16.7	
β-D-Fucose	97.3	72.8	74.0	72.5	71.9	16.7	
Methyl α-D-fucopyranoside	100.5	69.0	70.6	72.9	67.5	16.5	56.3
Methyl β-D-fucopyranoside	104.8	71.5	74.1	72.4	71.9	16.5	58.3
Allopyranoses[a]							
α-D-Allose	93.4	67.6	72.3	66.7	67.5	61.3	
β-D-Allose	94.0	71.9[b]	71.8	67.4	74.2[b]	61.8	
Amino Sugars[a,f]							
α-D-Glucosamine HCl	90.7	56.0	71.2	71.2	73.1	62.0	
β-D-Glucosamine HCl	94.3	58.5	73.6	71.2	77.6	62.0	
N-Acetyl-α-D-glucosamine	92.1	55.3	72.0	71.4	72.8	61.9	
N-Acetyl-β-D-glucosamine	96.2	58.0	75.2	71.2	77.2	62.0	
α-D-Mannosamine HCl	91.8	56.0	68.3	67.6	73.1	61.7	
β-D-Mannosamine HCl	92.4	57.1	70.9	67.4	77.4	61.8	
Pentopyranoses[a]							
α-D-Xylose	93.3	72.5[b]	73.9[b]	70.4	62.1		
β-D-Xylose	97.6	75.1	76.9	70.3	66.3		
Methyl α-D-xylopyranoside	100.6	72.3	74.3	70.4	62.0		56.0

(Continued)

TABLE I (Continued)

	C-1	C-2	C-3	C-4	C-5	C-6	OMe
Glucopyranoses[a]							
Methyl β-D-xylopyranoside	105.1	74.0	76.9	70.4	66.3		58.3
α-L-Arabinose[b,c]	97.8	73.0	73.5	69.6	67.5		
β-L-Arabinose[b,c]	93.7	69.6	69.8	69.8	63.6		
Methyl α-L-arabinopyranoside	105.1	71.8	73.4	69.4	67.3		58.1
Methyl β-L-arabinopyranoside	101.0	69.4	69.9[b]	69.96	63.8		56.3
Furanosides[a]							
Methyl α-D-galactofuranoside	103.1	77.4	75.5	82.3	73.7	63.4	56.1
Methyl β-D-galactofuranoside	109.2	81.9	77.8	84.0	72.0	63.9	56.1
Methyl α-D-arabinofuranoside	109.3	81.9	77.5	84.9	62.4		56.1
Methyl β-D-arabinofuranoside	103.2	77.5	75.7	83.1	64.2		56.3
Methyl α-D-lyxofuranoside	109.1	77.0	72.0	81.3	61.2		56.9
Methyl β-D-lyxofuranoside	103.2	72.9	70.7	81.9	62.4		56.5
Methyl α-D-xylofuranoside	103.0	77.7	76.0	79.3	61.5		56.6
Methyl β-D-xylofuranoside	109.6	80.9	76.0	83.5	62.1		56.2
Methyl α-D-ribofuranoside	104.2	72.1	70.8	85.5	62.2		56.5
Methyl β-D-ribofuranoside	109.0	75.3	71.9	83.9	63.9		56.3

[a] Gorin and Mazurek (6). Spectra in ^2H_2O solutions at 33°; referenced to external TMS = 0 ppm. [b] Reversed from Dorman and Roberts (11). [c] Reversed from Perlin et al. (12). [d] Resonance not reported. [e] Walker et al. (8); corrected to 33°. [f] Walker et al. (8), in H_2O at approximately 40°, referenced to external TMS = 0 ppm.

spaced resonances. All the resonances listed in Table I have been verified by isotopic labeling, and those values which differ from the original assignments are indicated in the table. These reassignments have been based on two approaches: the effects of deuterium substitution on the ^{13}C-chemical shifts, and the observation of ^{13}C–^{13}C coupling in 1-^{13}C-labeled sugars.

The use of deuteration for assigning ^{13}C resonances is based on the observation that conversion of C–H to C–D perturbs the substituted carbon atom so that it either disappears (4) from the nmr spectrum or appears as a triplet at higher field (5, 6). The chemical shift of the substituted carbon atom can thus be assigned unequivocally. In addition, carbon atoms β to the deuterium show small upfield shifts (0.02–0.17 ppm) due to the β-isotope effect (6, 7). The magnitude of the β-isotope effect is proportional to the number of protons exchanged for deuterium atoms on adjacent carbon atoms and is largest when all hydrogen atoms β to a carbon atom are replaced. The large (0.17 ppm) isotope shift observed for C-3 of β-D-[4,5-$^2H_3$]-arabinopyranose suggests the existence of an additional γ-isotope effect from the two γ-deuterium atoms. The β-isotope effect can best be observed by obtaining the spectrum of a mixture of the ^2H and ^1H compounds, particularly when the effect is very small. Thus, the spectrum of α-D-[4-^2H]-galactopyranose shows upfield shifts of 0.10 and 0.06 ppm, respectively, for C-3 and C-5 relative to α-D-galactopyranose; and the resonance due to C-4 is not observed (7). The resonance due to C-4 can therefore be assigned unequivocally, while those due to C-3 and C-5 can be identified and distinguished from one another on the basis of shifts observed for α-D-[6-^{2}H$_2$]-galactopyranose.

The enrichment of a specific carbon atom of a sugar with ^{13}C also permits unequivocal assignment of the resonances of the enriched carbon atom and neighboring carbon atoms to which it is coupled (8). Carbohydrates containing ^{13}C at 90+% enrichment in C-1 are prepared readily by an adaptation of the cyanohydrin synthesis (9, 10) and can be purified by established procedures (11). As shown by the data in Table II, 1-^{13}C-enriched carbohydrates show a pattern of coupling between C-1 and other carbon atoms of the ring which is highly predictable. One-bond coupling between C-1 and C-2 is large (42–47 Hz) and insensitive to structural alterations. Two-bond coupling between C-1 and C-3 is observed in β-D-pyranose derivatives and between C-1 and C-5 in all α-D-pyranoses. Three-bond coupling between C-1 and C-6 is apparent in both anomers; whereas, coupling between C-1 and C-4 has not been observed.

A typical [^{13}C] nmr spectrum of a [1-^{13}C]hexose (α,β-D-galactopyranose) is compared to a spectrum of the natural abundance form in Figure 1. In the 1-^{13}C-enriched compound, resonances due to C-2 are split by 46 Hz in the α-D anomer and by 46.6 Hz in the β-D anomer, allowing them to be dis-

TABLE II

Carbon-13 Coupling Constants of ^{13}C-Enriched Monosaccharides

| Compound | Coupling Constant (Hz)^a | | | | | |
	C-1,2	C-1,3	C-1,5	C-1,6	C-1, H-1	C-1, H-2
α-D-[1-^{13}C]Glucose	46.0	—	~1.8	*	169.8	—
β-D-[1-^{13}C]Glucose	46.0	~3.5	—	*	161.2	5.5
Methyl α-D-[1-^{13}C]glucopyranoside	46.4	—	~1.7	3.2	*	—
Methyl β-D-[1-^{13}C]glucopyranoside	46.8	~4.1	—	4.3	*	4.1 ± 1.0
α-D-[1-^{13}C]Mannose	46.8	—	1.7	*	170.4	—
β-D-[1-^{13}C]Mannose	42.4	4.3	—	*	160.7	b
Methyl α-D-[1-^{13}C]mannopyranoside	47.0	—	2.3	3.0	*	*
Methyl β-D-[1-^{13}C]mannopyranoside	43.8	3.4	—	4.0	*	*
α-D-[1-^{13}C]Galactose	46.6	—	2.1	*	168.6	—
β-D-[1-^{13}C]Galactose	46.0	3.7	—	*	162.7	5.7
α-L-[1-^{13}C]Fucose	45.8	—	2.3	3.5	*	*
β-L-[1-^{13}C]Fucose	46.0	4.1	—	3.5	*	*

^{a} Couplings designated by * were not measured; those designated by — refer to no observable coupling.
^{b} A broadening of ~1.6 Hz was observed in the proton coupled spectrum, which could be due to C-1–H-2 coupling.

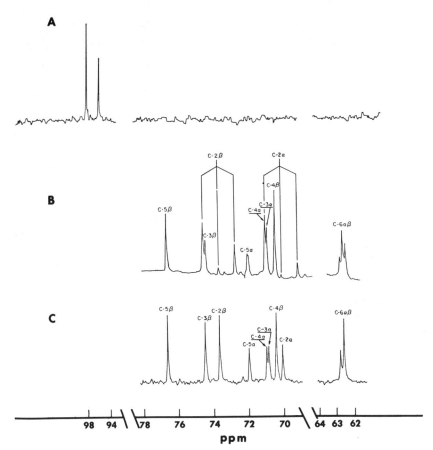

FIG. 1.—The proton decoupled, 25 MHz $[^{13}C]$ nmr spectra of α and β-D-galactopyranose with and without ^{13}C-enrichment at C-1. (A) A single scan of 0.2 M $[1-^{13}C]$ D-galactopyranose, 90% enrichment. Resonances are due to the β and α forms, present in the ratio of approximately 1.5:1. (B) The spectrum of the natural abundance region (carbon atoms 2, 3, 4, 5 and 6) of $[1-^{13}C]$ D-galactopyranose, 90% enrichment, 5000 scans at 2.05 sec per scan. Coupling between C-1 and C-2 is indicated. C-5α appears as a doublet and C-3β is a doublet overlapping the downfield resonance of C-2β. C-6$\alpha\beta$ appear as a complex. (C) The natural abundance spectrum of C-2 through 6 of D-galactopyranose.

tinguished from closely adjacent resonances of other carbon atoms. In the case of α-D-galactopyranose (Fig. 1C), the resonance at 70.0 ppm had been assigned to C-3 (12) or to C-4 (13), whereas the data presented in Figure 1B clearly demonstrate that this resonance is due to C-2. Thus, the resonances at 70.8 and 70.9 ppm are due to C-3 and C-4. Gorin (7) was able to assign the resonance at 70.9 ppm to C-4 on the basis of his studies with α-D-[4-^{2}H]-

galactopyranose. This distinction cannot be made using 1-^{13}C enriched galactose, since neither C-3 nor C-4 is coupled to C-1 in the α-D anomer. Earlier workers (12, 13) were able to observe only one resonance for C-3 and C-4 due to the inherently broader lines in the CW spectra. The chemical shifts for these carbon atoms are listed in Table I as being separated by 0.07 ppm. Although this implies an accuracy greater than that claimed for other assignments, the two resonances do appear less than 0.1 ppm apart within a single spectrum. The resonance due to C-5 can also be assigned unequivocally on the basis of the 0.1 ppm upfield shift observed for C-5 in the 6-^{2}H$_2$ derivative (7). In the case of 1-^{13}C-enriched monosaccharides, C-5 of the α-D anomer can be identified by the presence of a two bond coupling of about 1.5–2.5 Hz (8) as shown in Figure 1B.

The discussion above demonstrates that unequivocal assignment of resonances to carbon atoms 2 through 5 of hexoses is a difficult matter which may require considerable synthetic effort. Earlier assignments had been made using comparisons and empirical rules which the more recent studies demonstrate are not valid. For example, C-2 in the methyl α-D-glycopyranosides was assigned on the assumption that conversion of the reducing sugar to the corresponding methyl glycoside would result in a downfield shift of the C-2 resonance. As the corrected assignments indicate (6–8), however, C-2 shifts upfield about 0.5 ppm following methylation at C-1. Similarly, the original assignments for α-D-galactopyranose, in which OH-4 is axial, were based on comparison with α-D-glucopyranose chemical shifts. In D-glucose, the observation was made that the conversion of β- to α-D-glucopyranose resulted in nearly equivalent upfield shifts for C-3 and C-5 and a smaller upfield shift for C-2. Unfortunately, the assignments for C-2 and C-3 in α-D-glucopyranose were incorrect; consequently, the assignments for C-2 and C-3 in α-D-galactopyranose were reversed. There is no doubt that some of the errors originally made in assignment of chemical shifts stem from the use of continuous-wave spectrometers, which may be inadequate for the resolution of small chemical shift differences, particularly for comparisons between spectra. These difficulties are eliminated in most Fourier-transform spectrometers now in use.

It should be stressed that the correct assignment of chemical shifts is of utmost importance to all nmr studies of structure and conformation, and the effort that must be expended to establish them with accuracy is worthwhile. The use of ^{13}C-enriched derivatives is particularly useful in this regard, since the enriched carbon atom and adjacent carbon and hydrogen atoms can be identified unequivocally under all circumstances because of the intensity of the signal from the enriched carbon atom and the large one bond couplings it produces. Longer-range interactions must be interpreted with care, because, as is shown below, they appear to depend on the conformation of the molecule and on the position of substituents. These features will be changed in certain interactions of the monosaccharides, making the interpretation of coupling constants difficult, but of potentially high information content.

[^{13}C] NMR of Other Enriched Monosaccharides

Gorin and Mazurek (*14*) have examined a series of furanosyl derivatives specifically labeled with deuterium. Their spectra exhibit the same effects on chemical shift observed for the hexopyranoses, and therefore, the unequivocal assignments in Table I can be made. In theory, the resonances for any other saccharides can be assigned by the deuterium substitution technique provided appropriate derivatives can be synthesized. Gorin (*15*), for example, has assigned resonances from a phosphomannan isolated from *Rhodotorula glutinis* using appropriately deuterated precursors. Unfortunately, the small β-isotope effect was observed only for the C-5 resonance of the 6,6′-dideuterated polysaccharide, where the isotope shift was larger than the C-5 line width. Other polysaccharides can be examined provided there is no rearrangement of the label during incorporation of the monosaccharide into the polysaccharide.

Relatively few monosaccharides specifically enriched with ^{13}C other than the hexopyranoses have been examined; thus little is known about ^{13}C–^{13}C coupling patterns in other systems. Although Walker *et al.* (*16*) examined the ^{13}C nmr spectra of a series of precursors for methyl D-[5-^{13}C]ribofuranoside, the compounds did not include any furanosides labeled at C-1. The 5-^{13}C furanosides that were examined exhibited only one-bond coupling to C-4, and no long-range couplings to carbon atoms in the ring at the resolution of 2–5 Hz used in these studies. Further studies at higher resolution are required to evaluate the utility of long-range ^{13}C–^{13}C couplings in the furanosides.

Walker *et al.* (*17*) have examined the 1-^{13}C-labeled aldonic acids, salts and lactones leading to galactose and fucose. The one-bond couplings in all cases were 50–60 Hz, in the range expected for coupling to carbonyl groups. The open-chain aldonic acids and salts exhibited C-1–C-4 coupling of 0.7–2.4 Hz, consistent with the three-bond couplings observed for aliphatic carboxylic acids (*18*). The 1,4-lactones exhibit long range C-1–C-3 couplings ranging from 2.8 to 7.8 Hz. In contrast to the results obtained with the hexonic acid derivatives, no long-range couplings were observed in the pentose series (*16*). This apparent difference may be due either to the poor spectral resolution (2.5 Hz) used for the pentoses, or to real differences in the coupling patterns.

Carbon–Carbon and Carbon–Hydrogen Coupling
in 1-^{13}C-Monosaccharides

Coxon (Vol. VI [93]) has described the approaches generally used in the interpretation of coupling constants in terms of carbohydrate conformation with principal reference to proton magnetic resonance. Bundle and Lemieux (Vol. VII [17]), Perlin (Vol. VII [19]), Bock and Pedersen (19), and others (8, 20, 21) have discussed ^{13}C–^{1}H coupling in natural abundance and ^{13}C-enriched carbohydrates. With specifically enriched carbohydrates, it is possible to observe couplings that are otherwise difficult to ascertain. The data in Table II and Figure 1 show that values for coupling constants between C-1 and other carbon atoms are obtainable by examining the resonances due to unlabeled carbon atoms in specifically labeled ^{13}C-enriched monosaccharides with commonly available instruments operating in the Fourier transform mode. It is evident that the magnitude of carbon–carbon coupling constants generally varies with the configuration of the monosaccharides. All those listed are assumed to exist in the $^{4}C_{1}$ conformation in aqueous solution. Earlier workers (21, 22) utilizing uniformly ^{13}C-labeled carbohydrates, were unable to observe any long-range ^{13}C–^{13}C couplings, although they did observe that all resonances were somewhat broadened, indicating the presence of unresolved couplings. The cmr spectra of the 1-^{13}C-labeled sugars, however, have proven particularly useful as a means for examining long-range proton and carbon couplings to ^{13}C-1.

One-Bond Coupling

One-bond C-H coupling can be observed in proton-coupled ^{13}C nmr spectra of uniformly or specifically ^{13}C-enriched carbohydrates (8, 21, 22) or at high concentrations of natural abundance material (22). Pedersen et al. (19, 20, 23) examined a series of natural abundance hexo- and pentopyranoses and found a marked anomeric and conformational dependence of the C-1–H-1 coupling. For compounds which exist in the $^{4}C_{1}$ conformation, coupling constants of ~ 170 Hz for the α-D anomers and ~ 160 Hz for the β-D anomers are obtained. On the other hand, pyranoses in the $^{1}C_{4}$ conformation exhibit couplings of ~ 170 Hz for the β-D anomer and ~ 160 Hz for the α-D anomer. Bock and Pedersen (23) suggest that this effect might be caused by the proximity of H-1 to the lone pair of electrons on the ring oxygen atom, an effect shown to increase C-H coupling in other compounds (24). Thus, for α-D sugars in the $^{4}C_{1}$ conformation, where H-1 is equatorial, larger coupling is observed than in the β-D sugars where H-1 is axial and relatively farther from the lone pair. These observations provide a quantitative basis for calculating the percentage of $^{1}C_{4}$ conformation present in solution for a particular pyranose. Good agreement with ^{1}H nmr data has been observed for several pentopyranoses. β-D-Mannosamine, however, appears anomalous since the C-1–H-1 coupling constant of 165 Hz suggests considerable $^1C_4$ conformation for a structure thought to be $^4C_1$. Recently, Gorenstein (25) has suggested that the differences in C-1–H-1 couplings observed by Bock and Pedersen may be due to small changes in bond lengths and angles, rather than to changes in proximity to the lone pairs of electrons on oxygen atoms.

In the cases studied, one-bond coupling between C-1 and C-2 is large (41–47 Hz) and varies significantly with configuration at C-1 only in the *manno*-derivatives, in sharp contrast to the anomeric dependence observed for the C-1–H-1 coupling (Table II). The β-D-mannopyranosyl compounds, including the 2-amino-2-deoxy-D-mannopyranoses, have smaller couplings by 2.6–4.4 Hz than the α-D anomers, which may be due to the presence of an axial group at C-2 in the dominant conformer. In sugars having an equatorial group at C-2, differences in C-1–C-2 coupling between α- and β-D anomers is smaller, $\sim$0.5 Hz. The couplings for sugars with the *manno* configuration range both higher and lower than those for the other neutral sugars in which the C-1–C-2 coupling varies only from 45.8 to 46.8 Hz. Notably, the amino sugars show couplings $\sim$2 Hz smaller than those for the neutral sugars.

Two-Bond Coupling

Two-bond coupling between C-1 and H-2 can be observed readily in the coupled ^{13}C spectrum of 1-^{13}C-enriched sugars and in natural abundance samples of high concentration. These couplings have been studied extensively by Perlin *et al.* (21, 26–28), who found that coupling appears to depend on the conformation about the bonds through which coupling occurs and the dihedral angles subtended by substituents on the coupled atoms. Schwarcz and Perlin (26) propose that electronegative oxygen atoms *anti* to H-2 make a positive contribution to coupling between C-1 and H-2, whereas *gauche* oxygen atoms make a negative contribution. These observations are borne out by the data shown in Table II, in that the β-D-*gluco*- and *galacto* compounds (Figure 2a) display coupling of approximately 5 Hz, whereas the α-D anomers (Figure 2b,d) and the *manno*-isomers (Figure 2c,d) show only line broadening. Indeed, these smaller couplings are resolvable in the proton spectra (29), but only in those cases where the H-2 proton is shifted out of a broad envelope of resonances. The sign of the ^{13}C-1–H-2 coupling is negative in β-D-allopyranose (-5 Hz) (28), α- and β-D-glucosamine (-2.6 and -6.9 Hz), α- and β-D-mannosamine (-3.0 Hz) (29), supporting the proposal that the sign and magnitude of couplings are functions of the di-

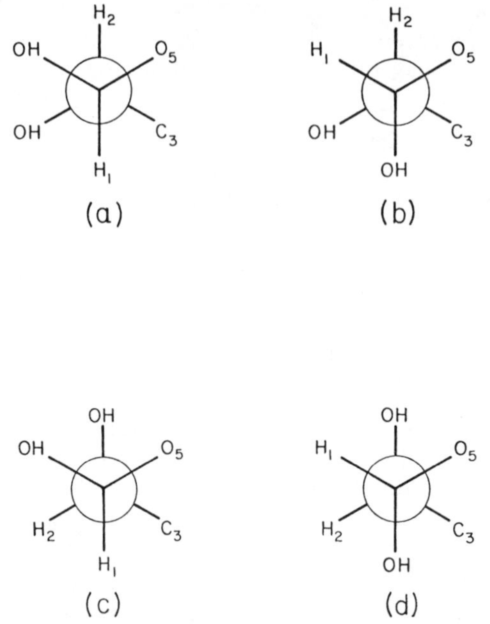

FIG. 2.—Projections along the C-1–C-2 bond of D-pyranose sugars in the $^4C_1$ conformation. (a) β-D anomers of gluco-, galacto-, allo- and gulopyranoses. (b) α-D anomers of the same. (c) β-D anomers of manno-, ido-, altro- and talopyranoses. (d) α-D anomers of the same.

hedral angles between the coupled atoms and their substituents in the hexopyranoses.

Similar considerations might be expected to apply to C-1–C-3 couplings if they were qualitatively similar to C-1–H-2 couplings. The data in Table II demonstrate, however, that C-1–C-3 coupling is observed only for the β-D sugars, in contrast to that expected from dihedral angle dependence.

The application of the dihedral angle dependence to the C-1–C-5 coupling is a fundamentally different problem for two reasons: (a) coupling is through an oxygen atom rather than a carbon atom, and (b) there is only one electronegative oxygen atom substituent to be considered at C-1. Thus, although the observation of C-1–C-5 coupling in the α-D anomers is consistent with the dihedral angle dependence, the agreement may be fortuitious. Clearly there are either additional factors involved in determining the two-bond C–C coupling or significant quantitative differences relative to the C-1–H-2 case.

Dorman proposed that the presence of axial hydroxyl groups on a carbon atom in cyclitols led to reduced two- and three-bond coupling to protons, whereas larger couplings were observed for carbon atoms with equatorial hydroxyl groups (*30*). This explanation is perhaps more consistent with the all or nothing effect observed for the two-bond ^{13}C–^{13}C couplings; however, it predicts no coupling for the α-D anomers where C-1–C-5 coupling is observed. Further data are required to determine whether other, as yet undetermined, factors may be of significant influence. Nevertheless, the couplings that have been established provide a basis for establishing configurations and conformations, and for exploring changes in these features with derivatization or with environmental changes that occur in complex formation, solvent perturbation, or carbohydrate–protein interactions.

Three-Bond Coupling

The relationship of three-bond coupling to the dihedral angle between ^{13}C and ^{1}H has been explored by Lemieux *et al.* (*31*) for uridine derivatives and by Schwarcz and Perlin (*26*) for a number of pyranosyl and furanosyl derivatives (Table III). Both studies demonstrate that the coupling constant is smallest when the dihedral angle is approximately 90° ($J \cong 0$ Hz) and greatest at 180° ($J \cong 6$–9 Hz). Over the range of dihedral angles from 40° to 180°, coupling between ^{13}C and H approximates a Karplus relationship. Other factors modify this relationship, such as the disposition of substituents on atoms in the coupling pathway, the hydridization state of carbon (sp^2 or sp^3), and the presence of higher order spin systems (*32*). The three-bond ^{13}C–^{13}C couplings in the hexopyranoses also obey a Karplus-type relationship. As shown in Figure 1, C-1 is coupled to C-6 (a *trans* relationship) but not to C-4 (a *gauche* relationship). There also appears to be a significant anomeric dependence of the C-1–C-6 coupling, coupling in the β-D anomer being about 1 Hz larger than that in the α-D anomer (*8*). This anomeric dependence suggests that the coupling is subject to other factors in addition to the dihedral angle effects discussed above.

Summary

The discussion above demonstrates that the application of "rules" for the assignment of chemical shifts to specific atoms is likely to lead to errors, even when the "rules" are applied with caution. Nevertheless, a number of valuable correlations exist that are useful in establishing chemical shifts and coupling constants, and from them, structural features. The following list of correlations is useful in this regard.

Enrichment with ^{13}C at a specific carbon atom allows unequivocal identification of carbon atoms α to the site of substitution because all α-carbon

TABLE III

$^{13}C-^1H$ Couplings in ^{13}C-Enriched Carbohydrates[a]

Compound	Coupling pathway	Dihedral angle (degree)	$^3J_{C-H}$ (Hz)	Coupling pathway	$^2J_{C-H}$ (Hz)
1,2,3,4,6-Penta-O-acetyl-α-D-[6-¹³C]glucopyranose	C-6–H-4	60	2.8	C-6–H-5	0
1,2,3,4,6-Penta-O-acetyl-β-D-[6-¹³C]glucopyranose	C-6–H-4	60	2	C-6–H-5	0
1,2,3,4,6-Penta-O-acetyl-β-D-[1-¹³C]glucopyranose	C-1–H-3	60	0	C-1–H-2	5.0
	C-1–H-5	60	2.5		
Methyl-α-D-[6-¹³C]glucopyranoside	C-6–H-9	60	3.3	C-6–H-5	0
α-D-[1-¹³C]Glucose	C-1–H-3			C-1–H-2	0
β-D-[1-¹³C]Glucose	C-1–H-3			C-1–H-2	5.7
β-D-[U-¹³C]Glucose	C-3–H-1	60	0		
	C-5–H-1	60	0		
α-D-[U-¹³C]Glucose	C-3–H-1	180	5.6		
1,2-O-Isopropylidene-3,5,6-tri-O-orthoformyl-α-D-[6-¹³C]glucofuranose	C-6–H-4	70	0	C-6–H-5	0
	C-6–H-formyl	160	4.6		
1,2-O-Isopropylidene-α-D-[1-¹³C]glucofuranose	C-1–H-3	140	5.5	C-1–H-2	5.5
	C-1–H-5	110	0		
1,2-O-Isopropylidene-α-D-[6-¹³C]glucofuranose	C-6–H-4		2.4	C-6–H-5	2
1,2-O-Isopropylidene-β-L-[6-¹³C]idofuranose	C-6–H-4		0	C-6–H-5	0
1,2,5,6-Di-O-isopropylidene-α-D-[6-¹³C]glucofuranose	C-6–H-4		2.7	C-6–H-5	0
1,2,5,6-Di-O-isopropylidene-α-D-[1-¹³C]glucofuranose	C-1–H-3	140	0	C-1–H-2	5.5
	C-1–H-5	110	0		
1,2-O-Isopropylidene-α-D-glucofuranurono-6,3-[6-¹³C]lactone	C-6–H-4	140	9	C-6–H-5	4.1
	C-6–H-3	100	0		
1,2-O-Isopropylidene-β-L-idofuranurono-6,3-[6-¹³C]lactone	C-6–H-4	140	6.6	C-6–H-5	0
	C-6–H-3	100	0		
D-Mannono-1,4-[1-¹³C]lactone	C-1–H-3	140	8.9	C-1–H-2	4.8
	C-1–H-5	110	0		

[a] Adapted from Schwarcz and Perlin (29); 0 Hz indicates ± 1 Hz.

atoms will show large splittings ($\sim$46 Hz in the case of 1-^{13}C-enriched pyranoses).

Similarly, protons attached to ^{13}C are split by approximately 160 Hz. The extent of coupling may be influenced by small changes in bond lengths and angles, or by associated structures, such as a ring oxygen atom. Characteristically, C-1–H-1 coupling is larger for the β-D anomer than for the α-D anomer in D-pyranoses with ^{4}C$_1$ conformations.

Two-bond coupling between C-1 and C-3 or C-5 has been observed in the β-D and α-D anomers, respectively, of the hexopyranosyl derivatives of several monosaccharides (Table II). These observations should not be generalized, however, since the basis for them is not well enough established to permit formulation of a rule.

Two-bond coupling between carbon and hydrogen atoms appears to be influenced by the type and position of substituents on the carbon atom, but firm correlations are not yet possible.

Replacement of a proton by a deuterium atom causes the resonances of substituted carbon atoms to be unobservable or observable as triplets of low intensity at higher field.

Carbon atoms β to a deuterium atom are shifted upfield by 0.02–0.17 ppm.

Conversion of an equatorial to an axial hydroxyl group, as occurs at C-4 when glucose is converted to galactose, causes displacements in chemical shifts; however, the direction of these shifts cannot be correlated with the change made (15).

Conversion of –CH$_2$OH at C-5 of a pyranose to –CH$_3$ causes C-4 to be displaced downfield by 2.0 to 5 ppm, and C-5 to be displaced upfield by 4 ppm. These correlations have been established in the D-gluco-, D-manno- and D-galactopyranosyl derivatives only and should be applied with caution to other systems (6).

Conversion of -CH$_2$OH at C-5 to –H causes C-5 to be shifted upfield by $\sim$ 10 ppm. This correlation has been established in the D-gluco- and D-galactopyranosyl systems only (6).

Conversion of CH$_2$OH at C-5 to –CO$_2$H or –CO$_2$CH$_3$ causes C-4 to be displaced upfield by 1.9 ppm and C-5 to be displaced downfield by 0.8 ppm in the D-glucopyranosyl system (6).

O-Alkylation displaces the resonance for the carbon atom downfield, whereas resonances of adjacent carbon atoms are generally shifted upfield by smaller amounts. O-Methylation has a greater effect than O-isopropylidination or O-glycosidation (14).

The correlations described in this section should be applied with caution to other systems. Assignments based on long-range couplings or on substituent effects should be checked by the use of isotopic replacements wherever possible.

Procedures

Earlier contributions to this series have described the procedures generally used to obtain nmr spectroscopic data (Vol. VI [93]; Vol. VII [17], [19]). For precise work, samples should be degassed or flushed with nitrogen or argon to remove paramagnetic oxygen, and if paramagnetic ions might be present, passed over Chelex 100 (Bio-Rad Laboratories, Richmond, California) to remove them. In ^{13}C nmr studies, the instrument is locked on, and D_2O contained in a capillary is used with samples dissolved in water. A suitable reference signal is that of C-1 of β-D-glucopyranose, which occurs at 97.4 ppm in 1 M solution in H_2O at 34° relative to external tetramethyl-silane. The instrument can be set to this value and checked occasionally to assure accurate estimation of chemical shift values.

When ^{13}C-enriched samples are used, it will often be valuable to examine the spectrum of the unenriched portion of the molecule using a spectral window which excludes the resonance signal from the enriched carbon atom.

References

(1) T. L. James, "Nuclear Magnetic Resonance in Biochemistry," Academic Press, New York (1975).

(2) S. E. Lasko and P. Milvy, eds., *Ann. N.Y. Acad. Sci.*, **222** (Part X, "Structural Studies of Carbohydrates"), 884–988 (1973).

(3) A. S. Perlin, "International Review of Science, Organic Chemistry, Series Two," Vol. 7, 1976, p. 1.

(4) H. J. Koch and A. S. Perlin, *Carbohydr. Res.*, **15**, 403 (1970).

(5) F. J. Weigert and J. D. Roberts, *J. Amer. Chem. Soc.*, **89**, 2967 (1967).

(6) P. A. J. Gorin and M. Mazurek, *Can. J. Chem.*, **53**, 1212 (1975).

(7) P. A. J. Gorin, *Can. J. Chem.*, **52**, 458 (1974).

(8) T. E. Walker, R. E. London, T. W. Whaley, R. Barker, and N. A. Matwiyoff, *J. Amer. Chem. Soc.*, **98**, 5807 (1976).

(9) A. S. Serianni, H. A. Nunez, and R. Barker, *Carbohydr. Res.*, **72**, 71 (1979).

(10) A. S. Serianni, E. L. Clark, and R. Barker, *Carbohydr. Res.* **72**, 79 (1979).

(11) J. K. N. Jones and R. A. Wall, *Can. J. Chem.*, **38**, 2290 (1960).

(12) D. E. Dorman and J. D. Roberts, *J. Amer. Chem. Soc.*, **92**, 1355 (1970).

(13) A. S. Perlin, B. Casu, and H. J. Koch, *Can. J. Chem.*, **48**, 2596 (1970).

(14) P. A. J. Gorin and M. Mazurek, *Carbohydr. Res.*, **48**, 171 (1976).

(15) P. A. J. Gorin, *Carbohydr. Res.*, **27**, 3 (1975).

(16) T. E. Walker, H. P. C. Hogenkamp, T. E. Needhaw, and N. A. Matwiyoff, *Biochemistry*, **13**, 2650 (1974).

(17) T. E. Walker, H. Nunez, and R. Barker, unpublished observation.

(18) M. Barfield, I. Burfitt, and D. Doddrell, *J. Amer. Chem. Soc.*, **97**, 2631 (1975).

(19) K. Bock and C. Pedersen, *Acta Chem. Scand.*, **B29**, 258 (1975).

(20) K. Bock, I. Lundt, and C. Pedersen, *Tetrahedron Lett.*, **13**, 1037 (1973).

(21) A. S. Perlin and B. Casu, *Tetrahedron Lett.*, **34**, 2921 (1969).

(22) R. E. London, V. H. Kollman, and N. A. Matwiyoff, *J. Amer. Chem. Soc.*, **97**, 3565 (1975).

(23) K. Bock and C. Pedersen, *J. Chem. Soc. Perkin Trans. 2*, 293 (1974).
(24) T. Yoneqawa and I. Marishima, *J. Mol. Spectrosc.*, **27**, 210 (1968).
(25) D. G. Gorenstein, *J. Amer. Chem. Soc.*, **99**, 2254 (1977).
(26) J. A. Schwarcz and A. S. Perlin, *Can. J. Chem.*, **50**, 3667 (1972).
(27) D. R. Bundle, H. J. Jennings, and I. C. P. Smith, *Can. J. Chem.*, **51**, 3812 (1973).
(28) J. A. Schwarcz, N. Cyr, and A. S. Perlin, *Can. J. Chem.*, **53**, 1872 (1975).
(29) T. E. Walker, R. E. London, R. Barker, and N. A. Matwiyoff, *Carbohydr. Res.*, **60**, 9 (1978).
(30) D. E. Dorman, *Ann. N.Y. Acad. Sci.*, **222**, 943 (1973).
(31) R. U. Lemieux, T. L. Nagabhushan, and B. Paul, *Can. J. Chem.*, **50**, 773 (1972).
(32) J. A. Pople, W. G. Schnieder, and H. J. Bernstein, "High Resolution Nuclear Magnetic Resonance," McGraw-Hill, New York (1959).

Section II. Preparation of Mono-, Oligo-, and Polysaccharides and Their Derivatives

MONOSACCHARIDES AND THEIR DERIVATIVES
Articles 22 through 28

POLYSACCHARIDES
Article 29 ..

UNSATURATED SUGARS
Article 30

DEOXYHALO SUGARS
Article 31

GLYCOSIDES AND GLYCOSYLAMINES
Articles 32 through 38

ESTERS AND AMIDES
Articles 39 through 46

ETHERS
Article 47

CYCLIC ACETALS
Articles 48 and 49

OXIDIZED PRODUCTS
Article 50

MONOSACCHARIDES AND THEIR DERIVATIVES

[22] Methyl α-D-Altropyranoside and Its 4,6- and 3,4-O-Isopropylidene Derivatives

By Michael E. Evans

The Australian Wine Research Institute, Glen Osmond, South Australia

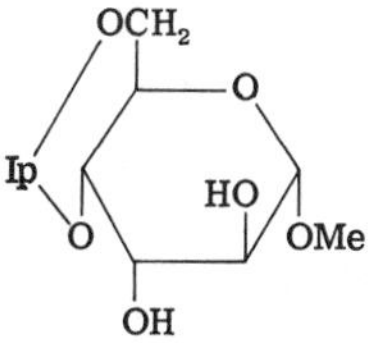

(I) 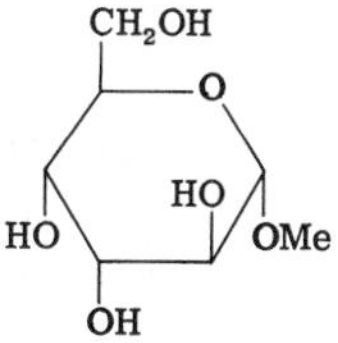   (II) 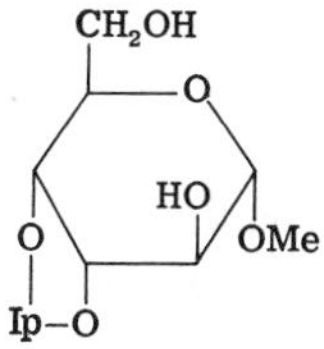   (III)

Introduction

Methyl α-D-altropyranoside (II) is usually prepared from methyl α-D-glucopyranoside by successive benzylidenation, p-toluenesulfonylation, epoxide formation, alkali-catalyzed hydrolysis, and acid-catalyzed hydrolysis. The procedure requires 8–20 days and the yield is 40–50%, depending on the exact method used (Vol. I [30] and references therein). The synthesis may be improved greatly by employing monomolar methanesulfonylation of methyl 4,6-O-isopropylidene-α-D-glucopyranoside (1), followed by generation, in situ, of an epoxide and its hydrolysis to methyl 4,6-O-isopropylidene-α-D-altropyranoside (I). Methyl 4,6-O-isopropylidene-α-D-altropyranoside (I) is obtained in 55% yield, based on methyl α-D-glucoside; and only 2–3 days are required for the preparation. It may be converted by hydrolysis to methyl α-D-altropyranoside (II) or by isomerization to methyl 3,4-O-isopropylidene-α-D-altropyranoside (III), each obtained in almost quantitative yield.

Procedure

Methyl 4,6-O-Isopropylidene-α-D-altropyranoside (I)

Methyl α-D-glucopyranoside (4.0 g), 0.04 g of p-toluenesulfonic acid monohydrate, 16 ml of 2,2-dimethoxypropane, and 16 ml of N,N-dimethyl-

169

formamide (DMF) are stirred together for 16 h at 23°. Sodium hydrogen carbonate solution (1.2 M, 10 ml), and 30 ml of water are added successively; and the solution is extracted continuously with chloroform for 6 h in a liquid–liquid extractor. The extract is dried with anhydrous magnesium sulfate and evaporated under reduced pressure; then, the bath temperature is raised to 70° to remove the DMF. Xylene (six 4-ml portions) is evaporated from the residue until the residue forms a brittle gel which is then dried for 5 min at 70°/0.01 torr.

The product is dissolved in 12 ml of pyridine, and the solution is stirred at <10° while 1.45 ml of methanesulfonyl chloride is added. After 10 min, the solution is examined by tlc (Vol. VI [6]) (silica gel G plates irrigated with 2:1 v/v ethyl acetate–hexane). If methyl 4,6-O-isopropylidene-α-D-gluco-pyranoside (R_f value 0.1–0.2) remains, more acid chloride is added in 0.1 ml portions until the glucoside has all reacted. [Note: The bismethanesulfonate of methyl 4,6-O-isopropylidene-α-D-glucopyranoside is resistant to the alkaline hydrolysis conditions used in the next step; use of too much meth-anesulfonyl chloride results in a reduced yield of I.] Water (40 ml) and 40 ml of chloroform are added; the mixture is stirred for 10 min, and the aqueous layer is separated and extracted with two 20-ml portions of chloroform. All the chloroform layers are combined, dried with anhydrous magnesium sulfate, and evaporated. Toluene (30 ml) is evaporated from the residue, which is then dried 30 min at 30°/0.01 torr. M Sodium hydroxide solution (60 ml) is added; and the mixture is boiled for 16 h, cooled, and extracted continuously for 6 h with chloroform. The extract is dried with anhydrous magnesium sulfate and evaporated. The residue is dissolved in 50 ml of hot ethyl acetate, and the hot solution is passed through 10 g of 1:1 w/w carbon–celite, then through 6 g of silica gel packed in the same column (2 cm i.d.). The product is eluted from the column with 100 ml of ethyl acetate. Pressure should be used to drive the solution and the first 20 ml of the washing through the column, otherwise I tends to crystallize and block the flow. The eluate is concentrated, and the product is dissolved in 5 ml of n-butyl acetate. The solution is cooled slowly to 3° to give a crystalline product which is collected by filtration, washed with 1:1 v/v ethyl acetate–hexane, and dried at 40°; yield of I 2.57 g (57%), mp 167°–169° (sublimes from 160°), [α]D +125° (c 0.61, water).

Methyl α-D-Altropyranoside (2)

A solution of 1.5 g of I in 10 ml of 0.1 M hydrochloric acid is kept for 2 h at 23°; 2 ml of Dowex 1X8(HCO$_3$⁻) anion-exchange resin is added, and the solution is aerated for 10 min. The resin is removed by filtration and washed with four 5-ml portions of water. The filtrate and washings are combined and concentrated to a syrup which crystallizes rapidly when kept at 50°. The product is dried 1 h at 23°/0.1 torr to give II; yield 1.2 g (97%), mp 106°–107°. Methyl α-D-altropyranoside is recrystallized from 1-propanol; yield 1.11 g (90%), mp 107°–108°, [α]D +125° (c 3.0, water).

Methyl 3,4-O-Isopropylidene-α-D-altropyranoside (III)

Compound I (0.5 g) is stirred for 10 min with 5 ml of acetone containing 5 μl of conc. sulfuric acid; 20 μl of conc. ammonium hydroxide is added and the mixture is concentrated to dryness. The residue is suspended in 2 ml of ethyl acetate and applied to a column of 5 ml of silica gel; the column is eluted with 30 ml of ethyl acetate. Concentration of the eluate gives III; yield 0.49 g (98%). Although methyl 3,4-O-isopropylidene-α-D-altropyranoside (III) forms large crystals, no satisfactory conditions for recrystallization were found. Analysis of the crude product by acetylation and glc (Vol. VI [5]) showed it to contain ~98% of III (2). The product is dissolved in 1 ml of ethyl acetate; 1.3 ml of hexane is added, and the solution is nucleated, then kept 4 days at 3° to give pure III, mp 61°–62°, [α]D +103° (c 1.66, water).

References

(1)   M. E. Evans and F. W. Parrish, *Carbohydr. Res.*, **3**, 453 (1967).
(2)   M. E. Evans, *Carbohydr. Res.*, **30**, 215 (1973).

[23] L-Gulose and Methyl β-L-Gulofuranoside

By Michael E. Evans

The Australian Wine Research Institute, Glen Osmond, South Australia

AND

Frederick W. Parrish

Eastern Regional Research Center, Science and Education Administration, U.S. Department of Agriculture, Philadelphia, Pennsylvania

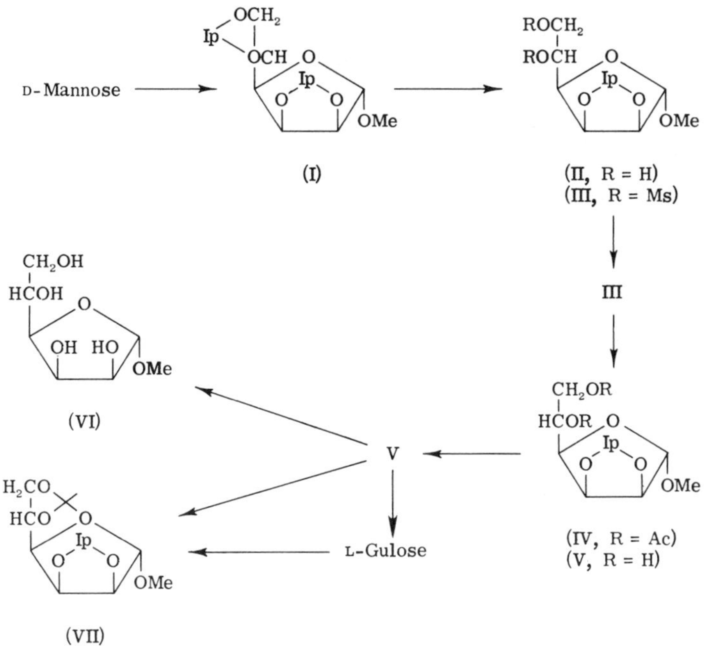

D-Mannose ⟶ (I) ⟶ (II, R = H) (III, R = Ms)

(III)

(VI)

(VII)

(IV, R = Ac) (V, R = H)

V

L-Gulose

Introduction

D-Gulose may be prepared by reduction of D-gulono-1,4-lactone (Vol. I [39]), by reduction of 3-O-acetyl-1,2:5,6-di-O-isopropylidene-α-D-*erythro*-hex-3-enofuranose (*1*, Vol. VI [17]), by oxidation and reduction of 4,6-O-ethylidene-1,2-O-isopropylidene-α-D-galactopyranose (*2*), or by reduction of 2,3,5,6-tetra-O-benzoyl-D-gulono-1,4-lactone (Vol. VI [18]). L-Gulose is

173

METHODS IN CARBOHYDRATE
CHEMISTRY, VOL. VIII

accessible from 2,4-*O*-benzylidene-D-glucitol (Vol. I [40]) or from D-mannose (*3*) by successive glycosidation/isopropylidenation, acid-catalyzed hydrolysis, methylsulfonylation, acetate exchange, saponification, and acid-catalyzed hydrolysis. Methyl *β*-L-gulofuranoside may be obtained by using milder conditions for the final hydrolysis.

Treatment of L-gulose with acidified 2,2-dimethoxypropane–acetone–methanol gives methyl 2,3:5,6-di-*O*-isopropylidene-*β*-L-gulofuranoside which crystallizes well and is suitable for characterization of L-gulose.

Procedure

Methyl 2,3-*O*-Isopropylidene-5,6-di-*O*-methylsulfonyl-α-D-mannofuranoside (III)

A solution of 1 g of D-mannose, 3.4 ml of 2,2-dimethoxypropane, 3.3 ml of acetone, 3.3 ml of methanol, and 0.1 ml of conc. hydrochloric acid is boiled under reflux for 2 h. The solution is cooled, diluted with 10 ml of water, and concentrated to ~10 ml at <30°. Methanol (10 ml) and 0.25 ml of conc. hydrochloric acid are added, and the solution is kept 200 min at 23°. At this stage tlc (Vol. VI [6]) (silica gel G plates developed in 1:1 v/v ethyl acetate–hexane) should show that almost no I is left and that the product is essentially pure compound II. Sodium hydrogen carbonate solution (1 *M*, 7.5 ml) is added; the mixture is concentrated to remove the methanol and then extracted continuously with chloroform for 3 h in a liquid–liquid extractor. The chloroform extract is dried with anhydrous magnesium sulfate and concentrated to a syrup. This syrup is dissolved in 5 ml of pyridine, and the solution is stirred at <35° while methanesulfonyl chloride (1.5 ml) is added, then kept 2 h at 20°. Excess acid chloride is decomposed by cautious addition of water, keeping the temperature at <50°; then 150 ml of water is added. The product is collected by filtration, washed thoroughly with water, and dried to constant weight; yield of III 1.67 g (77%), mp 143°–144°. The product is recrystallized from 25 ml of ethanol; yield 1.64 g, mp 144.5°–146°,[1] [α]D +33° (*c* 0.50, chloroform).

Methyl 2,3-*O*-Isopropylidene-*β*-L-gulofuranoside (V)

A mixture of 5.0 g of III, 5.0 g of sodium acetate, and 40 ml of *N,N*-dimethylformamide is boiled under reflux for 6 h; 7.5 ml of acetic anhydride

---

[1] Lerner reported (*4*) that III is more soluble in methanol than in ethanol and that recrystallization from methanol gives material melting at 147°–148°. We found that III has similar solubilities in methanol (1 g in ~17 ml) and in ethanol (1 g in ~15 ml). When the material with mp at 144.5°–146° was recrystallized from methanol, the product melted at 145.5–146.5°.

is added cautiously to the mixture and boiling is continued for 1 h. The solution is cooled, and the solvents are removed at 65°/30 torr with the aid of a short path evaporation adaptor (5). The residue is dried 30 min at 95°/30 torr. The product is dissolved in 6 ml of methanol, and after 5 min, the solution is stirred rapidly while 25 ml of sodium hydrogen carbonate solution and 70 ml of water are added successively. The solution is extracted continuously with hexane for 6 h. The extract is dried with anhydrous magnesium sulfate and concentrated to dryness. The residue is dissolved in 10 ml of 0.1 M sodium methoxide in methanol. The solution is kept at 23° for 5 h, then neutralized with carbon dioxide, and concentrated to dryness. The product is dissolved in 30 ml of ethyl acetate and passed successively through 6 g of 1:1 w/w carbon–Celite and 20 g of silica gel packed in the same 22-mm (i.d.) column. The product is eluted with 150 ml of ethyl acetate. The eluate is concentrated to a syrup, which is dissolved in 4 ml of hot ethyl acetate; 12 ml of hexane is added, and the solution is nucleated and cooled slowly to 3°; yield of V 1.94 g (64.5%), mp 76.5°–77° (unchanged by recrystallization), [α]D +82° (c 1.17, methanol).

Methyl β-L-Gulofuranoside (VI)

A solution of 1.0 g of V in 12 ml of methanol and 28 ml of 0.033 M hydrochloric acid is kept at 45° for 3 days, then cooled, and applied to a 2.5 × 35-cm column of Dowex 1X2(OH⁻) anion-exchange resin (200–400 mesh). The column is eluted with water, and 10-ml fractions are collected and examined polarimetrically.

Fractions 10–13 (positive rotations) give V; yield 0.18 g (18%).

Fractions 25–28 (negative rotations) give a product tentatively identified as methyl α-L-gulofuranoside; yield 0.013 g.

Fractions 42–59 (positive rotations) give VI; yield 0.53 g (64%), mp 101° (unchanged on recrystallization from 1-propanol), [α]D +108° (c 1.0, water); reported (6) for the D enantiomer), mp 101°, [α]D −108° (water).

L-Gulose

A solution of 0.50 g of V in 5 ml of 0.01 M hydrochloric acid is boiled under reflux for 150 min, then cooled to 30°, 0.5 ml of Dowex 1X8(HCO₃⁻) anion-exchange resin is added, and the mixture is aerated for 10 min. The resin is removed by filtration and washed with four 5-ml portions of water. The filtrate and washings are combined, concentrated to a syrup, and dried for 3 h at 40°/30 torr to give L-gulose; yield 0.395 g (100%), [α]D +21° (c 4.58, water); reported (Vol. I [40]) [α]D +20° (water). The product gives a phenylhydrazone melting at 143°; reported for the D enantiomer (Vol. I [39]) mp 143°.

When the hydrolysis is allowed to proceed for 24 h, the product contains a considerable amount of a second compound, probably 1,6-anhydro-β-L-gulose.

Methyl 2,3:5,6-di-O-isopropylidene-β-L-gulofuranoside (VII)

A. From V. A solution of 0.105 g of V and 1.5 mg of p-toluenesulfonic acid monohydrate in 2 ml of 2,2-dimethoxypropane is stirred at 23° for 30 min. Pyridine (0.05 ml) is added, and the mixture is washed through 1.5 ml of silica gel with 15 ml of 1:1 v/v of ethyl acetate–hexane. The eluate is concentrated to give VII; yield 0.12 g (97%), mp 76°–77°. The product is recrystallized from 0.5 ml of hexane to give material melting at 76°–77°; [α]D +42° (c 1.13, chloroform). A sample of the D enantiomer of VII, prepared (7) by Kuhn methylation (Vol. II [39], Vol. V [72]) of 2,3:5,6-di-O-isopropylidene-β-D-gulose melted at 75°–76°, [α]D −45° (chloroform).

B. From L-Gulose. A mixture of 0.14 g of L-gulose, 0.12 g of p-toluenesulfonic acid monohydrate, 1 ml of acetone, 1 ml of 2,2-dimethoxypropane, and 1 ml of methanol is boiled under reflux for 2 h, then cooled, and treated with 5 ml of M sodium hydrogen carbonate solution. The product is concentrated to dryness, then kept 15 min at 70°/30 torr. The residue is extracted with three 6-ml portions of hexane, and the extracts are combined and evaporated. The product is recrystallized from 0.25 ml of methanol; yield of VII 0.05 g, mp 76°–77° alone or mixed with the product from V.

References

(1) K. N. Slessor and A. S. Tracey, Can. J. Chem., 47, 3989 (1969).
(2) G. J. F. Chittenden, Carbohydr. Res., 15, 101 (1970).
(3) M. E. Evans and F. W. Parrish, Carbohydr. Res., 28, 359 (1973).
(4) L. M. Lerner, Carbohydr. Res., 36, 392 (1974).
(5) M. E. Evans, Carbohydr. Res., 21, 473 (1972).
(6) H. G. Fletcher, Jr., H. W. Diehl, and R. K. Ness, J. Amer. Chem. Soc., 76, 3029 (1954).
(7) S. J. Angyal, personal communication.

[24] L-Idose and L-Iduronic Acid

By Michèle Blanc-Muesser and Jacques Defaye

Centre de Recherches sur les Macromolécules Végétales,
CNRS, Grenoble, France

AND

Derek Horton and Ji-Hsiung Tsai

Department of Chemistry, The Ohio State University, Columbus, Ohio

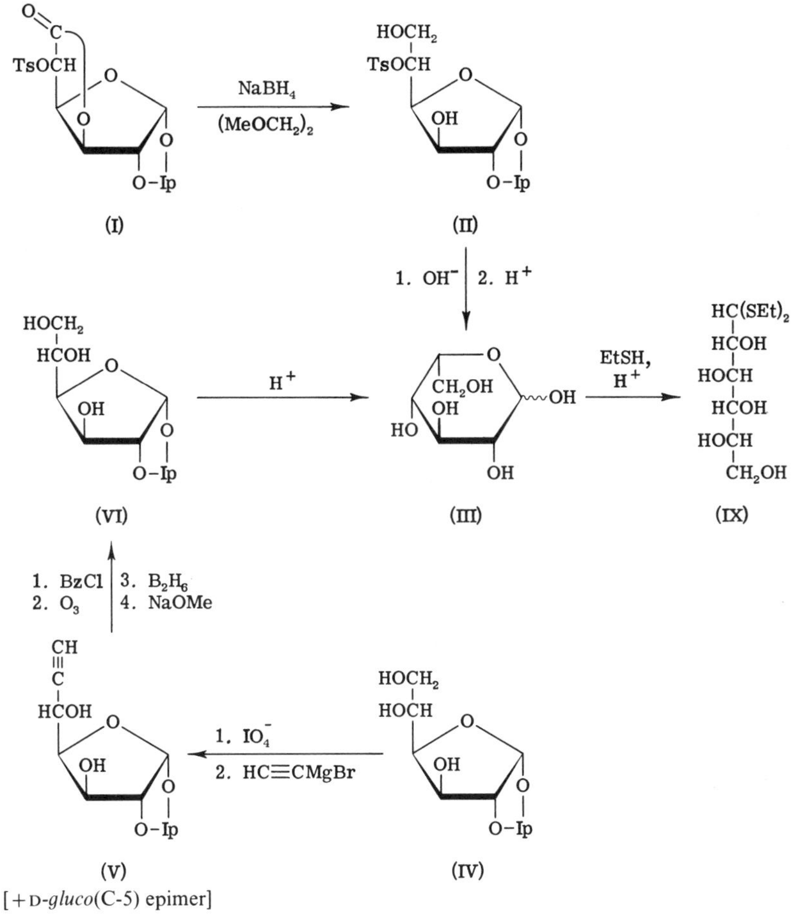

(I) $\xrightarrow[\text{(MeOCH}_2\text{)}_2]{\text{NaBH}_4}$ (II)

$\xdownarrow$ 1. OH⁻ | 2. H⁺

(VI) $\xrightarrow{\text{H}^+}$ (III) $\xrightarrow[\text{H}^+]{\text{EtSH,}}$ (IX)

1. BzCl   3. B₂H₆
2. O₃   4. NaOMe

(V) $\xleftarrow[\text{2. HC≡CMgBr}]{\text{1. IO}_4^-}$ (IV)

[+D-gluco(C-5) epimer]

177

(V)

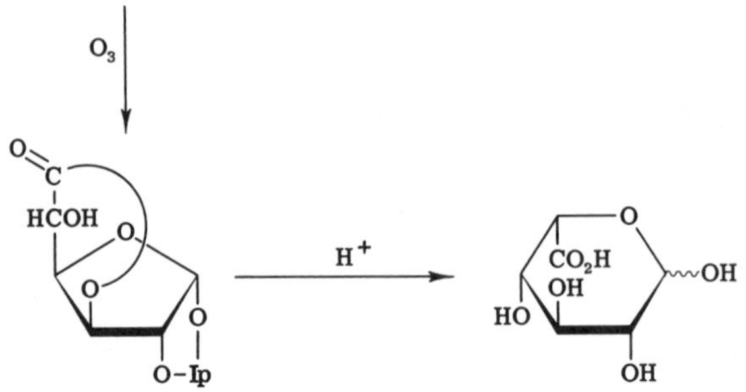

O_3

(VII)

H$^+$

(VIII)

Introduction

The preparation of L-idose (III) (see also Vol. I [42]) is achieved in a sequence of high-yielding steps (1) starting from the commercially available D-glucurono-6,3-lactone, and the syrupy product (III) is characterized as its crystalline diethyl dithioacetal (IX). An alternative route (2) that begins with 1,2-O-isopropylidene-α-D-glucofuranose (IV) utilizes an acetylenic intermediate (V) having the L-*ido* configuration as a common precursor for either the crystalline 1,2-isopropylidene acetal (VI) of L-idose or the crystalline 1,2-isopropylidene acetal (VII) (see also Vol. II [11]) of L-iduronic acid, from which the syrupy free acid (VIII) may be obtained.

Procedures

1,2-O-Isopropylidene-5-O-p-tolylsulfonyl-α-D-glucofuranurono-6,3-lactone (I) (3)

A solution of 30 g (170 mmoles) of D-glucurono-6,3-lactone in 600 ml of dry acetone is stirred with 10 g of Amberlite IR-120(H$^+$) cation-exchange resin for 15 h at the boiling point (56°) of the mixture, and then the supernatant solution is decanted off and evaporated under reduced pressure to dryness. The residue is triturated with 100 ml of ether, collected by filtration, washed with ether, and dissolved in 100 ml of hot benzene containing 25 ml of ethyl acetate; the solution is cooled to 50°, whereupon some unreacted starting material precipitates and is removed by filtration. The filtrate is then heated to the boiling point, and ether is added cautiously to give, after refrigeration, 1,2-O-isopropylidene-α-D-glucofuranurono-6,3-lactone (4, 5) as large needles; yield 17.66 g (48%), mp 120°, [α]D +70° (c 1, acetone).

The foregoing product (5 g, 23.1 mmoles) is dissolved in 50 ml of pyridine, and 5 g (26.2 mmoles) of p-tolylsulfonyl chloride is added at 20°–25°. The mixture, which becomes warm at the outset, is kept for 1 day at 20°–25° and is then poured onto 200 g of ice. The crystalline sulfonate I (3) that separates (7.75 g) is collected by filtration, washed with two 100-ml portions of cold water, and recrystallized from ∼700 ml of boiling methanol; yield 6.18 g (72%), mp 189°–194°, [α]D +84° (c 0.4, acetone).

1,2-O-Isopropylidene-5-O-p-tolylsulfonyl-α-D-glucofuranose (II) (1)

A solution of 3 g (8.1 mmoles) of I in 90 ml of 1,2-dimethoxyethane is cooled in ice; and 9 ml of acetic acid is added to the magnetically stirred solution, followed by 6 g (157 mmoles) of sodium borohydride added portionwise during 45 min. The mixture is kept for 8 h at 0° and then poured into 600 ml of ice and water. The product is extracted with two 150-ml portions of chloroform, and the extract is washed with two 150-ml portions of water. The extract is dried with anhydrous sodium sulfate and then evaporated under diminished pressure to a crystalline solid (2.45 g) that is dissolved in the minimum volume of hot benzene (∼60 ml) with addition of ∼3 ml of petroleum ether (bp 40°–65°), not allowing the onset of turbidity, to give after 5–10 min 2.12 g (70%) of pure II; mp 122°–124°, [α]^{22}D +6° (c 1, chloroform).

L-Idose (III)

The sulfonate II (3 g, 8.1 mmoles) is dissolved in 20 ml of alcohol-free chloroform (Vol. I [144]) cooled to −15°; and 14 ml of methanol, in which 280 mg of sodium has been dissolved, is added. The mixture is maintained for 3 h between −5° and 0° by means of an ice–salt bath. At the end of this time, the solution is neutralized by passing carbon dioxide through it; the mixture is then filtered, and the filtrate is evaporated to dryness. The residue is then dissolved in ∼100 ml of acetone, and the slightly turbid mixture is filtered through Celite. The filtrate is then evaporated to give 1.64 g (quantitative) of crude, crystalline 5,6-anhydro-1,2-O-isopropylidene-β-L-idofuranose (6), mp 73°–76°. The compound crystallizes readily when it is obtained by evaporation from solution, but it proves difficult to crystallize from a solvent.

The foregoing anhydro derivative (1.5 g, 7.42 mmoles) is dissolved in 30 ml of 50 mM sulfuric acid, and the solution is kept for 2.5 h at 45°. Progress of the reaction is conveniently monitored by tlc observation of the disappearance of starting material (Vol. VI [6]). The solution is then cooled and made neutral with barium carbonate. Barium salts are removed by centrifugation, and the supernatant is then evaporated under diminished pressure to give III as a colorless syrup; yield 1.5 g ($\sim 100\%$), $[\alpha]^{22}_D$ $-13°$ (c 1.4, water).

L-Idose Diethyl Dithioacetal (IX) (*1*)

A mixture of 2.5 g of L-idose (III), 5 ml of conc. hydrochloric acid (d 1.19), and 7.5 ml of ethanethiol is agitated vigorously with a mechanical shaker for 30 min at $\sim 20°$. The mixture is then diluted with 50 ml of methanol and stirred with an excess of lead carbonate until it becomes neutral. The mixture is centrifuged, and the supernatant is evaporated under diminished pressure. The resultant oil is purified by passing it through a 1-m × 1.5-cm column of 70 g of silica gel (Merck No. 60, 70–230 mesh) with 5:1 v/v ethyl acetate–ethanol as the eluent. The product is crystallized from ethyl acetate (about 25 ml); yield of IX 2.36 g (59%), mp 96°–97°, $[\alpha]^{22}_D$ $-7°$ (c 1, methanol).

6,7-Dideoxy-1,2-*O*-isopropylidene-β-L-*ido*-(and α-D-*gluco*-)-hept-6-ynofuranose and Isolation of the L-*ido* Epimer (V) (*2*)

To a solution of 30 g (13.7 mmoles) of 1,2-*O*-isopropylidene-α-D-gluco-furanose (Vol. II [83]) in 350 ml of 37% aqueous formaldehyde cooled in ice is added slowly with stirring a solution of 30 g (14 mmoles) of sodium meta-periodate in 250 ml of water. After 3 h, 1.5 ml of ethylene glycol is added to decompose any unreacted periodate. The solution is then concentrated to one half of the original volume, and the sodium iodate that precipitates is removed by filtration. The solution is extracted with five 100-ml portions of chloroform. The chloroform extract is dried with anhydrous magnesium sulfate, then evaporated to a syrup that contains mainly a mixture of 1,2-*O*-isopropylidene-3,5-*O*-methylene-α-D-*xylo*-pentodialdo-1,4-furanose aldehy-drol and 1,2-*O*-isopropylidene-α-D-*xylo*-pentodialdo-1,4-furanose dimer; yield 20.65 g (69.5%).

A solution of 48 g of ethyl bromide in 300 ml of dry tetrahydrofuran is added slowly to 10 g of magnesium dust that is covered with 50 ml of dry tetrahydrofuran. The mixture is stirred vigorously to initiate the exothermic reaction, which is allowed to continue until the exotherm subsides. Acetylene is then passed slowly through 500 ml of dry tetrahydrofuran in a 2-liter flask by means of a gas-dispersion tube, and the liquid is stirred magnetically. After 1 h, the solution of ethylmagnesium bromide is added dropwise to the acetylene solution through an addition funnel. The resultant solution is stirred for 1 h while the stream of acetylene gas is maintained continuously.

To the dark-brown solution, a solution of 20 g (101 mmoles) of the foregoing periodate-oxidized product in 100 ml of dry tetrahydrofuran is added dropwise with stirring at 20°–25°. A slow stream of acetylene is passed through the solution throughout the reaction period. Stirring and passage of acetylene gas are continued for an additional 2 h. The solution is evaporated, and 200 ml of ethyl ether is added to the residual syrup. The mixture is shaken with a cold, aqueous, saturated ammonium chloride solution, and the aqueous phase is extracted with four 200-ml portions of ether. The combined ethereal extracts are washed with water, dried with anhydrous magnesium sulfate, and evaporated to give a dark-brown syrup that is a mixture of the title epimers; yield 14.4 g (73.5%). The epimers have R_f 0.20 (D-*gluco*) and 0.10 (V), respectively (tlc, 1:1 v/v ether–petroleum ether, Silica Gel G, Merck, activated at 110°).

To separate the epimers, a suspension of 450 g of silica gel (Merck, No. 7734, 0.05–0.2 mesh) in 1:1 v/v petroleum ether–ether is poured into a 1.25-m × 5-cm column and allowed to settle. The syrupy mixture of epimers, dissolved in 20 ml of 1:1 v/v petroleum ether–ether, is placed on the top of the column. Elution with the same solvent gives the two components resolved completely from each other. The first product to be eluted, the D-*gluco* epimer, is a solid; yield 2.58 g (18%); it may be purified by vacuum distillation at 140°/0.05 torr, whereupon it melts at 87°–89°, $[\alpha]^{25}_D$ −7° (c 1.0, chloroform). The second component to be eluted, the L-*ido* epimer (V), is also obtained as a solid; yield 3.80 g (26%). Recrystallization from ether–petroleum ether gives pure V, mp 126°–128°, $[\alpha]^{25}_D$ −23° (c 0.7, chloroform).

1,2-*O*-Isopropylidene-*β*-L-idofuranurono-6,3-lactone (VII)

The crystalline, epimerically pure acetylene V (5.0 g) is dissolved in 1 L of 85:15 v/v carbon tetrachloride–glacial acetic acid at 0°. The solution is maintained at 0° while ozone[1] is passed through it for 2 h, and then pure oxygen is passed through it until the blue color is discharged. The solution is evaporated to a syrup; and several portions of toluene are added to, and evaporated from, the residual syrup. The latter is dissolved in 200 ml ether, and the solution is passed through a small column of silica gel. Crystallization of the effluent gives pure VII; yield 2.1 g (42%); mp 134°–135°, $[\alpha]^{25}_D$ +102° (c 0.5, acetone), +109° (c 0.8, chloroform).

---

[1] A Welsbach ozonator, model T-408 (The Welsbach Company, Philadelphia, Pennsylvania) is suitable for this purpose.

L-Iduronic Acid (VIII)

The procedure described (Vol. II [11]) for acid-catalyzed hydrolysis of VII to VIII is used.

1,2-*O*-Isopropylidene-β-L-idofuranose (VI)

To a solution of 6.0 g (2.8 mmoles) of pure, crystalline V in 50 ml of dry pyridine (Vol. II [43], [53], [63], [73]; Vol. IV [73]; Vol. VII [2]) at 0° is added slowly 10 ml of benzoyl chloride dissolved in 30 ml of dichloromethane. The solution is stirred overnight. The mixture is poured into 300 ml of cold, aqueous sodium hydrogen carbonate, and the product is extracted with two 300-ml portions of dichloromethane. The extract is concentrated, and toluene is evaporated several times from the residue to remove pyridine. The resultant dibenzoate of V is recrystallized from abs. ethanol; yield 13.0 g (87%), mp 195°–196°, $[\alpha]^{25}$D −15° (c 1, chloroform). Similar benzoylation of the crude, unseparated mixture of acetylenic epimers, followed by recrystallization of the product from absolute ethanol, may also be used to give this product in about 34% yield.

To an ice-cold solution of 200 mg (0.48 mmoles) of the foregoing dibenzoate in 85 ml of carbon tetrachloride is added 15 ml of acetic acid. Ozone is passed through the solution for 40 min at 0°, and then pure oxygen is passed through until the blue color of ozone is discharged. The solution is evaporated to a syrup, and several small portions of toluene are evaporated from the residue to remove acetic acid. To 15 ml of tetrahydrofuran containing 15 mmoles of diborane under nitrogen is added 15 ml of a solution in tetrahydrofuran of 500 mg (1.19 mmoles) of the crude product prepared by ozonolysis of V dibenzoate. After keeping the solution under nitrogen for 3 h at 0°, 1.5 ml of water is slowly added dropwise. The solvent is removed by evaporation under diminished pressure; and several portions of methanol are added to, and evaporated from, the syrupy residue to remove residual boric acid. The syrup is dissolved in 1:1 v/v ether–petroleum ether, and the solution is passed through a small column packed with silica gel. Recrystallization of the eluted product from ether–petroleum ether gives the pure 3,5-dibenzoate of VI; yield 193 mg (38%), mp 75°–77°, $[\alpha]^{25}$D +8.5° (c 0.3, chloroform). This product (113 mg) is dissolved in 10 ml of methanol, and 3 ml of 1 *M* methanolic sodium methoxide (Vol. II [53]) is added. The solution is stirred for 3 h at ∼25° and then neutralized with 20 ml of Amberlite IR-120(H^{+}) cation-exchange resin for 10 min. The resin is removed by filtration, and the filtrate is evaporated to give compound VI as a solid that is recrystallized from methanol–ether to give pure VI; yield 50 mg (87%), mp 112°–113°.

Acid-catalyzed hydrolysis of VI (Vol. I [42]) gives syrupy L-idose (III), which may be characterized as already described in the alternative route via I and II.

References

(*1*)   M. Blanc-Muesser and J. Defaye, *Synthesis*, 568 (1977).
(*2*)   D. Horton and J.-H. Tsai, *Carbohydr. Res.*, **58,** 89 (1977).
(*3*)   J. K. N. Jones, *Can. J. Chem.*, **34,** 310 (1956).
(*4*)   H. Weidmann, *Ann. Chem.*, **679,** 178 (1964).
(*5*)   L. N. Owen, S. Peat, and W. J. G. Jones, *J. Chem. Soc.*, **339** (1941).
(*6*)   R. E. Gramera, T. R. Ingle, and R. L. Whistler, *J. Org. Chem.*, **29,** 878 (1964).

[25] Applications of the 1,3-Dithiane Procedure for the Synthesis of Branched-Chain Carbohydrates

L-Streptose, Methyl β-D-Hamameloside, Methyl 2,3,6-Trideoxy-2-C-(2-hydroxyacetyl)-α-L-*threo*-hexopyranosid-4-ulose, Methyl-4-C-(2-Benzoxyacetyl)-2,3,6-trideoxy-α-L-*threo*-hexopyranoside

By Hans Paulsen, Volker Sinnwell, and Joachim Thiem

*Institute of Organic Chemistry and Biochemistry,
University of Hamburg, Hamburg, West Germany*

Introduction

The synthesis of simple methyl-branched carbohydrates by reaction of selectively protected ketoses with methyllithium or methylmagnesium halides causes comparatively little problems (*1, 2*). However, the preparation of complex, branched-chain carbohydrates with a functionalized side chain represents a more complicated task. In this connection, the 1,3-dithiane procedure of Seebach (*3*) proved to be particularly advantageous and is preferred to other methods which make use of corresponding anions.

By release of a proton, 1,3-dithiane is transformed to its anion with *n*-butyl-lithium (*3*). The easy nucleophilic addition of this 1,3-dithiane anion to carbonyl compounds results in the formation of a side chain containing a potential aldehyde or keto group (*4–6*). In contrast, only rather reactive epoxides (*7*) and halides (*8*) react with this anion. Thus, the main application of the dithiane method resides in addition reactions to carbonyl compounds (*9*).

One of the best examples represents the synthesis of L-streptose (*4*). Reaction of the 3-keto compound I with the anion of 1,3-dithiane results in the stereoselective formation of the adduct II. A general rule was derived, according to which, in case of a neighboring isopropylidene group, nucleophilic addition to the keto group occurs always in a way that the newly introduced side chain adopts a trans-position to the isopropylidene group (*4, 6*).

For the cleavage of the dithiane ring, various methods are at hand. The most favorable proved to be either the reaction with boron trifluoride etherate with mercury(II) oxide in aqueous acetone (*10*) or methylation using methyl iodide and barium carbonate in aqueous acetone (*11, 12, 13*). After

185

cleavage of the dithiane ring to III and hydrolysis of the isopropylidene group, free L-streptose (IV) can be obtained easily (4).

For the synthesis of hamamelose (4) a stereoselective addition of the 1,3-dithiane anion to ketose V forms the adduct VI. By dethioacetalation the aldehyde VII is obtained, and subsequent reduction and hydrolysis yields methyl β-D-hamameloside (VIII).

In addition to 1,3-dithiane itself corresponding anions of 2-alkyl-1,3-dithiane react similarly with carbonyl compounds. In this way a series of branched-chain carbohydrates having extended alkyl side chains could be prepared (5, 6, 14, 15).

Furthermore, two particularly valuable variants of the dithiane procedure are known. With enones of type IX as substrates, the usual dithiolane reaction will always yield a product of 1,2-addition to the carbonyl group. However, by application of the anion prepared from 2-ethoxy-1,3-dithiane in a stereoselective 1,4-addition, compound XI results (16), which is formed by hydrolysis of the primary adduct X. The introduction of the side chain occurs stereoselectively "trans" in relation to the anomeric methoxy group (16, 17). This method is favorably applied to the synthesis of compounds that do not carry a hydroxyl group at the branching point.

The reduction of the carbethoxy group of the primarily formed lithium enolate X can be directly performed using lithium aluminum hydride. Following hydrolysis, the branched-chain carbohydrate XII can be obtained; it readily forms the hemiacetal XIII which predominates in the equilibrium. After subsequent dethioacetalation, the open-chain form XIV is again favored (16). The structure of compound XIV having a glycoloyl side chain at C-2 was formerly supposed to be the formula of pillarose.

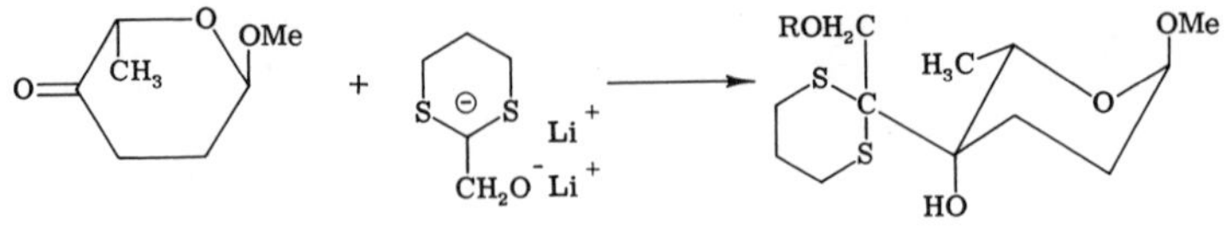

O ODMe CH₃ + S ⊖ S Li⁺ CH₂O⁻Li⁺ ROH₂C H₃C OMe O S C S HO

(XV)

(XVIa R = H)
(XVIb R = Bz)

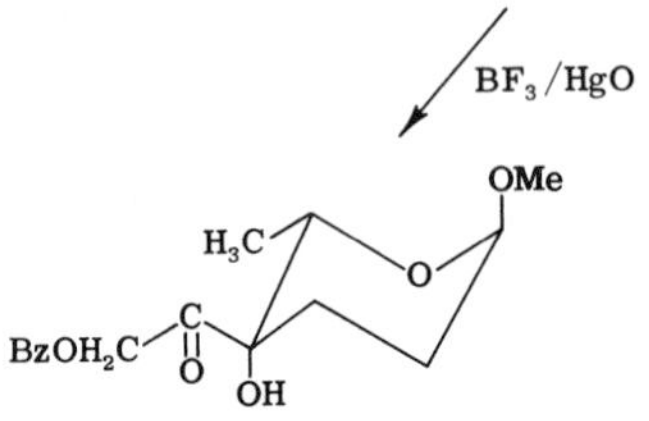

BF₃/HgO

OMe
H₃C O
BzOH₂C–C OH
O

(XVII)

A further variation of the dithiane method represents the application of a dianion of 2-hydroxymethyl-1,3-dithiane (*18*). Anions resulting from dithanes with a neighboring alkoxy group give rise to elimination and degradation. However, the corresponding unprotected alcohols react with one mole of *n*-butyllithium to form the lithium alcoholates, that without elimination, can be transformed to dianions with a second mole of *n*-butyllithium. Nucleophilic additions of these dianions proceed similar to those of mono-anions (*18*).

Thus, for the synthesis of pillarose (*19*), the ketose XV could be reacted with the dianion and gives rise to the predominant formation of the adduct XVIa. Following benzoylation to compound XVIb and subsequent dethio-acetalation, the monobenzoate of methyl α-L-pillaroside (XVII) can be characterized (*19*).

A generally applicable method for the transformation of complex branched-chain carbohydrates having a hydroxyl group at the branching point towards the corresponding deoxy branched-chain derivatives has been developed. Following benzoylation or *p*-cyanobenzoylation at the branching hydroxyl group, the ester can be removed by way of a radical deoxygenation using tri-*n*-butyltin hydride (*20*).

Procedures

L-Streptose

5-Deoxy-3-C-(1,3-dithiane-2-yl)-1,2-O-isopropylidene-
β-L-lyxofuranose (II) (4)

A solution of 3.4 g (28.3 mmole) of freshly sublimed 1,3-dithiane (*3*) in

25 ml of anhydrous tetrahydrofuran is treated under nitrogen with 18 ml (29.7 mmoles) of n-butyllithium in hexane. After 2 h at $-20°$ to $-30°$, metallation is complete. The mixture is cooled to $-78°$, and with efficient stirring, 4.5 g (26.2 mmole) of 5-deoxy-1,2-O-isopropylidene-β-L-$threo$-pentofuranos-3-ulose (I) (4) dissolved in 25 ml of tetrahydrofuran is added dropwise. Within 2 h, the mixture is gradually warmed to room temperature. Completion of the reaction is checked by thin-layer chromatography on silica gel using 5:1 v/v benzene–acetone (Vol. VI [6]). The mixture is poured into 130 ml of water and extracted several times with chloroform. The combined organic phases are washed with water, dried over potassium carbonate, and evaporated under diminished pressure. Excessive 1,3-dithiane is removed by sublimation at $42°$ and 10^{-2} Torr. The residue is recrystallized from ethyl acetate–petroleum ether; yield 6.8 g (89%), mp $141°$, $[\alpha]^{20}_D$ $-3.0°$ (c 1, acetone).

5-Deoxy-3-C-formyl-1,2-O-isopropylidene-β-L-lyxofuranose (III) (4)

To a suspension made of red mercury(II) oxide (3.5 equiv) and boron trifluoride etherate (3 equiv) in 35 ml of 17% aqueous acetone are added dropwise 1.5 g of compound II dissolved in 10 ml of tetrahydrofuran with efficient stirring under nitrogen. Stirring is continued for 20 h until, by thin-layer chromatography, no residual compound II can be detected. The mixture is treated with 15 ml of water and 30 ml of acetone and neutralized with sodium hydroxide. The precipitates are filtered and centrifuged; the filtrate is evaporated under diminished pressure to 10 ml, and extracted several times with chloroform. In order to enhance yields, the aqueous phase is evaporated to dryness, and the residue is extracted with warm chloroform. The combined chloroform extracts are dried and evaporated under diminished pressure. Thin-layer chromatography of the syrup obtained in 3:1 v/v ethyl acetate–petroleum ether shows a double spot of compound III and its hydrate; yield 770 mg (70%). For the complete transformation to the hydrate, the syrup is treated with a little water. The hydrate is crystallized from dry ether; it is recrystallized from acetone; mp $80°–82°$, $[\alpha]^{20}_D$ $+43°$ (c 1, dioxane).

L-Streptose (IV) (5-Deoxy-3-C-formyl-α,β-L-lyxose) (4)

Compound III (700 mg) dissolved in 8 ml of 1:1 v/v water–dioxane is stirred for 50 h with Amberlite IR-120(H$^+$) cation-exchange resin until thin-layer chromatography in 7:2:1 v/v 2-propanol–ethyl acetate–water shows no residual compound II. After filtration, the mixture is treated with charcoal and evaporated under diminished pressure. L-Streptose (IV) is obtained as a colorless, rather hygroscopic glass. In aqueous solvents the compound adopts the hydrated form completely; yield 520 mg (85%), $[\alpha]^{20}$D $-18°$ (c 0.65, water).

Methyl β-D-Hamameloside

Methyl-2-C-(1,3-dithiane-2-yl)-3,4-O-isopropylidene-
β-D-ribopyranoside (VI) (4)

1,3-Dithiane (14.8 g, 123.5 mmole) dissolved in 200 ml of anhydrous tetrahydrofuran is treated under nitrogen with 80 ml of a 20% solution of n-butyllithium in n-hexane. To achieve complete metallation, the mixture is stirred for 1.5 h at $-25°$ to $-30°$, then cooled to $-78°$. To this solution is added dropwise, over a few minutes, a solution of 25 g (123.7 mmole) of methyl 3,4-O-isopropylidene-β-D-*erythro*-pentopyranosid-2-ulose (V) (4) dissolved in 200 ml of anhydrous tetrahydrofuran. During this addition, the temperature rises to $-60°$. The reaction mixture is warmed to room temperature within 1 h, after which thin-layer chromatography in 4:6 v/v ethyl acetate–hexane shows complete reaction of compound V. The reaction mixture is poured into 100 ml of water and extracted with five 50-ml portions of chloroform. The combined chloroform extracts are washed with 20 ml of 10% aqueous potassium hydroxide and then with 50 ml of water, dried over potassium carbonate, and evaporated under diminished pressure. The resulting syrup is dissolved in 3 ml of chloroform, and 20 ml of n-hexane is added. Crystallization occurs after 10 min at $-20°$. After 1 h, another 60 ml of n-hexane is added, and the flask is kept at $-20°$, which results in a complete crystallization after 2 days. The crystals are filtered, washed with n-hexane, and dried *in vacuo*; yield 31 g (78%), mp 103°, $[\alpha]^{24}$D $+74°$ (c 0.1, chloroform).

Methyl 2-C-formyl-3,4-O-isopropylidene-β-D-ribopyranoside (VII) (4)

Tetrahydrofuran (90 ml), 10 ml of water, and 15.5 g of red mercury(II) oxide are mixed under stirring at 0°. The pH of the suspension is 9 which changes to 1 after addition of 2.8 ml of boron trifluoride etherate. To the cold mixture, 3.5 g (10.9 mmole) of compound VI is added, which causes a short rise of the pH to 2.5 and subsequently a gradual decline to 1 again within the next 2 h. After 30 min, the cooling bath is removed and stirring is continued at room temperature. By thin-layer chromatography in 6:4 v/v ethyl acetate–n-hexane, no starting material can be detected after 4.5 h. The suspension is neutralized with sodium hydroxide causing the precipitation of grayish-green material which is centrifuged and extracted with three 50-ml portions of chloroform. The combined organic solutions are evaporated under diminished pressure and dissolved in 100 ml of chloroform. The chloroform solution is stirred for 15 min with charcoal, filtered, and evaporated under diminished pressure to yield a syrup; yield 2 g (79%). An analytical sample is obtained by column chromatography on silica gel; $[\alpha]^{24}$D − 192° (c 1, methanol).

Methyl β-D-Hamameloside (VIII) (Methyl 2-C-Hydroxymethyl-β-D-ribopyranoside) (4)

Compound VII (1.5 g, 6.4 mmole) is refluxed in 70 ml of anhydrous tetrahydrofuran with 1.8 g of lithium aluminum hydride for 4 h. Under cooling, water is added slowly to decompose excessive lithium aluminum hydride. The white precipitate is filtered and washed twice with 50 ml of ether. The organic solutions are combined and evaporated under diminished pressure to yield 1.2 g (79%) of methyl 2-C-hydroxymethyl-3,4-O-isopropylidene-β-D-ribopyranoside. The syrupy compound proved to be pure by tlc in 4:1:5 v/v 1-butanol–ethanol–water (organic phase) and can be used directly in the next step.

The syrup is dissolved in 20 ml of water and stirred for 5 h with Amberlite IR-120(H⁺) cation-exchange resin. The resin is removed by filtration and washed twice with 20 ml of water; the water solutions are combined and stirred with charcoal for 15 min. After filtration and evaporation, the resulting syrup is tituated with ethyl acetate, which causes spontaneous crystallization; yield 820 mg (72%). An analytical sample is obtained by two sublimations at 120° and 10⁻³ Torr; mp 131°–133°, $[\alpha]^{24}$D − 149° (c 1, methanol).

Methyl 2,3,6-Trideoxy-2-C-(2-hydroxyacetyl)-α-L-threo-hexopyranosid-4-ulose

Methyl 2-C-[2-carbethoxy-1,3-dithiolane-2-yl)-2,3,6-trideoxy-α-L-threo-hexopyranosid-4-ulose (XI) (16)

Diisopropylamine (0.91 ml, 6.6 mmole) is treated under nitrogen at 4° with 6 mmole of n-butyllithium in 3 ml of n-hexane. After 20 min, the mixture is diluted with 5 ml of anhydrous tetrahydrofuran and cooled to − 78°; to this solution is added 1.07 g (6 mmole) of 2-carbethoxy-1,3-dithiolane in 4 ml of anhydrous tetrahydrofuran. Methyl 2,3,6-trideoxy-α-L-glycero-hex-2-enopyranosid-4-ulose (IX) (16) (650 mg 4.6 mmole) dissolved in 10 ml of anhydrous tetrahydrofuran is slowly added dropwise. Within 3 h, the temperature is allowed to rise to − 10°; the mixture is then poured into 100 ml of water, extracted with three 100-ml portions of chloroform, washed with 50 ml of water, dried over magnesium sulfate, and evaporated to give a syrup. The raw material is cleaned by column chromatography on silica gel using 1:1 v/v ether–n-hexane as eluent; yield 1.1 g (74%), $[\alpha]^{20}_D$ −132° (c 2.3, methanol).

Methyl 2,3,6-Trideoxy-2-C-(2-hydroxymethyl-1,3-dithiolane-2-yl)-α-L-threo-hexopyranosid-4-ulose (XIII) (16)

Under nitrogen, 2.34 ml (17 mmole) of diisopropylamine is treated at 4° with 7.7 ml of n-butyllithium (15.4 mmole) in n-hexane and left for 20 min. After dilution with 10 ml of anhydrous tetrahydrofuran, the mixture is cooled to −78° and 2.7 g (15.7 mmole) of 2-carboethoxy-1,3-dithiolane dissolved in 5 ml of anhydrous tetrahydrofuran is added dropwise. After 20 min, 2 g (14.1 mmole) of compound IX in 10 ml of anhydrous tetrahydrofuran is slowly introduced with efficient stirring. The mixture is allowed to rise to −15° within 3 h, then cooled to −40° and transferred under nitrogen to a precooled solution (−40°) of 400 mg of lithium aluminum hydride in 10 ml of anhydrous tetrahydrofuran. The cooling bath is removed, and stirring is continued for 1 h at room temperature and another hour at 45°. Excessive lithium aluminum hydride is decomposed first with ethyl acetate, then with water; and the mixture is diluted with 700 ml of ether. The precipitate is removed by filtration and washed with 50 ml of water; the washings and filtrates are evaporated. The residue is dissolved in ether; the ether solution is dried with magnesium sulfate, filtered, and evaporated under diminished pressure. The remaining syrup is crystallized from ether–n-hexane; yield 2.1 g (53%), mp 83°, $[\alpha]^{20}_D$ +43° (c 2.5, methanol).

Methyl 2,3,6-trideoxy-2-C-(2-hydroxyacetyl)-α-L-threo-hexopyranosid-4-ulose (XIV) (16)

Compound XIII (500 mg, 1.8 mmole) is dissolved in 25 ml of 80% aqueous acetonitrile and treated under efficient stirring and cooling at 0° with 1.2 g of mercury(II) oxide and 1 ml of boron trifluoride etherate. After 1 h, the mixture is neutralized with sodium carbonate, filtered, and evaporated. The remaining material is dissolved in acid-free chloroform; the resulting solution is washed with a little water, dried over magnesium sulfate, and chromatographed on a column of silica gel using 4:1 v/v ether–n-hexane as the eluent; yield 230 mg (63%); $[\alpha]^{20}_D$ −221° (c 1.1, chloroform).

Methyl 4-C-(2-Benzoxyacetyl)-2,3,6-trideoxy-α-L-*threo*-hexopyranoside

Methyl 2,3,6-Trideoxy-4-C-(2-hydroxymethyl-1,3-dithiane-2-yl)-α-L-threo-hexopyranoside (XVIa) (19)

2-Hydroxymethyl-1,3-dithiane (1.66 g, 11.1 mmole) is dissolved in 40 ml of anhydrous tetrahydrofuran; the resulting solution is cooled to $-70°$, treated with 11.1 ml of a 2 N n-butyllithium solution in n-hexane, and stirred for 30 min. Methyl 2,3,6-trideoxy-α-L-*glycero*-hexopyranosid-4-ulose (XV) (*19*) (500 mg, 3.5 mmole) dissolved in 20 ml of anhydrous tetrahydrofuran are added dropwise, and the mixture is stirred for 1.5 h at $-70°$ and another 2 h at $-40°$. After a thin-layer chromatographic check in ethyl acetate shows complete reaction, the mixture is hydrolyzed with water at $-40°$. The product is extracted with 200 ml of chloroform; the chloroform layer is washed with water, dried over sodium sulfate, evaporated under diminished pressure, and repeatedly codistilled with toluene to dryness. Column chromatography on 100 g of silica gel using 1:2, 1:1, and 3:1 v/v ethyl acetate–toluene as eluents yields 460 mg (44%) of raw product. After recrystallization from ether 300 mg (30%) of compound XVIa are obtained: mp 123°–127°, $[α]^{20}_D$ $-108°$ (c 1, chloroform).

Methyl 4-C-(2-Benzoxymethyl-1,3-dithiane-2-yl)-2,3,6-trideoxy-
α-L-threo-hexopyranoside (XVIb) (19)

Compound XVIa (382 mg, 1.28 mmole) is treated with 0.5 ml of benzoyl chloride in 5 ml of anhydrous pyridine and left at room temperature for 24 h. The mixture is hydrolyzed with iced water for 4 h and extracted with chloroform. The chloroform extract is dried over sodium sulfate and evaporated under diminished pressure. Recrystallization from ether–n-hexane yields 450 mg (87%) of compound XVb; mp 123°–124°, $[α]^{20}_D$ $-81°$ (c 0.9, chloroform).

Methyl 4-C-(2-Benzoxyacetyl)-2,3,6-trideoxy-α-L-threo-hexopyranoside
(Methyl 4²-O-benzoyl-α-L-pillaroside) (XVII) (19)

Compound XVIb (100 mg, 0.25 mmole) is stirred for 1 h with 300 mg of mercury(II) oxide and 0.2 ml of boron trifluoride etherate in 5 ml of 4:1 v/v acetonitrile–water at 0°. The reaction mixture is neutralized using a saturated solution of sodium hydrogen carbonate and filtered, and the precipitate is washed carefully with acetonitrile. The filtrate is concentrated under diminished pressure; the residue is dissolved in chloroform; and the chloroform solution is washed with water, dried over sodium sulfate, and evaporated under diminished pressure to give 45 mg of raw crystalline material. Recrystallization from ether–n-hexane yields 20.8 mg (27%) of XVII; mp 104°–106°, $[α]^{20}_D$ $-96°$ (c 0.51, chloroform).

References

(*1*) A. A. Feast, W. G. Overend, and N. R. Williams, *J. Chem. Soc. C*, 303 (1966).
(*2*) J. S. Burton, W. G. Overend, and N. R. Williams, *J. Chem. Soc. C*, 3433 (1966).
(*3*) D. Seebach, *Synthesis*, 17 (1969).

(4)   H. Paulsen, V. Sinnwell, and P. Stadler, *Chem. Ber.*, **105,** 1978 (1972).
(5)   H. Paulsen and H. Redlich, *Chem. Ber.*, **107,** 2992 (1974).
(6)   H. Paulsen, B. Sumfleth, and H. Redlich, *Chem. Ber.*, **109,** 1362 (1976).
(7)   A. M. Sepulchre, G. Lukacs, G. Vass, and S. D. Gero, *Bull. Soc. Chim. Fr.*, 4000 (1972).
(8)   A. M. Sepulchre, G. Vass, and S. D. Gero, *Tetrahedron Lett.*, 3619 (1973).
(9)   H. Paulsen, *Staerke*, **25,** 389 (1973).
(10) E. Vedejs and P. Fuchs, *J. Org. Chem.*, **36,** 366 (1971).
(11) H. Paulsen, P. Stadler, A. Banaszek, and F. Toedter, *Chem. Ber.*, **110,** 1908 (1977).
(12) M. Fetizon and M. Jurion, *Chem. Commun.*, 382 (1972).
(13) H.-L. Wang Chang, *Tetrahedron Lett.*, 1989 (1972).
(14) H. Paulsen and V. Sinnwell, *Angew. Chem.*, **88,** 476 (1976); *Angew. Chem. Intern. Ed. Engl.*, **15,** 438 (1976); *Chem. Ber.*, **111,** 869, 879 (1978).
(15) A. M. Sepulchre, A. Gateau-Olesker, G. Vass, and S. D. Gero, *Biochimie*, **55,** 613 (1973).
(16) H. Paulsen and W. Koebernick, *Carbohydr. Res.*, **56,** 53 (1977).
(17) H. Paulsen and W. Koebernick, *Chem. Ber.*, **110,** 2127 (1977).
(18) H. Paulsen, K. Roden, V. Sinnwell, and W. Koebernick, *Angew. Chem.*, **88,** 477 (1976); *Angew. Chem. Intern. Ed. Engl.*, **15,** 439 (1976).
(19) H. Paulsen, K. Roden, V. Sinnwell, and W. Koebernick, *Chem. Ber.*, **110,** 2146 (1977).
(20) H. Redlich, H. J. Neumann, and H. Paulsen, *Chem. Ber.*, **110,** 2911 (1977).

[26] 2,6-Dideoxy-D-*ribo*-hexose (Digitoxose)

By Derek Horton, Tak-Ming Cheung, and Wolfgang Weckerle

Department of Chemistry, The Ohio State University, Columbus, Ohio

Introduction

Digitoxose widely occurs as the sugar component of such cardiac glyco-sides (*1*, *2*) as digitoxin, gitoxin, and digoxin, from any of which it is readily obtained (*3*) by very mild, acid hydrolysis (Vol. I [62]). More recently,

195

digitoxose has been found to be a constituent of two antibiotic substances, α-lipomycin (4) and oleficin (5).

Digitoxose was first synthesized by Iselin and Reichstein (6) by the glycal method (Vol. I [55] and [57]) and later by Gut and Prins (7), and by Bollinger and Ulrich (8), in moderate yields from a 2,3-anhydro-D-*allo* precursor through an epoxide ring-opening reaction and subsequent deoxygenation at C-6.

The procedure (9, 10) described here provides a convenient, preparative-scale conversion of methyl 4,6-O-benzylidene-2-deoxy-α-D-*ribo*-hexopyranoside (I) into crystalline digitoxose (VI) by a five-step sequence involving treatment of the benzoylated[1] derivative II with N-bromosuccinimide (NBS) (Vol. VI [28]), followed by reductive debromination, catalytic transesterification (Vol. II [54]), and acid-catalyzed hydrolysis; the net, overall yield is ~58%. The 6-bromo derivative III constitutes (9, Vol. VIII [27]) a suitable precursor to the 5-epimer of digitoxose, 2,6-dideoxy-L-*lyxo*-hexose (2-deoxy-L-fucose), a sugar that is a component of several biologically important compounds.

Procedure

Methyl 4,6-O-Benzylidene-2-deoxy-α-D-*ribo*-hexopyranoside (I) (12–14)

In a 2-liter, three necked flask equipped with a heating mantle, magnetic stirrer-bar, and a reflux condenser protected by a drying tube, is placed a suspension of 30 g (113.5 mmole) of methyl 2,3-anhydro-4,6-O-benzylidene-α-D-allopyranoside (Vol. I [30], Vol. II [45]) in anhydrous ether (or tetrahydrofuran) (800 ml). Lithium aluminum hydride (10 g, 263.5 mmole) is added in small portions during a period of 10 min, and the mixture is boiled for 4 h under reflux. The flask is then fitted with a dropping funnel and a paddle-type mechanical stirrer and placed into an ice–water bath. The excess of lithium aluminum hydride is decomposed by successive addition of 10 ml of water, 10 ml of 15% (w/v) aqueous sodium hydroxide, and 30 ml of water. The granular precipitate formed is filtered off by suction, resuspended in 300 ml of dichloromethane, and again removed by filtration. The combined organic filtrates are evaporated to dryness to afford crude methyl 4,6-O-benzylidene-2-deoxy-α-D-*ribo*-hexopyranoside (I), sufficiently pure for most purposes; yield 27.4 g (91%). Analytically pure I is obtained by recrystallization from ~100 ml of ethanol; yield 25 g (83%), mp 127°–128°, $[\alpha]^{20}_D$ +146° (c 1, chloroform).

---

[1] Treatment of I with N-bromosuccinimide leads (11) to a syrupy product.

Methyl 3-*O*-Benzoyl-4,6-*O*-benzylidene-2-deoxy-α-D-*ribo*-hexopyranoside (II) (*9*, *10*)

To a solution of 20 g (75 mmole) of I in 250 ml of dry pyridine, cooled in an ice–water bath, is added 10 ml (85.7 mmole) of benzoyl chloride. The mixture is kept for 15 h at ∼25° and then poured with vigorous stirring onto 600 ml of ice–water containing 25 g of potassium carbonate. After stirring for 1 h at 0°, the product is extracted successively with 150-, 100-, and 100-ml portions of dichloromethane. The combined organic phase is washed with aqueous sodium hydrogencarbonate, and then dried with magnesium sulfate and evaporated. To remove residual pyridine, three 20-ml portions of toluene are successively added to and then evaporated from the residue. The resulting crystalline solid is recrystallized from ∼80 ml of ethanol to give pure II; yield 24.7 g (89%), mp 94°–95°, $[\alpha]^{22}$D $+185°$ (*c* 1, chloroform).

Methyl 3,4-Di-*O*-benzoyl-6-bromo-2,6-dideoxy-α-D-*ribo*-hexopyranoside (III) (*9*, *10*)

A mixture of 25.7 g (69.4 mmole) of the benzylidene acetal II, 13.55 g (75 mmole) of *N*-bromosuccinimide (NBS) and 20 g of barium carbonate in 500 ml of dry carbon tetrachloride is placed in a 1-liter, round-bottomed flask, equipped with a magnetic stirrer-bar and a reflux condenser protected by a drying tube. The mixture is boiled for 5 h under reflux, during which time the color of the solution, originally colorless, becomes successively yellow, red, and finally pale yellow. The solvent then is evaporated off on a rotary evaporator, and the residue is extracted with three 150-ml portions of dichloromethane. The combined organic phase is washed successively with 250 ml of 10% sodium hydrogensulfite and 250 ml of 10% sodium hydrogencarbonate, and then dried with magnesium sulfate and evaporated. The resulting crystalline solid is recrystallized from ∼80 ml of ethanol to give analytically pure III; yield 24.7 g (79%), mp 96°–98°, $[\alpha]^{22}$D $+192.5°$ (*c* 0.6, chloroform).

Methyl 3,4-Di-*O*-benzoyl-2,6-dideoxy-α-D-*ribo*-hexopyranoside (IV) (*9*)

A mixture of 10.2 g (22.7 mmole) of the bromide III, ∼8 g of Raney nickel,[2] and 8 ml of triethylamine in 200 ml of methanol is shaken under hydrogen at 4.2 kg. cm^{-2} (60 lb. in.$^{-2}$) for 18 h at ∼25°. Celite filter-aid (10 g) is added,[3] and the mixture is filtered through a medium-porosity,

---

[2] The amount of catalyst is not very crucial; if a more accurate weight determination is required, see Vol. II [23].

[3] This precaution prevents the catalyst from igniting when exposed to air.

sintered-glass funnel. The filter cake is thoroughly washed with 100 ml of methanol; the combined filtrate is evaporated, and the residue is dissolved in 180 ml of dichloromethane. The solution is washed twice with water to remove triethylammonium bromide, dried with magnesium sulfate, and and evaporated to give a crystalline residue that is recrystallized from ~70 ml of hexane, affording pure IV; yield 7.9 g (94%), mp 80°–82°, $[\alpha]^{23}$D +228° (c 1, chloroform).

Methyl 2,6-Dideoxy-α-D-ribo-hexopyranoside (Methyl α-Digitoxoside, V)

This method is a modification of the methods described by Horton et al. (9) and by Haga et al. (10). To a solution of 5 g (13.5 mmole) of the methyl glycoside IV in 50 ml of absolute methanol (Vol. VII [3]) is added 1 ml of 1 M sodium methoxide (Vol. II [54]), and the mixture is kept for 15 h at 25°. The solution then is treated for 30 min at 0° with 5 ml of a weakly acidic, ion-exchange resin [such as Amberlite IRC-50(H⁺)] prewashed with methanol. The resin is removed by filtration, and the filtrate is evaporated. Water (three 15-ml portions) is added to and evaporated (rotary evaporator, water aspirator, bath temperature ~45°) from the syrupy residue for azeotropic removal of the methyl benzoate formed. The last traces of water are removed by distilling ethanol from the residue, affording V as a mobile syrup; yield 2.1 g (96%). An analytically pure sample is obtained by distillation under diminished pressure (40 mtorr, bath temperature 60°–80°), $[\alpha]^{21}$D +192° (c 1.5, methanol) and +169° (c 2.3, chloroform).

2,6-Dideoxy-D-ribo-hexose (Digitoxose, VI) (9)

A solution of 1 g (6.2 mmole) of the glycoside V in 40 ml of 3:1 v/v water–acetic acid is heated for 45 min on a boiling-water bath. The solvent is then removed by distillation, and last traces of acetic acid are removed by repeated evaporation of small portions of water from the residue. The product is kept in a desiccator over potassium hydroxide and phosphorus pentaoxide for 24 h. The glassy residue readily crystallizes upon addition of a little acetone. The crystals are collected and recrystallized from acetone by addition of ether; yield 850 mg (93%), mp 105°–108°, $[\alpha]^{22}$D +42.5° (5.20 min) → +51.7° (7.50 min) → +47.5° (20 min) → +47.8° (1 h, equil, c 1, water) and +36.5° (c 0.7, methanol). For physical constants reported in the literature for this compound, see (9) and Vol. I [62].

References

(*1*) T. Reichstein and E. Weiss, *Advan. Carbohydr. Chem.*, **17,** 65 (1962).

(*2*) T. Reichstein, *Naturwissenschaften*, **54,** 53 (1967).

(*3*) R. C. Elderfield, *Advan. Carbohydr. Chem.*, **1,** 147 (1945).

(*4*) B. Kunze, K. Schabacher, H. Zähner, and A. Zeek, *Arch. Mikrobiol.*, **86,** 147 (1972).

(*5*) G. Horváth, J. Gyimesi, and Z. Méhesfalvi-Vajna, *Tetrahedron Lett.*, 3643 (1973).

(*6*) B. Iselin and T. Reichstein, *Helv. Chim. Acta*, **27,** 1203 (1944).

(*7*) M. Gut and D. A. Prins, *Helv. Chim. Acta*, **30,** 1223 (1947).

(*8*) H. R. Bolliger and P. Ulrich, *Helv. Chim. Acta*, **35,** 93 (1952).

(*9*) T. M. Cheung, D. Horton, and W. Weckerle, *Carbohydr. Res.*, **58,** 139 (1977).

(*10*) M. Haga, M. Chonan, and S. Tejima, *Carbohydr. Res.*, **16,** 486 (1971).

(*11*) S. Hanessian and N. R. Plessas, *J. Org. Chem.*, **34,** 1045 (1969).

(*12*) D. A. Prins, *J. Amer. Chem. Soc.*, **70,** 3955 (1948).

(*13*) A. Rosenthal and P. Catsoulacos, *Can. J. Chem.*, **46,** 2868 (1968).

(*14*) A. C. Richardson, *Carbohydr. Res.*, **4,** 422 (1967).

[27] 2,6-Dideoxy-α-L-*lyxo*-hexose (2-Deoxy-α-L-fucose)

By Derek Horton, Tak-Ming Cheung, and Wolfgang Weckerle

Department of Chemistry, The Ohio State University, Columbus, Ohio

Introduction

2,6-Dideoxy-L-*lyxo*-hexose (VI) was prepared by Iselin and Reichstein (*1*) from L-fucose by the glycal[1] method (Vol. I [55] and [57]) long before it was found in nature (*4–8*) as a constituent of rhodomycins (*9–11*), cinerubins A (*9–12*) and B (*9–11, 13*), the macrolide antibiotic azalomycin-B (*14*) and, apparently, as a component of a cardiac glycoside (*15*) from *Pentopetia androsaemifolia*.

The D enantiomer[2] of VI is known as D-oliose (*16–18*) and occurs in a number of antitumor antibiotics, as does its 4-methyl ether [D-olivomose,

---

[1] For a somewhat controversial (reaction temperature) improvement in the synthesis of the required di-*O*-acetyl-L-fucal (3,4-di-*O*-acetyl-1,5-anhydro-1,2,6-trideoxy-L-*lyxo*-hex-1-enitol), see (*2*) and (*3*).

[2] Except for a brief mention in a preliminary communication (*16–18*), it appears that this sugar has never been prepared synthetically.

METHODS IN CARBOHYDRATE
CHEMISTRY, VOL. VIII

D-chromose A (*16–19*)] and its 3-acetate [D-chromose D (*19*)]. Both VI and its enantiomer are often encountered (*4–8*) as their 3-methyl ethers (D- and L-diginose) as components of cardiac and pregnane glycosides.

The procedure (*20*) described here provides a convenient, preparative-scale conversion of the readily available methyl 4,6-O-benzylidene-2-deoxy-α-D-*ribo*-hexopyranoside (I) (*21–23*, Vol. VIII [26]) into crystalline 2,6-dideoxy-α-L-*lyxo*-hexose (VI) via methyl 3,4-di-O-benzoyl-6-bromo-2,6-dideoxy-α-D-*ribo*-hexopyranoside (II) by a four-step sequence from II. Elimination of hydrogen bromide (see also Vol. II [105]), followed by catalytic hydrogenation, introduces the terminal C-methyl group with concomitant inversion at C-5 to generate the required L-*lyxo* stereochemistry. Catalytic transesterification (see Vol. II [54]), followed by acid-catalyzed hydrolysis, then leads to 2,6-dideoxy-L-*lyxo*-hexose (VI) as the crystalline, slightly hygroscopic, α anomer; the net, overall yield, based on II, exceeds 50%.

Procedure

Methyl 3,4-Di-O-benzoyl-2,6-dideoxy-α-D-*erythro*-hex-5-enopyranoside (III) (*20*)

A mixture of 20 g (44.5 mmole) of methyl 3,4-di-O-benzoyl-6-bromo-2,6-dideoxy-α-D-*ribo*-hexopyranoside (II; Vol. VIII [25]) and 15 g of dry, technical-grade silver fluoride[3] in 200 ml of dry pyridine (Vol. II [43], [53], [63], [73]; Vol. IV [73]; Vol. VII [2]) is thoroughly stirred for 48 h at 25° with the exclusion of moisture. The dark solution then is poured with vigorous stirring into 1 liter of ether. After 15 min, the inorganic precipitate is removed by filtration, and the filtrate is evaporated on a rotary evaporator at a bath temperature of ∼40°. To remove all the pyridine, three 20-ml portions of toluene are successively added to and then evaporated from the residue. The resulting, syrupy residue, first colorless but turning dark on being kept in the open air, is freed from residual silver salts by filtration through a small column (600 × 30 mm) of silica gel with acetone as eluant and monitor-

---

[3] Best results are achieved with silver fluoride prepared in the laboratory: Commercial silver carbonate (60 g) is carefully added (*24*) (CO_2 evolution) to 40 ml of 48% hydrofluoric acid in a platinum bowl. Under continuous stirring (a hardwood rod may be used instead of a platinum spatula), the mixture is evaporated by use of a bunsen burner in a well-ventilated hood. Toward the end of the evaporation (∼4 h for a 60-g batch of silver carbonate), an infrared lamp (surface evaporator) may be used for thorough drying of the granular, dark brown, extremely hygroscopic product, which may then be kept for an extended period in a desiccator over potassium hydroxide and phosphorus pentaoxide.

ing the effluent by the potassium permanganate test.[4] Evaporation of the solvent affords syrupy III, sufficiently pure for the following step; yield 15.1 g (92%), $[\alpha]^{23}$D +123° (c 2.9, chloroform).

Methyl 3,4-Di-*O*-benzoyl-2,6-dideoxy-β-L-*lyxo*-hexopyranoside (IV) (*20*)

A mixture of 10 g (27.2 mmole) of the unsaturated sugar III and 1 g of 10% palladium on barium sulfate in 150 ml of methanol is shaken under hydrogen at atmospheric pressure for 1 h at 25°. Celite filter-aid (∼1 g) is then added,[5] and the mixture is filtered through a medium-porosity, sintered-glass funnel. The filter cake is thoroughly washed with methanol (30 ml), and the combined filtrates are evaporated. The resulting, crystalline residue is recrystallized from ethanol (∼70 ml) to furnish analytically pure IV as fine needles; yield 8.8 g (87%), mp 122°–123°, $[\alpha]^{22}$D +64° (c 0.9, chloroform).

Methyl 2,6-Dideoxy-β-L-*lyxo*-hexopyranoside (V)

This procedure is a modification of the one described (*20*). To a solution of 8.5 g (23 mmole) of the glycoside IV in 50 ml of absolute methanol (Vol. VII [3]) is added 1 ml of 1 M sodium methoxide (Vol. II [54]), and the mixture is kept for 12 h at 25°. The solution then is treated for 30 min at 0° with 5 ml of a weakly acidic, ion-exchange resin [such as Amberlite IRC-50(H$^+$), prewashed with methanol]. The resin is removed by filtration, and the filtrate is evaporated. Three 15-ml portions of water are added to and evaporated (rotary evaporator, water aspirator, bath temperature ∼45°) from the residue to remove azeotropically the methyl benzoate formed. The last traces of water are removed by distilling ethanol from the product; and the resulting, crystalline solid is recrystallized from 3:1 v/v isopropyl ether (or ether)–dichloromethane (∼150 ml) to afford pure V; yield 3.3 g (89%), mp 110°, $[\alpha]^{22}$D +44° (c 0.8, methanol) and +50° (c 1.2, chloroform).

2,6-Dideoxy-α-L-*lyxo*-hexose (2-Deoxy-α-L-fucose, VI) (*20*)

A solution of 2 g (12.3 mmole) of the methyl glycoside V in 3:1 v/v water–acetic acid is heated for 45 min on a boiling-water bath. The solvent then is removed by distillation and three successive 10-ml portions of water are added to the concentrated solution (∼3 ml) and removed by evaporation after each addition to remove all the acetic acid. Finally, the solution is

---

[4] The product is detected in aliquants of the effluent by spotting the aliquants on tlc plates (silica gel) and streaking the plates, by means of a capillary, with aqueous KMnO$_4$. The violet line turns yellow in zones containing the product.

[5] This precaution prevents the catalyst from igniting when exposed to air.

evaporated to dryness to afford a colorless, hygroscopic syrup that is kept over phosphorus pentaoxide and potassium hydroxide for several days, affording crude 2,6-dideoxy-L-*lyxo*-hexose (VI) as a glass; yield 1.79 g (98%). Addition of a little acetone effects crystallization, and the crystals are collected with the aid of a little hexane in an inert atmosphere (dry box) and dried to give analytically pure VI; yield 1.3 g (71%). A second crop of crystals may be obtained (*1*) by evaporating the mother liquor and subliming VI from the residue (50 mtorr, bath temperature 120°). The crystalline, slightly hygroscopic α anomer (VI) melts at $102°-105°$, $[\alpha]^{23}$D $-97°$ (initial, extrapolated)→ $-81.2°$ (4 min)→ $-73.8°$ (6 min) → $-64.8°$ (10 min)→ $-51.5°$ (2 h, equil, c 1, water) and $-134°$ (initial, extrapolated)→ $-131.1°$ (7.5 min)→ $-128.3°$ (15 min)→ $-122.7°$ (30 min)→ $-75.6°$ (10 h, equil, c 1, acetone).

References

(*1*)   B. Iselin and T. Reichstein, *Helv. Chim. Acta*, **27**, 1200 (1944).
(*2*)   P. J. Garegg, B. Lindberg, and T. Norberg, *Acta Chem. Scand.*, **B28**, 1104 (1974).
(*3*)   H. S. El Khadem, D. L. Swartz, J. K. Nelson, and L. A. Berry, *Carbohydr. Res.*, **58**, 230 (1977).
(*4*)   S. Hanessian, *Advan. Carbohydr. Chem.*, **21**, 143 (1966).
(*5*)   R. F. Butterworth and S. Hanessian, *ibid.*, **26**, 279 (1971).
(*6*)   S. Hanessian and T. H. Haskell, *in* W. Pigman and D. Horton, eds., "The Carbohydrates," Academic Press, New York, 1970, Vol. IIA, p. 139.
(*7*)   T. Reichstein and E. Weiss, *Advan. Carbohydr. Chem.*, **17**, 65 (1962).
(*8*)   T. Reichstein, *Naturwissenschaften*, **54**, 53 (1967).
(*9*)   H. Brockmann and T. Waehneldt, *Naturwissenschaften*, **48**, 717 (1961).
(*10*)   H. Brockmann, *Fortschr. Chem. Org. Naturst.*, **21**, 121 (1963).
(*11*)   H. Brockmann, B. Scheffer, and C. Stein, *Tetrahedron Lett.*, 3699 (1973).
(*12*)   W. Keller-Schierlein and W. Richle, *Antimicrob. Agents Chemother.*, 68 (1970).
(*13*)   W. Richle, E. K. Winkler, D. M. Hawley, M. Dobler, and W. Keller-Schierlein, *Helv. Chim. Acta*, **55**, 467 (1972).
(*14*)   S. Takahashi, M. Kurabayashi, and E. Ohki, *Chem. Pharm. Bull.*, **15**, 1657 (1967).
(*15*)   E. Wyss, H. Jäger, and O. Schindler, *Helv. Chim. Acta*, **43**, 664 (1960).
(*16*)   Yu. A. Berlin, S. E. Esipov, M. N. Kolosov, and M. M. Shemyakin, *Tetrahedron Lett.*, 1431 (1966).
(*17*)   Yu. A. Berlin, O. A. Kiseleva, M. N. Kolosov, M. M. Shemyakin, V. S. Soifer, I. V. Vasina, and I. V. Yartseva, *Nature*, **218**, 193 (1968).
(*18*)   Yu. A. Berlin, M. N. Kolosov, and I. V. Yartseva, *Khim. Prir. Soedin.*, **9**, 539 (1973) [*Chem. Abstr.*, **80**, 27439p (1974)] and earlier papers in this series, cited therein.
(*19*)   M. Miyamoto, Y. Kawamatsu, M. Shinohara, Y. Nakadaira, and K. Nakanishi, *Tetrahedron*, **22**, 2785 (1966); compare J. S. Brimacombe and D. Portsmouth, *Carbohydr. Res.*, **1**, 128 (1965).
(*20*)   T. M. Cheung, D. Horton, and W. Weckerle, *Carbohydr. Res.*, **58**, 139 (1977).
(*21*)   D. A. Prins, *J. Amer. Chem. Soc.*, **70**, 3955 (1948).

(22) A. Rosenthal and P. Catsoulacos, *Can. J. Chem.*, **46**, 2868 (1968).
(23) A. C. Richardson, *Carbohydr. Res.*, **4**, 422 (1967).
(24) W. Kwasnik, in G. Brauer, ed., *Handbuch der präparativen anorganischen Chemie*, Vol. I, F. Enke Verlag, Stuttgart, 1960, p. 224; *Handbook of Preparative Inorganic Chemistry*, Vol. I, Academic Press, New York, 1963, p. 240.

[28] Methoxymercuration in Synthetic Carbohydrate Chemistry

Methyl 2,6-Dideoxy-α-L-*arabino*-hexopyranoside

By Leon Goodman

Department of Chemistry, University of Rhode Island,
Kingston, Rhode Island

AND

Edward M. Acton, Carol W. Mosher, and John P. Marsh, Jr.

Department of Bio-organic Chemistry, Stanford Research Institute,
Menlo Park, California

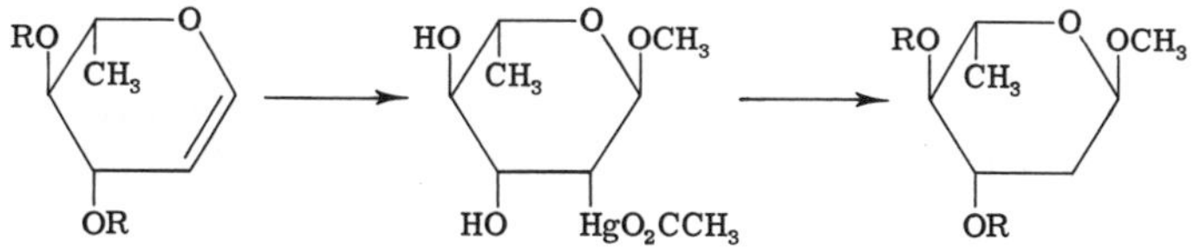

(I, R = CH₃CO)
(II, R = H)

(III)

(IV, R = H)
(V, R = CH₃CO)

Introduction

Addition reactions of unsaturated sugars, especially electrophilic additions, provide a large number of useful carbohydrate derivatives. Electrophilic additions initiated by mercuric salts provide a recent example of such reactions and possess great versatility because the nucleophile that completes the reaction can be varied and because, in a secondary reaction, demercuration can lead to other sugar derivatives of which the deoxy sugars prepared by reduction demercuration with borohydrides are probably the most important. The first examples of such reactions in the carbohydrate area were the methoxymercuration studies of Inglis (*1*) and Manolopoulos (*2*) with D-glucal and its triacetate. Methoxymercuration of a glycal provided a key intermediate in the first synthesis of daunosamine (*3*); a careful study of the stereochemistry of the methoxymercuration of various glycal acetates has been reported recently (*4*). Alkoxymercurations of glycals have been noted (*5*) as have oxymercurations and acetoxymercurations (*6*). Unsaturated sugars other than glycals have been subjected to mercuration reactions. In a linco-

207

METHODS IN CARBOHYDRATE
CHEMISTRY, VOL. VIII

mycin synthesis, the terminal vinyl group of 7,8-dideoxy-1,2:5,6-di-O-isopro-pylidene-D-*glycero*-α-D-*galacto*-oct-7-eno-pyranose was oxymercurated, then treated with sodium borohydride to give 8-deoxy-1,2:3,4-di-O-isopropyl-idene-D-*erythro*-α-D-*galacto*-octopyranose (*7*) and azidomercuration of a *C*-methylene furanose sugar has been reported (*8*) in a novel approach to aminosugars.

Procedure

Methyl 2-Acetoxymercuri-2,6-dideoxy-α-L-mannopyranoside (III) (*3*)

To a solution of 100 g (0.467 mole) of 3,4-di-O-acetyl-6-deoxy-L-glucal (I) (Vol. II [102]) in 1800 ml of absolute methanol (Vol. VII [3]) is added, with vigorous stirring, a solution of 2.0 g of sodium methoxide in 200 ml of absolute methanol. The resulting solution of II, protected from moisture, is allowed to stand at 20°–25° for 1 h; then to it is added 150 g (0.470 mole) of mercuric acetate, and stirring is continued until a complete solution is obtained. After it stands an additional hour at 20°–25°, the solution is evaporated to dryness under diminished pressure. The white residual solid is redissolved in 500 ml of methanol; the solution is concentrated to a slurry to effect crystallization (volume of about 70 ml), and 300 ml of ether is added. The mixture is chilled overnight, and the white crystalline product is collected and washed with two 50-ml portions of cold ether; yield 160 g (81%); mp 145°–150°; $[\alpha]^{23}$D −2.3° (*c* 1.0, methanol); NMR (δ, DMSO-d_6) 1.32,d-(CH$_3$,J$_{5,6}$ = 6.0 Hz), 2.05,s(CH$_3$CO), 3.32,s(OCH$_3$), 2.8–4.2,m(H-2,3,4,5), 4.2,s(HOD), 5.0,s(H-1); λ^{Nujol}_{max} (μm) 2.88,2.98(OH), 6.83(CH$_3$CO), 9.0,9.55-(C—O—C).

Methyl 2,6-Dideoxy-α-L-*arabino*-hexopyranoside (IV) (*3*)

A solution of 4.9 g (0.13 mole) of sodium borohydride in 140 ml of 1 *M* sodium hydroxide is added dropwise to a solution of 196 g (0.467 mole) of the 2-acetoxymercuri sugar (III) in 2.6 liters of absolute methanol (Vol. VII [3]) and 558 ml of 1 *M* sodium hydroxide at 20°–25° (the addition is mildly exothermic). The reaction mixture is stirred at 20°–25° for 1 h (free mercury noted); then 3.5 g more of solid sodium borohydride is added in three portions at 20-min intervals. After the final addition, the mixture is stirred 20 min more and is filtered through a Celite pad; then the filter cake is washed with a minimum volume of water and methanol. Concentration of the filtrate leaves a pasty, white residue which is dried by repeated addition and evaporation of methanol to give 161 g of a crude product still containing inorganic salts.

Purification is achieved by acetylation to give V which can be separated from the salts. The above crude product is heated at 50°–55° for 45 min with a solution of 700 ml of pyridine (dried over calcium hydride) and 1040 ml of acetic anhydride. The reaction mixture is evaporated to dryness under diminished pressure, and the residue is partitioned between 890 ml of water and 2450 ml of chloroform. The chloroform extract is washed with three 900-ml portions of water, dried with magnesium sulfate, and concentrated under diminished pressure to give about 120 g of the diacetate (V) as a yellow oil. The oil is deacetylated by dissolving it in a solution of 3.0 g of sodium methoxide and 1650 ml of absolute methanol and heating the resulting solution at reflux for 45 min. The basic (to pH paper) methanolic solution is neutralized with Amberlite IRC-50(H^+ form, previously washed with methanol), using approximately 100 g. The resin is removed by filtration and washed with methanol, and the combined filtrates are evaporated to leave a pale yellow oil; 70 g (92%); n^{24}D 1.4651; $[\alpha]^{21}$D $-112.6°$ (c 0.7, water); NMR (δ,CDCl$_3$) 1.28,d(CH$_3$,J$_{5,6}$ = 6.0 Hz), 1.55–2.35,m(H-2), 3.3,s(OCH$_3$), 2.85–4.15,m(H-3,4,5), 4.7,d(H-1,J$_{1,2}$ = 3.5 Hz); λ_{max}^{film} (μm), 2.95(broad,OH), 8.9(OCH$_3$), 9.5(C—O—C), 11.1,11.5,12.0,13.15 (characteristic for α-anomers). The infrared and NMR spectral data reported are in excellent agreement with those of methyl chromoside C, the enantiomer of IV, prepared from a sugar isolated from the hydrolysis of chromomycin A$_3$ (10).

The syrup can be distilled at 95° (0.35 mm).

References

(1) G. R. Inglis, J. C. P. Schwarz, and L. McLaren, *J. Chem. Soc.*, 1014 (1962).

(2) P. T. Manolopoulos, M. Mednick, and N. N. Lichtin, *J. Amer. Chem. Soc.*, **84**, 2203 (1962).

(3) J. P. Marsh, Jr., C. W. Mosher, E. M. Acton, and L. Goodman, *Chem. Commun.*, 973 (1967).

(4) K. Takiura and S. Honda, *Carbohydr. Res.*, **21**, 379 (1072).

(5) J. H. Leftin and N. N. Lichtin, *Israel J. Chem.*, **3**, 107 (1965).

(6) K. Takiura and S. Honda, *Carbohydr. Res.*, **23**, 369 (1972).

(7) D. G. Lance, W. A. Szarek, J. K. N. Jones, and G. B. Howarth, *Can. J. Chem.*, **47**, 2871 (1969).

(8) J. S. Brimacombe, J. A. Miller, and U. Zakir, *Carbohydr. Res.*, **49**, 233 (1976).

(9) M. Miyamoto, Y. Kawamatsu, M. Shinohara, Y. Nakadaira, and K. Nakanishi, *Tetrahedron*, **22**, 2785 (1966).

POLYSACCHARIDES

[29] Isolation and Purification of Carbohydrate Antigens

BY JOHN H. PAZUR AND L. SCOTT FORSBERG

*Department of Biochemistry and Biophysics,
The Pennsylvania State University, University Park, Pennsylvania*

Introduction

The discovery that the carbohydrates in bacterial cell walls and capsules are immunogenic substances was made by Avery and Heidelberger (*1*) who observed that the immunogenicity of various pneumonoccal capsules was attributable to the carbohydrates in the capsules. These initial observations led to the development of the serological method for classifying the pneumococci and streptococci based on the type of immunogenic carbohydrate present in the cell wall matrix. Such methods have been invaluable for the detection and identification of pathogenic organisms and, in turn, the clinical management of infectious and inflammatory diseases and related maladies (*2*).

A number of procedures have been used for extracting the carbohydrate antigen from bacterial capsules and cell walls. The extraction agents which have been employed include trichloroacetic acid (*3*), dilute potassium chloride–hydrochloric acid solution (*4, 5*), formamide (*6*), and dilute acetic acid (*7*). The extractions have been performed under a variety of conditions including cold room (4°) to elevated temperature (180°) and for short (5 min) and extended periods (48 h). A description of the older methods is presented in an immunological methods series (*8*). Two procedures used recently in our laboratory for obtaining carbohydrate antigens will be described. One is based on the use of 10% trichloroacetic acid (*3*), and the other on the use of 0.05 *M* potassium chloride and 0.01 *M* hydrochloric acid of pH 2 (*9*) for extracting immunogenic glycans from group D streptococci. Also described are methods for the purification of the glycans to homogeneity and the criteria of purity for the antigens. These procedures can be applied to the isolation of carbohydrate containing antigenic substances from other bacterial groups and from different types of cells and tissues including mammalian tumors and virally transformed cells.

211

METHODS IN CARBOHYDRATE
CHEMISTRY, VOL. VIII

Procedure

Extraction of Carbohydrate Antigens from *Streptococcus bovis* by the KCl–HCl Method

Streptococcus bovis, strain C3, was the organism used to isolate the antigenic glycans by this procedure. The glycans are type-specific antigens located in the cell wall. *S. bovis* is grown in a 3% Todd–Hewett media until a culture of 18 L is obtained. The cells are harvested in a refrigerated Sharples centrifuge (Sharples Corp., Philadelphia, Pennsylvania) yield of cells, 40 g wet weight. The wet cells are suspended in 200 ml of HCl–KCl solution of pH 2 (10.6 ml of 0.2 M HCl and 50 ml of 0.2 M KCl diluted to a volume of 200 ml) and heated in a boiling water bath successively for specified periods of time (5, 10, 20, 40, and 60 min). At the end of each heating period, the residue is removed by centrifugation and resuspended in 200 ml of new HCl–KCl solution. The supernatants after each heating period are saved for isolation of solubilized products. After the final extraction, all supernatants are neutralized individually with a few drops of 1 N NaOH and dialyzed separately for 48 h against distilled water in order to remove low-molecular-weight products. The solutions from the dialysis tubing are transferred to a beaker and mixed with an equal volume of 10% trichloroacetic acid to precipitate protein and nucleic acids which are also extracted from the cells. The precipitates are removed by centrifugation, and the supernatants are dialyzed against distilled water for 48 h and then taken to dryness by lyophilization. The yield of lyophilized product range from 0.1 to 0.3 g for the various periods. These lyophilized samples constitute the crude preparations of antigenic glycans.

Extraction of Carbohydrate Antigens from *Streptococcus faecalis* by the Trichloroacetic Acid Method

Streptococcus faecalis, strain N was the test organism used in this method. This organism also contains two type-specific carbohydrate antigens in its cell wall. *Streptococcus faecalis* is grown on 0.5% nutrient broth containing 0.1% D-glucose essentially by the procedure described above. The cells from 18 L are collected in a Sharples centrifuge; yield of cells 25 g wet weight. The cell wall components are extracted by stirring the cells in 200 ml of 10% trichloroacetic acid with a magnetic stirrer at 4° for 48 h. The residue is removed by centrifugation, and the supernatant is used for the isolation of the glycans. The supernatant is concentrated under reduced pressure to 30 ml, and the protein components are precipitated by the addition of an equal volume of 1:20 v/v 2 N HCl–absolute ethanol. The mixture is kept at 4°

for 24 h, and the precipitate is removed by centrifugation. The glycans in the supernatant are then precipitated by addition of 5 volumes of acetone. The mixture is maintained at 4° for 24 h, and the precipitated material is collected by centrifugation. The residue is redissolved in water, dialyzed against distilled water for 48 h, and lyophilized to dryness; yield of white amorphous product, 0.2 g.

Separation of the Carbohydrate Antigens

In the initial studies with the antigens from *S. bovis*, a separation of the antigens was achieved by filtration through columns of BioGel P-60 poly-(acrylamide) gel beads (*5*, see Vol. VIII [5]). Later, an improved separation method based on DEAE-cellulose chromatography was devised (*9*) and the procedure is now used routinely for the preparation of antigens from this organism. In the latter procedure, a 35 × 3-cm column of 25 g of DEAE-cellulose (medium mesh, 0.86 meq/g, Sigma Chemical Co., St. Louis, Missouri) is equilibrated with 0.001 M potassium phosphate buffer of pH 7.5 in 0.01 M sodium chloride. A sample of 0.1 g of the lyophilized carbohydrate antigens is introduced onto the column, and elution of the glycan is effected with an additional 200 ml of the above salt solution followed by 600 ml of a salt gradient solution of pH 7.5 (300 ml of 0.001 M potassium phosphate of pH 7.5 in 0.01 M sodium chloride mixed slowly with 300 ml of 0.001 M potassium phosphate buffer in 0.4 M sodium chloride). Fractions of approximately 6 ml of the eluates· are collected, and aliquots of 0.1 ml of these fractions are analyzed for carbohydrate by the phenol–sulfuric acid method (*11*, Vol. II [115], [116]).

Figure 1 shows data on the separation of glycans in the 40-min extract from the *S. bovis* cells. Two major and one minor carbohydrate component were present in this extract. One major component was eluted at low ionic strength salt-buffer in fractions 10 to 30, and the other at high ionic strength in fractions 90 to 120. The minor component was eluted in fractions 60 to 88. From the 5- and 10-min extracts, only one major component was obtained, and this was eluted with the salt buffer solution of high ionic strength. From the 20- and 60-min extracts, the two major peaks and the minor peak of carbohydrate material were obtained corresponding to the peaks from the 40-min extract but in somewhat different ratios. The fractions containing the individual components were combined, dialyzed against distilled water, and taken to dryness by lyophilization. To check for purity, a rechromatography on DEAE-cellulose can be performed following the above directions.

Separations of the two antigens from *S. faecalis* can be achieved by fractional precipitation with ehtanol. A sample of 0.2 g of the trichloroacetic acid-soluble material from the cell walls is dissolved in 9 ml of a 10% solution

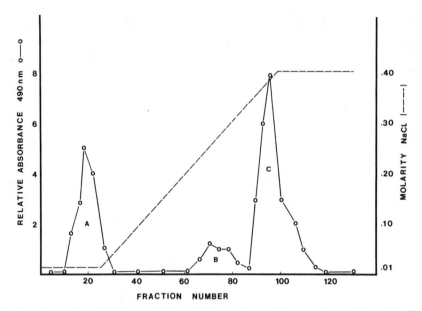

FIG. 1.—Elution pattern of the heteroglycans of *S. bovis* from a DEAE-cellulose column with a sodium chloride gradient: (A) diheteroglycan of D-glucose and L-rhamnose; (B) structure unknown; (C) tetraheteroglycan of 6-deoxy-L-talose, L-rhamnose, D-galactose, and D-glucuronic acid.

of trichloroacetic acid. After removal of a slight amount of insoluble material by centrifugation, 1 volume of absolute alcohol is added to the clear supernatant, and the mixture is maintained at 4° for 18 h. The white precipitate which appears in this time period is isolated by centrifugation. The supernatant is saved for subsequent work. The precipitate is washed with 3 ml portions of ethanol and with diethyl ether, collected by centrifugation, and dried in a desiccator; yield, 0.07 g. This material is redissolved in 3 ml of a 10% solution of trichloroacetic acid, and the precipitation with alcohol is repeated; yield after the second precipitation, 0.06 g.

The supernatant from the first precipitation is stirred with four additional volumes of ethanol, and the solution is stored at 4° for 18 h. A second precipitate appears and is collected by centrifugation, washed with ethanol and diethyl ether (Vol. V [16]), and dried in a desiccator; yield, 0.07 g. This glycan was also subjected to a second fractionation by the procedure described above; yield after the second fractionation, 0.05 g.

Determination of Purity

Agar Diffusion and Immunoelectrophoresis

Antisera directed against the glycans are obtained from rabbits immunized with a vaccine of non-viable cells of *S. faecalis*. The vaccines are prepared by the following procedure. Freshly grown cells from 500 ml of 18-h cultures are suspended in 30 ml of 0.1 M phosphate buffer of pH 7.2, containing 0.2% formaldehyde, and stirred at 4° (cold room) for 16 h. The cells are then washed several times with the saline–phosphate buffer. After the final washing, viability tests are performed on the samples by innoculating a tube of media with a loop of the cell suspension. If growth is obtained, the cell suspension is again treated with 0.2% formaldehyde and rewashed thoroughly with buffer. When the suspension of cells is no longer viable, the sample is diluted two fold with the buffer solution to yield a slightly turbid solution. Rabbits are immunized by intravenous injection of 0.3 ml of the vaccine daily for 4 d followed by a rest period of 3 d. This schedule is repeated for 4 additional weeks. After a rest period of 1 week, a second administration of vaccine is used following the above schedule. Blood samples are obtained from the rabbits weekly, and the sera from the samples are tested for potency utilizing the appropriate glycan as antigen. Generally, serum samples after the 9th week of immunization exhibit a high titer against the antigenic glycans.

The purified glycans are tested with appropriate antisera for reactivity by the double diffusion in agar method and by immunoelectrophoresis. Diffusion tests are performed in a standard manner in 0.1% agarose gel in 0.1 *M* phosphate buffer of pH 7.2 with several concentrations of the glycan. Immunoelectrophoresis is performed in 0.1% agarose in 0.1 *M* Veronal buffer of pH 8.3 on microscope slides at voltage of 60 V and amperage of 15 ma for 1–2 h. Subsequently, antisera are placed in the center trough, and diffusion is allowed to occur overnight. The bands of antigen–antibody complex are visualized under indirect light. In these tests, the crude antigen preparation yields two bands of precipitin complex with the homologous antisera, but the purified antigenic glycans yield single bands.

Ultracentrifugation

Samples of 2 mg of the pure glycans and reference glycans of known molecular size are introduced on different density gradient solutions (5 to 25% glycerol). All samples are then centrifuged at the same time and for 14 h at 65,000 rpm in a centrifuge such as a Spinco model L-65 (Spinco Div.,

Beckman Instruments, Palo Alto, California). At the end of this time, the density gradient columns are fractionated into 0.2-ml samples by means of an automatic fractionator (*11*). Aliquots of the samples obtained from the tubes are analyzed for carbohydrate content (*10*), and the carbohydrate concentration is plotted versus the fraction number. With the glycans from *S. faecalis*, symmetrical peaks were indeed obtained indicating homogeneity in molecular size. These two glycans sedimentated at different rates. These data and an empirical formula can be used to calculate the molecular weights of the glycans. The diheteroglycan of glucose and galactose had a molecular weight of 15,000, while the tetraheteroglycan of L-rhamnose, D-glucose, D-galactose, and 2-acetamido-2-deoxy-D-glucose had a molecular weight of 5,000 (*11*). The molecular weight of the glycans from *S. bovis* was estimated to be 6000 for the tetraheteroglycan and 12,000 for the diheteroglycan (*9*).

Polyacrylamide Gel Filtration

Samples of the glycans ranging in amounts from 10 to 50 mg are used for polyacrylamide gel filtration. A 94 cm × 2.5-cm column containing polyacrylamide gel (100–200 mesh BioGel P-30, Bio-Rad Laboratories,

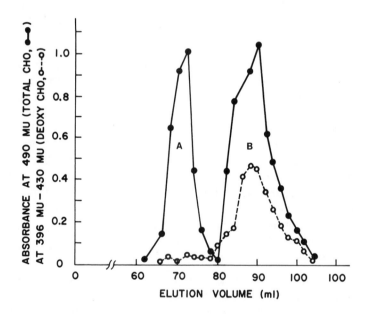

FIG. 2.—Elution pattern from a BioGel P-30 column of the diheteroglycan of (A) D-glucose and D-galactose from *S. faecalis* and the diheteroglycan (B) of D-glucose and L-rhamnose from *S. bovis* (*13*).

Richmond, Califormia) was prepared and washed thoroughly with saline and 0.1 M phosphate buffer of pH 7.0. The carbohydrate sample was introduced in the column and washing was continued with the above buffer. Fractions of 2 ml were collected from the column and analyzed for carbohydrates by the phenol sulfuric acid method (*10*) and by the cysteine-sulfuric acid method (*12*). Glycans which are homogeneous will yield symmetrical elution peaks. BioGel filtration patterns for the diheteroglycan from *S. faecalis* and tetraheteroglycan from *S. bovis* are reproduced in Figure 2. It will be noted that both the glycans yield symmetrical peaks indicating homogeneity in molecular size for both preparations.

References

(*1*) M. Heidelberger and O. T. Avery, *J. Exptl. Med.*, **40**, 301 (1924).

(*2*) M. McCarty, "Streptococcal Infections," Columbia University Press, New York, 1954.

(*3*) J. H. Pazur, J. S. Anderson, and W. W. Karakawa, *J. Biol. Chem.*, **246**, 1793 (1971).

(*4*) R. C. Lancefield, *J. Exptl. Med.*, **47**, 91 (1928).

(*5*) J. H. Pazur, J. A. Kane, D. J. Dropkin, and L. M. Jackman, *Arch. Biochem. Biophys.*, **150**, 382 (1972).

(*6*) R. M. Krause and M. McCarty, *J. Exptl. Med.*, **114**, 127 (1961).

(*7*) T. Liu and E. C. Gotschlich, *J. Biol. Chem.*, **238**, 1928 (1963).

(*8*) R. Krause, *in* "Methods in Immunology and Immunochemistry," C. A. Williams and M. W. Chase, Eds., Academic Press, New York, N.Y., Vol. I, 1967, p. 34.

(*9*) J. H. Pazur, D. J. Dropkin, K. L. Dreher, L. S. Forsberg, and C. S. Lowman, *Arch. Biochem. Biophys.*, **176**, 257 (1976).

(*10*) M. Dubois, K. A. Gilles, J. K. Hamilton, P. A. Rebers, and F. Smith, *Anal. Chem.*, **28**, 350 (1956).

(*11*) J. H. Pazur, K. Kleppe, and J. S. Anderson, *Biochim. Biophys. Acta*, **65**, 369 (1962).

(*12*) Z. Dische and L. B. Shettles, *J. Biol. Chem.*, **175**, 595 (1948).

(*13*) J. H. Pazar, D. J. Dropkin, and L. S. Forsberg, *Carbohydr. Res.*, **66**, 155 (1978).

UNSATURATED SUGARS

[30] 3-Deoxyglycals (1,5-Anhydro-1,2,3-trideoxyhex-1-enitols) from Hex-2-enopyranosides

By Bert Fraser-Reid, Bruno Radatus, and Steve Yik-Kai Tam

*Guelph-Waterloo Centre for Graduate Work in Chemistry,
Waterloo Campus, University of Waterloo,
Waterloo, Ontario, Canada*

Introduction

The need for ready routes to 3-deoxysugars was recently brought to the fore by the convincing demonstration of Umezawa and co-workers that 3-deoxykanamycin exhibited inhibition against bacteria normally resistant to kanamycin (*1*). The long history of glycals in synthetic carbohydrate chemistry (*2*), particularly in relation to the Lemieux–Nagabhushan reaction (Vol. VI [89]), suggests that 3-deoxyglycals would provide access to a wide variety 3-deoxysaccharides. Indeed this aspect is already being explored (*3*, *4*), and the value of 3-deoxyglycals has been shown for the synthesis of specifically labeled 2-deoxyriboses (*5*) and 2-deoxyribonucleotides (*6*). In this article a simple route to 3-deoxyglycals[1] is described (*7–9*).

The transformations of particular interest are shown in Table I. They involve the lithium aluminum hydride induced formation of 3-deoxyglycals (VI–XI) from hex-2-enopyranosides (I→V). The latter are well known compounds which are readily prepared by the Tipson–Cohen (*11*) and the Ferrier (Vol. VI [54]) procedures, both of which can be conducted on a 0.5 molar scale. In the Tipson–Cohen reaction (*11*) the elimination is cleaner and more reproducible (*12*), especially on a large scale, when zinc–copper couple replaces zinc as the reducing agent.[2] As an example, the preparation of the less well known hex-2-enopyranoside, II, from methyl 4,6-di-*O*-acetyl-2,3-di-*O*-methylsulfonyl-β-D-glucopyranoside is described. The results in Table I are shown for lithium aluminum deuteride to emphasize the stereochemical aspects of the transformations. The latter have been given detailed analysis elsewhere, and anyone interested in the preparation

---

[1] See also the work of Achmatowicz and Szechner (*10*).

[2] When this reaction is done on a very large scale (∼12 liters of solvent), emergency cooling should be readily available in case the reaction becomes too vigorous.

METHODS IN CARBOHYDRATE
CHEMISTRY, VOL. VIII

TABLE I

Reductive Rearrangement of Hex-2-enopyranosides with Lithium Aluminum
Hydride (Deuteride) in Refluxing Dioxane (isolated yields)

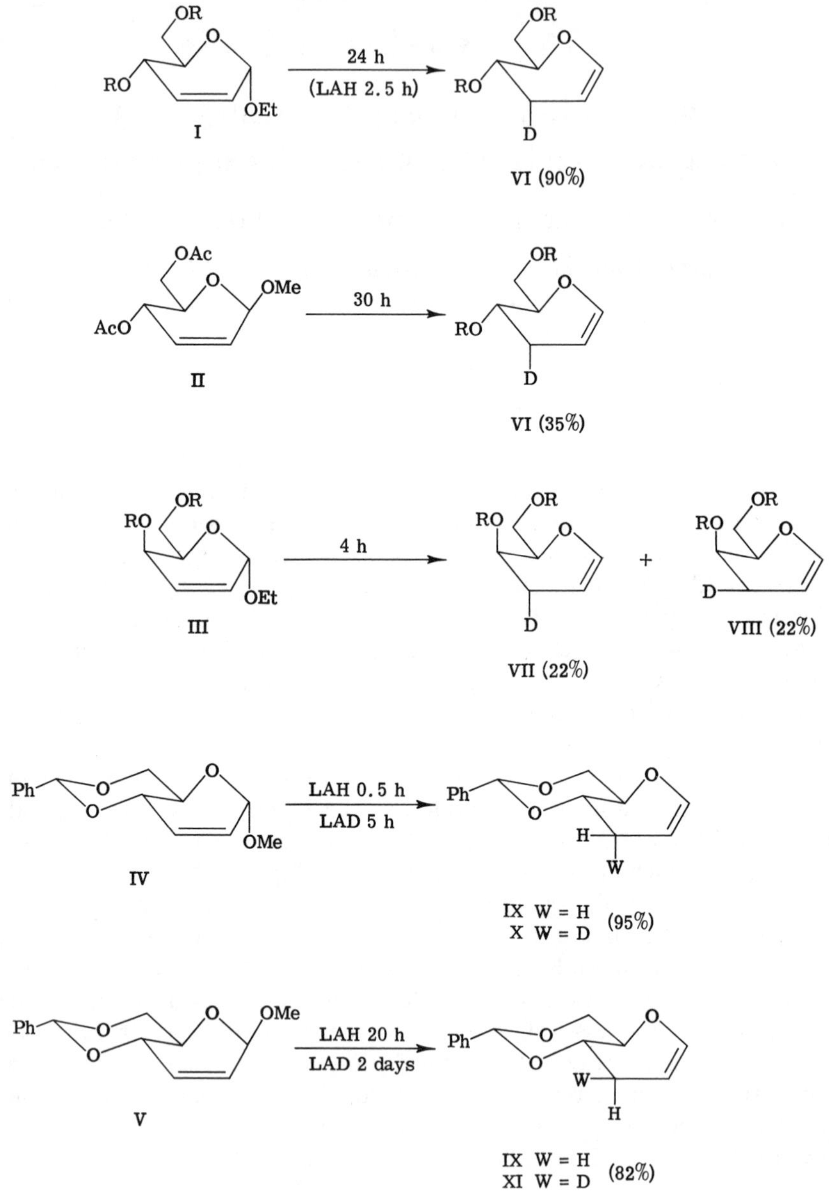

I
24 h
(LAH 2.5 h)
VI (90%)

II
30 h
VI (35%)

III
4 h
VII (22%)
+
VIII (22%)

IV
LAH 0.5 h
LAD 5 h
IX W = H
X   W = D   (95%)

V
LAH 20 h
LAD 2 days
IX W = H
XI W = D   (82%)

a series R = H
b series R = Ac

---

[a] Determined at 220 MHz in CDCl$_3$ (TMS). J values were read directly from spectra run at 250 Hz sweep width.

of specifically labeled derivatives should consult the pertinent publications (8, 9). However the results suggest that the rearrangements take place via one of the mechanisms expressed in XII and XIII, and there are four perti-

(classical S_n2')   (abnormal S_n2')
(XII)   (XIII)

nent conclusions: (A) The hex-2-enopyranosides react whether the geometry of the oxygens is cis-1,4 or trans-1,4. (B) If there is an hydroxyl or an ester group on C-4 (both of which would be immediately converted to alkoxide under the reaction conditions), the abnormal S_N2' transition state XIII may be brought into operation. Etherification of O-4 (as in V) apparently prevents the attainment of transition state XIII. (C) An axial allylic oxygen permits easier attainment of the S_N2' transition state than does its equatorial counterpart. This conclusion follows from the greater reactivity of IV as compared to V. (D) Considered in relation to the operative transition state, the reactions are always *stereo*specific but may or may not be *regio*specific.

Acylic vinyl ethers (such as XIV), and 2-deoxyhex-3-enopyranosides (such as XV) are sometimes obtained as by-products in these reactions. The amounts range from 0 to 15 percent, and their occurrence depends strongly upon the structure of the substrate (8, 9).

(D or H)   (D or H)
(XIV)   (XV)

Characterization of the Products

The progress of the reaction is readily followed on thin-layer chromatograms (Vol. VI [6]) sprayed with sulfuric acid, whereupon the starting materials (I–V) char immediately in the cold to a black residue, while the products (VI–XI, XIV, XV) require heating in an oven before charring occurs. On the other hand, the vinyl ether products, but not the starting materials, are readily visualized by iodine vapor.

Some definitive nmr data are shown in Table II. Typically, H-1 appears as a doublet ($J_{12} = 6.0$ Hz) at 7.37 ± 0.06 ppm. Infrared absorptions at 1645, 1233 and 1080 cm^{-1} are indicative of the vinyl ether structures.

TABLE II

Proton Magnetic Resonance Parameters for Some 3-Deoxyglycals[a]

Compound	Chemical shifts (δ)				J values (Hz)		
	H-1	H-2	H-3	H-4	$J_{1,2}$	$J_{1,3}$	$J_{2,3}$
VIa	6.44	4.70	2.33	~4.1	6.5	1.0	4.0
VIb	6.37	4.70	2.40	5.03	6.0	1.0	5.0
VIIa	6.44	4.69	2.36	~4.3	6.0	1.0	5.0
VIIb	6.43	4.70	2.40	5.19	6.0	1.0	4.6
VIIIa	6.44	4.9	2.02	~4.3	6.0	2.1	~1.0
VIIIb	6.43	4.70	2.40	5.19	6.0	2.1	2.5
IX	6.35	4.73	2.25	3.89	6.0	2.7	2.5 & 5.5
X	6.33	4.72	2.33	3.92	6.0	1.0	5.5
XI	6.31	4.71	2.24	3.92	6.0	2.7	2.0

Procedures

Methyl 4,6-Di-O-acetyl-2,3-dideoxy-β-D-erythro-hex-2-enopyranoside (II)

Methyl 4,6-di-O-acetyl-2,3-di-O-methylsulfonyl-β-D-glucopyranoside (9) (5.7 g, 0.013 mole) is added to an efficiently stirred mixture of potassium iodide (11.0 g, 0.066 mole) and zinc–copper couple (13) (4.16 g) in 40 ml of N,N-dimethylformamide. The reaction is complete within 1 h (tlc), and the hot solution is poured into ice and water. Dichloromethane (methylene chloride) is added, and the mixture is stirred well and then filtered through Celite. The filter-cake is washed well with dichloromethane. The combined organic layers are washed with sodium thiosulfate solution and then water, dried with anhydrous sodium sulfate, and evaporated under diminished pressure. The residue (2.5 g) shows two components by tlc on silica gel using 1:1 v/v petroleum ether (30°–60°)–ethyl acetate of R_f 0.34 and R_f 0.52, the former being the 6-O-monoacetate (12). Acetylation converts the slower into the faster moving substance and purification by passage through a silica gel column gives pure II; yield 2.5 g (78%).

General Experimental Procedure for Reductive Rearrangements

The following procedure is given for reactions using lithium aluminum deuteride (LAD) in refluxing dry dioxane. However lithium aluminum hydride (LAH), diethyl ether, or tetrahydrofuran[3] can be substituted without changing the product composition. In general, reactions with LAH go

---

[3] Dioxane, diethyl ether, and tetrahydrofuran are dried with benzophenone ketyl (14).

faster than those with LAD; on the other hand, the lower boiling solvents cause the reaction to be much slower than with dioxane.

In a typical experiment, the substrate (3 mmole) is dissolved in 25 ml of dry dioxane and (0.38 g (9 mmole) of LAD is added with efficient stirring with a magnetic stirrer–hot plate.[4] The mixture is heated to gentle reflux and the progress of the reaction is followed by tlc (for details, see sample procedures below). Upon completion, the mixture is allowed to cool; and the method of work-up depends upon the solubility of the product and on the scale of the reaction.

Small Scale Reactions

For *water-soluble products* (for example, the diols VIa–VIIIa), excess LAD is destroyed by dropwise addition of ethyl acetate. The salts are removed by filtration through Celite, and the filter-cake is washed several times with ethyl acetate. The filtrate is evaporated to dryness, and the residue (which contains some aluminum salts) is acetylated with acetic anhydride and pyridine in the usual way (Vol. II [53]) affording the diacetates (VIb–VIIIb).

For *water-insoluble products* (IX–XI), the excess hydride (deuteride) is destroyed by dropwise addition of the minimum of water, and the aluminum hydroxide is removed by filtration through a bed of Celite. The filtrate is then dried with anhydrous sodium sulfate and evaporated under reduced pressure, whereupon the product crystallizes.

Large Scale Reactions

In large scale reactions (those requring >2 g LAH), the "usual" work-up procedure gives a gelatinous precipitate of aluminum salts which causes the filtration to be painfully slow. The following alternative work-up gives a granular precipitate which is easier to filter. For *n* grams of LAH (LAD), *n* ml of water is added dropwise, followed by *n* ml of 15% sodium hydroxide solution, followed in turn by 3 *n* ml of water (*15*).

1,2,3-Trideoxy-3-deuterio-D-*ribo*-hex-1-enopyranose (VIa) and Its Diacetate (VIb)

(1) Compound Ia (1.74 g; 0.01 mole) or Ib (2.58 g; 0.01 mole) is treated with 1.20 g of LAD in 50 ml of dry dioxane under reflux for 24 h when tlc on silica gel using ethyl acetate as the developer indicates the complete

---

[4] For large scale reactions, a mechanical stirrer and heating mantle must be used.

absence of starting material, and the presence of a single new substance. The product VIa (VI, R = H or D) is isolated by the procedure for water-soluble products (see above) and is converted directly to the diacetate VIb (VI, R = Ac). The material is purified by passage through a 10 × 1.9-cm silicic acid column. Compound VIb is a syrup; $[\alpha]^{23}_D$ +116° (c 4.05, CHCl$_3$), m/e = 215. Some nmr data for VIb are shown in Table II.

(2) Compound II (1 g, 0.004 mole) is treated with 425 mg (0.010 mole) of LAD in 20 ml of dry dioxane under reflux for 30 h. The reaction is then worked up and acetylated as described above for water-soluble products. Two products are indicated by tlc on silica gel using 4:1 v/v petroleum ether (30°–60°)–ethyl acetate as developer. The one with R_f 0.32 is VIb while the one with R_f 0.25 is an acyclic vinyl ether (c.f. XIV). Separation is readily effected by silicic acid column chromatography, affording 305 mg (35%) of VIb.

1,2,3-Trideoxy-3-deuterio-D-*xylo*-hex-1-enopyranose (VIIa).
1,2,3-Trideoxy-3-deuterio-D-*lyxo*-hex-1-enopyranose (VIIIa),
and Their Acetates VIIb and VIIIb

Compound IIIa (400 mg, 2.3 mmole) of IIIb (640 mg, 2.3 mmole) (9) is treated with 240 mg (5.7 mmole) of LAD in refluxing dioxane for 4 h, and the reaction is worked up by the procedure for water-soluble products. Three zones [R_f = 0.45, 0.20 and 0.14] are revealed by tlc on silica gel using ethyl acetate as developer. The material of R_f 0.20 is found to be a mixture of the epimers VIIa and VIIIa and acetylation affords 201 mg (43.5%) of a mixture, R_f = 0.31 using 1:1 v/v petroleum ether (30°–60°)–ethyl acetate.

The nmr data for VIIb and VIIIb shown in Table II were obtained by analysis of the 220 MHz spectrum of the mixture (9).

The compounds with R_f 0.45 and 0.14 are congeners of XV and XIV, respectively (9).

4,6-O-Benzylidene-1,2,3-trideoxy-D-*erythro*-hex-1-enopyranose (IX)

(1) Methyl 4,6-O-benzylidene-2,3-dideoxy-α-D-*erythro*-hex-2-enopyranoside (IV) (Vol. VI [52]) (2.48 g, 0.01 mole) is treated with 0.15 g of LAH in refluxing dioxane for 0.5 h. The reaction is followed by tlc on silica gel using 4:1 v/v petroleum ether (30°–60°)–ethyl acetate as the developer. Isolation by the procedure for water-insoluble products affords material containing a small amount (~4%) of an acyclic vinyl ether (compare XII). One recrystallization from ethanol affords 2.08 g of IX.

(2) Methyl 4,6-O-benzylidene-2,3-dideoxy-β-D-*erythro*-hex-2-enopyrano-

side (V) (*18*) treated as described for the α-anomer in part (1) required 20 h for completion. Compound IX is isolated by fractional crystallization from ethanol; yield 1.8 g (82%), mp 113°–114°, $[\alpha]^{23}_D$ +66° (*c* 10.7, chloroform). Some nmr data are shown in Table II.

4,6-O-Benzylidene-1,2,3-trideoxy-3-deuterio-D-ribo-hex-1-enopyranose (*X*)

When compound IV is treated as described above with LAD instead of LAH, the reaction requires 5 h for completion. The product X is isolated as described above for IX and recrystallized from ethanol; mp 113°–114°, $[\alpha]^{23}_D$ +42.5° (*c* 7.70, chloroform). Some nmr data are shown in Table II.

4,6-O-Benzylidene-1,2,3-trideoxy-3-deuterio-D-arabino-hex-1-enopyranose (*XI*)

When compound V is treated as described above with LAD instead of LAH, the reaction requires 2 days for completion. The product (XI) is isolated as described for IX; mp 117°–119°, $[\alpha]^{23}_D$ +43.5° (*c* 6.6, chloroform). Some nmr data are shown in Table II.

References

(*1*) S. Umezawa, Special Lectures, XXIII Intern. Cong. Pure Appl. Chem., **2**, 173 (1971).

(*2*) R. J. Ferrier, *Adv. Carbohyd. Chem.*, **20**, 67 (1965); **24**, 199 (1969).

(*3*) Smithline Corp., Belgian Patent 14334, 17.08.1973.

(*4*) M. Kugelman and A. K. Mallams, *J. Chem. Soc., Perkin I*, 113 (1976).

(*5*) B. Radatus, M. B. Yunker, and B. Fraser-Reid, *J. Amer. Chem. Soc.*, **93**, 3086 (1971).

(*6*) B. Fraser-Reid and B. Radatus, *J. Amer. Chem. Soc.*, **93**, 6342 (1971).

(*7*) B. Fraser-Reid and B. Radatus, *J. Amer. Chem. Soc.*, **92**, 6661 (1970).

(*8*) B. Fraser-Reid and S. Y.-K. Tam, *Tetrahedron Lett.*, 489 (1973).

(*9*) B. Fraser-Reid, S. Y.-K. Tam, and B. Radatus, *Can. J. Chem.*, **53**, 2005 (1975).

(*10*) O. Achmatowicz and B. Szechner, *Tetrahedron Lett.*, 1205 (1972).

(*11*) R. S. Tipson and A. Cohen, *Carbohyd. Res.*, **1**, 338 (1965–1966).

(*12*) B. Fraser-Reid and B. Boctor, *Can. J. Chem.*, **47**, 393 (1969).

(*13*) Organic Syntheses, **41**, 72 (1961).

(*14*) M. S. Kharash and O. Reinmuth, "Grignard Reactions of Nonmetallic Substances," Prentice-Hall, Inc., New York, 1954, p. 25.

(*15*) V. M. Mićović and M. L. J. Mihailović, *J. Org. Chem.*, **18**, 1190 (1953).

(*16*) R. U. Lemieux, E. Fraga, and K. A. Watanabe, *Can. J. Chem.*, **46**, 61 (1968).

DEOXYHALO SUGARS

[31] Preferential Halogenation of the Primary Hydroxyl Group

By Roy L. Whistler and Abul Kashem M. Anisuzzaman

Department of Biochemistry, Purdue University, Lafayette, Indiana

227

(VII)   (VIII)

Introduction

Deoxyhalogenosugars constitute an important and versatile class of compounds for potential application in the synthesis of other carbohydrate derivatives. The carbon–halogen bond is susceptible to attack by a nucleophile; and products arising from inter- or intramolecular halogen displacement and elimination may be processed further giving rise to useful compounds (1–6). A variety of reagents (7–14) are available for the direct replacement of a hydroxyl group by a halogeno substituent. Methanesulfonyl chloride in N,N-dimethylformamide (15, Vol. VI [30]) or triphenylphosphine-N-halosuccinimide in N,N-dimethylformamide (16, Vol. VII [11]) are reported to effect halogenation only of the primary hydroxyl group. Edwards and co-workers (17), however, found that with MsCl–DMF, halogenation of the secondary hydroxyl groups also takes place. The use of Ph$_3$P-N-halosuccinimide–DMF reagent gives low yield in many instances (16, 18), and for a number of nucleosides, the method is not selective (16). Moreover, acetal migration (16) can occur with this procedure, and the purification step may involve further chemical modifications (16).

Triphenylphosphine–carbon tetrahalide (19, 20) is a useful reagent for the replacement of both primary and secondary hydroxyl groups with halogens. However, it is found (21) that, if halogenation is performed in pyridine, only the primary hydroxyl group of a polyhydroxy compound reacts and almost quantitative yield of halogeno derivatives is obtained. The reaction can be performed over the wide temperature range of 5°–70° and has been applied to a variety of compounds. Stereoselectivity of halogenation under those conditions may be due to the initial formation of a bulky halogenating complex arising from triphenylphosphine, carbon tetrahalide, and pyridine.

Procedure

Methyl 6-Deoxy-6-halogeno-α-D-glucopyranoside (II)

Methyl 6-Chloro-6-deoxy-α-D-glucopyranoside (IIa)

To a cold solution of 1.94 g (10 mmole) of methyl α-D-glucopyranoside (I) (Sigma Chemical Co.) and 5.2 g (20 mmole) of triphenylphosphine (Aldrich Chemical Co., Inc., Milwaukee, Wisconsin) in 70 ml of anhydrous pyridine (Vol. II [43], [53], [63], [73]; Vol. IV [73], Vol. VII [2]) is added 10 ml of carbon tetrachloride. The resulting solution is heated at 50° for 20 min. Thin-layer chromatography (tlc) on silica gel (E. Merk, Darmstadt, Germany) coated glass plates (irrigant A: 6:1 v/v $CHCl_3$–MeOH) indicates complete conversion of I to a single product. Methanol (10 ml) is added, and after keeping the resulting solution at 50° for another 20 min, solvent is removed under reduced pressure. The residue is chromatographed on silica gel (J. T. Baker Chemical Co.). Elution first with neat chloroform and then with 20:1 v/v chloroform–methanol gives IIa; yield 2.06 g (97%), mp 113° after crystallization from hexane, $[\alpha]^{25}$D +152.8° (*c* 1, MeOH). The 6-bromo compound (IIb) is similarly prepared in 98% yield by using carbon tetrabromide (Aldrich Chemical Co., Inc.) in place of carbon tetra-chloride, mp 127° after crystallization from chloroform–hexane; $[\alpha]^{25}$D +125.2° (*c* 0.8, MeOH).

Methyl 6-Deoxy-6-iodo-α-D-glucopyranoside (IIc)

A solution of 0.97 g (0.5 mmole) of I and 4 g (1.5 mmole) of triphenyl-phosphine in 55 ml of anhydrous pyridine is cooled to 0°, and to this cooled solution is gradually added at 0° 4 g (0.77 mmole) of carbon tetraiodide (Ventron, Denver, Massachusetts). An organge-colored precipitates separates out, and the mixture is stirred at 25° for 18 h. The dark yellow solution, on examination with irrigant A, reveals quantitative conversion of I to IIc. Working up and final chromatographic purification as described above gives pure IIc; yield 1.44 g (95%), mp 148° after crystallization from chloroform–hexane, $[\alpha]^{25}$D +107.8° (*c* 1, MeOH).

6-Chloro-6-deoxy-1,2-*O*-isopropyliene-α-D-glucofuranose (IVa)

Carbon tetrachloride (10 ml) is gradually added to a solution of 1.1 g (0.5 mmole) of III (Pfanstiehl Laboratories, Inc., Waukegan, Illinois) and

2.6 (1 mmole) of triphenylphosphine in 50 ml of anhydrous pyridine at 0°. After holding the resulting solution at 5° for 18 h, methanol is added, and the mixture is evaporated to a crystalline residue which is chromatographed on a silica gel column. Elution, first with chloroform and then with 6:1 v/v chloroform–acetone gives pure IVa; yield 1.13 g (95%), mp 80° after crystallization from ether–hexane, $[\alpha]^{25}$D $-11°$ (c 2, CHCl$_3$). Hydrolysis of IVa with aqueous sulfuric acid gives 6-chloro-6-deoxy-D-glucose; mp. 137°.

The 6-bromo compound IVb is obtained in 90% yield by using carbon tetrabromide and working up the reaction mixture as described above; mp 89° after crystallization from ether–hexane, $[\alpha]^{25}$D $-10.9°$ (c 2, CHCl$_3$).

6,6′-Dichloro-6,6′-dideoxysucrose (VI)

Sucrose (4.3 g) is dissolved in 250 ml of pyridine by heating, and the solution is cooled to 25°. Triphenylphosphine (20 g) and 20 ml of carbon tetrachloride are added, and the mixture is heated at 65°–70° for 0.5 h. Methanol (20 ml) is added, and after 1 h at 50°, the mixture is evaporated to a residue. Tlc on silica gel with solvent B (45:5:3 v/v EtOAc–EtOH–H$_2$O) indicates complete conversion of sucrose to a single product. The residue is then chromatographed on a silica gel column. Elution first with neat chloroform and then with irrigant B gives the 6,6′-dichlorosucrose VI; yield 4.36 g (92%), mp 84°–88° after crystallization from 2-butanone; $[\alpha]^{25}$D $+60°$ (c 1, H$_2$O). The hexa-O-benzoate of VI melts at 88°–89° after crystallization from ethanol; $[\alpha]^{25}$D $+5.7°$ (c 1, CHCl$_3$).

5′-Chloro-5′-deoxyinosine (VIII)

Inosine (1.3 g) is dissolved in 200 ml of boiling pyridine. The solution is cooled to 25°, and 3.5 g of triphenylphosphine and 2 ml of carbon tetrachloride is added. The mixture is heated at 60°–65° for 30 min, and 20 ml of methanol is added. After keeping the mixture at 50° for another hour, solvent is removed and the residue is chromatographed on a silica gel column. Elution first with neat chloroform and then with irrigant B gives pure VIII as white crystals; yield 1.3 g (95%, mp 191° after crystallization from methanol [α]^{25}D $-47.5°$ (c 1, DMSO). 5′-Chloro-5′-deoxy-2′,3′-O-isopropylidene-inosine, mp 196°, is obtained by treating VIII with acetone containing 2,2′-dimethoxypropane and toluene-p-sulfonic acid.

References

(1)   C.-W. Chiu and R. L. Whistler, *J. Org. Chem.*, **28**, 832 (1973).

(2)   B. T. Lawton, W. A. Szarek, and J. K. N. Jones, *Chem. Commun.*, 787 (1969).

(*3*) B. T. Lawton, W. A. Szarek, and J. K. N. Jones, *Carbohydr. Res.*, **14**, 255 (1970).

(*4*) R. G. Almquist and E. J. Reist, *Carbohydr. Res.*, **46**, 33 (1976).

(*5*) M. J. Robins, Y. Fouron, and R. Mengel, *J. Org. Chem.*, **39**, 1564 (1974).

(*6*) B. J. Megerlein, *Tetrahedron Lett.*, 33 (1970).

(*7*) B. Helferich, *Ber.*, **54**, 1082 (1921).

(*8*) J. P. H. Verheyden and J. G. Moffatt, *J. Org. Chem.*, **35**, 2868 (1970).

(*9*) K. Haga, M. Yoshikawa, and T. Kato, *Bull. Chem. Soc. Japan*, **43**, 3922 (1970).

(*10*) A. Klemer and G. Mersmann, *Carbohydr. Res.*, **22**, 425 (1972).

(*11*) G. A. Wiley, R. L. Hershkowitz, B. M. Rein, and B. C. Chung, *J. Amer. Chem. Soc.*, **86**, 964 (1964).

(*12*) R. F. Dods and J. S. Roth, *J. Org. Chem.*, **34**, 1627 (1969).

(*13*) H. B. Sinclair, *J. Org. Chem.*, **30**, 1283 (1965).

(*14*) I. M. Downie, J. B. Lee, and M. S. Matough, *Chem. Commun.*, 1350 (1968).

(*15*) M. E. Evans, L. Long, Jr., and F. W. Parrish, *J. Org. Chem.*, **33**, 1074 (1968).

(*16*) S. Hanessian, M. M. Ponpipom, and P. Lavalle, *Carbohydr. Res.*, **24**, 45 (1972).

(*17*) R. G. Edwards, L. Hough, A. C. Richardson, and E. Tarelli, *Carbohydr. Res.*, **35**, 111 (1974).

(*18*) S. Hanessian and P. Lavalle, *Carbohydr. Res.*, **28**, 303 (1973).

(*19*) J. B. Lee and T. J. Nolan, *Tetrahedron*, **23**, 2789 (1967).

(*20*) J. P. H. Verheyden and J. C. Moffatt, *J. Org. Chem.*, **37**, 2289 (1972).

(*21*) A. K. M. Anisuzzaman and R. L. Whistler, *Carbohydr. Res.*, **61**, 511 (1978).

GLYCOSIDES AND GLYCOSYLAMINES

[32] Complexes of Carbohydrates with Metal Cations

Methyl β-D-Mannofuranoside

By S. J. Angyal, M. E. Evans, and R. J. Beveridge

*School of Chemistry, The University of New South Wales,
Kensington, N.S.W., Australia*

Introduction

In this preparation (*1*) complex formation between the glycoside and calcium chloride (*2*) is used three times. First, the equilibrium of the methyl D-mannofuranosides is shifted in favor of the β-isomer by the presence of calcium chloride; in its absence the yield is only 2.5% (*3*). Second, the required compound is separated from other products and starting material, which do not form complexes with cations, by the use of a cation-exchange column in its calcium form (*4*). Thirdly, the product is isolated as its addition compound with calcium chloride. Because of its low mp (46°–47°), methyl β-D-mannofuranoside is not readily obtained crystalline unless seed crystals are available; the calcium chloride complex, however, crystallizes well (*3*) and provides a convenient means of storing the compound as a solid.

Procedure

Preparation of the Calcium Ion Column

Dowex AG 50W-X2(H^+) resin (200–400 mesh, 500 ml; Bio-Rad Laboratories, Richmond, California) is gently stirred with water in a 5-liter beaker. After settling, the water is decanted and the process is twice more repeated to remove fine particles. To the resin, 300 ml of a 25% solution of calcium acetate in water is added; the mixture is gently stirred and then poured into a 60 × 3.7-cm glass column and allowed to settle.[1] The column is washed

---

[1] The resin contracts considerably when converted from the hydrogen ion form to the calcium ion form. If the conversion is conducted in a glass tube, the column is likely to break up into an uneven mass.

METHODS IN CARBOHYDRATE
CHEMISTRY, VOL. VIII

first with a further 700 ml of 25% calcium acetate solution,[2,3] then with about 15 bed volumes of water until the effluent is free of calcium ions (test with sodium carbonate solution), and finally with 3 liter of 3:7 v/v methanol–water.[4,5]

Methyl β-D-Mannofuranoside

A mixture of 10.0 g of D-mannose, 12 g of anhydrous calcium chloride dried at 220° for 12 h, 150 ml of anhydrous methanol, and 0.5 ml of acetyl chloride is boiled under reflux for 2 h.[6] The solution is cooled, neutralized to pH 8–9 with 2 M sodium hydroxide solution, and evaporated on a rotary evaporator under reduced pressure. The residue is dissolved in 20 ml of 30% methanol and chromatographed on the above column with 30% methanol as eluent, 25-ml fractions being collected. Fractionation is followed by spotting the fractions on tlc plates of silica gel, spraying with 50% sulfuric acid solution and heating; should the two fractions not be completely separated, their contents can be identified by tlc with 5:1:4 v/v acetone–water–ethyl acetate as developer. Fractions 10–26 contain calcium chloride and the other methyl D-mannosides; fractions 36–100 are combined and evaporated under reduced pressure to give methyl β-D-mannofuranoside as a syrup; yield 4.3 g (40%).[7] Addition of 13 ml of a 70% (w/v) aqueous solution of calcium chloride causes immediate crystallization (6); 10 ml of 1-propanol is added to the mixture which is then kept at 0° overnight. The crystals of the calcium chloride complex ($C_7H_{14}O_6 \cdot CaCl_2 \cdot 3H_2O$) are collected by filtration, washed with anhydrous ethanol and dried in a desiccator over anhydrous calcium chloride; yield 6.1 g, $[\alpha]^{20}D$ −56° (c 1.45, water).

---

[2] The capacity of the resin is quoted as 0.7 meq/ml; the indicated quantity of calcium acetate represents a 4–5 fold excess.

[3] The column of resin in the calcium ion-form is usually prepared by treatment of a resin column with calcium chloride (4). However, the column then retains some hydrogen ions which may catalyze the hydrolysis of the very acid-sensitive furanoside. Calcium acetate gives a completely neutral column.

[4] Aqueous methanol as an eluant gives a better separation than water (5).

[5] The column can be used repeatedly; once all the fractions of a run have been eluted, it is ready for another separation. If it is to be totally regenerated, it is first converted to the hydrogen ion form by using an excess of 2 M hydrochloric acid and then treated with calcium acetate as described above.

[6] The composition of the mixture is 53.5% methyl β-D-mannofuranoside (R_t 5.8 min), 37.5% of the other mannosides (unresolved, R_t 6.8 min) and 9% of D-mannose (R_t 12.0 and 13.0 min) as determined after acetylation by glc on a 0.45 × 115-cm column containing 3% SP 2401 (Supelco, Inc., Bellefonte, Pennsylvania) on Chromosorb W at 230°. The yield of the furanosides is at a maximum after 2 h; longer heating results in increasing amounts of methyl pyranosides being formed.

[7] The compound is chromatographically pure and suitable for further reactions.

References

(*1*) S. J. Angyal, C. L. Bodkin, and F. W. Parrish, *Aust. J. Chem.*, **28**, 1541 (1975).
(*2*) S. J. Angyal, *Tetrahedron*, **30**, 1695 (1974).
(*3*) A. Scattergood and E. Pacsu, *J. Amer. Chem. Soc.*, **62**, 903 (1940).
(*4*) V. E. Felicetta, M. Lung, and J. L. McCarthy, *Tappi*, **42**, 496 (1959).
(*5*) S. J. Angyal, G. S. Bethell, and R. J. Beveridge, *Carbohydr. Res.*, **73**, 9 (1979).
(*6*) M. H. Randall, *Carbohydr. Res.*, **11**, 173 (1969).

[33] 1,2-Orthoester Formation and Glycosidation with Amide Acetals

By S. Hanessian and J. Banoub

Department of Chemistry, University of Montreal,
Montreal, Quebec, Canada

1,2-Orthoester Formation

CH_2OBz — O, Cl + $Me_2N-CH(OR)_2$ $\xrightarrow{CH_2Cl_2}$ CH_2OBz — O, O, O$-$C$\sim$OR, Ph

BzO OBz   BzO O$-$C$\sim$OR

Where R = methyl, 2-propyl, cyclohexyl

CH_2OAc — O, OAc + $Me_2N-CH(OR)_2$ $\xrightarrow[CH_2Cl_2]{CF_3SO_3Ag}$ CH_2OAc — O, OAc, O$-$C$\sim$OR, Me

AcO Br   AcO
OAc

Where R = methyl, propyl, 2-propyl, 1-butyl, 2,2-dimethylpropyl

Glycosidation

CH_2OBz — O, OAc + $Me_2N-CH(OR)_2$ $\xrightarrow[CH_2Cl_2]{SnCl_4}$ CH_2OBz — O, OR

BzO OBz   BzO OBz

Where R = methyl, 2-propyl, benzyl, cyclohexyl, 2,2-dimethylpropyl

237

METHODS IN CARBOHYDRATE
CHEMISTRY, VOL. VIII

Introduction

The orthoester method (*1*) has been used extensively for the synthesis of glycosides derived from simple and complex alcohols, including sugar derivatives (*1, 2,* Vol. VI [88]). Glycoside formation by this method takes place with a high degree of stereocontrol and leads, in general, to 1,2-*trans*-glycosides. Some secondary reactions, which are known to occur (*3–5*), result in variations in the anomeric composition.

1,2-Orthoesters

There are a number of well-established procedures for the preparation of carbohydrate 1,2-orthoesters (*6–9*), and several modifications have been made in recent years (*10*). This transformation can now be effected rapidly and efficiently by the treatment of per-*O*-acylglycosyl halides with amide acetals, which are the sources of the alkoxyl group in the resulting ortho-esters (*11*). Thus, treatment of 2,3,5-tri-*O*-benzoyl-β-D-ribofuranosyl chlo-tide (Vol. II [29], see also *12*) with *N*,*N*-dimethylformamide dialkyl acetals in dichloromethane leads to the corresponding orthoesters. In the case of 2,3,4,6-tetra-*O*-acetyl-α-D-glucopyranosyl bromide (Vol. II [55]) an analo-gous transformation takes place, but only in the presence of silver trifluoro-methanesulfonate (triflate).

Glycosidation with Amide Acetals

In the presence of stannic chloride and various *N*,*N*-dimethylformamide dialkyl acetals, 1-*O*-acetyl-2,3,5-tri-*O*-benzoyl-β-D-ribofuranose (*13*), is converted into the corresponding alkyl β-glycosides (*14*). The aglycon originates from the alkoxyl group of the amide acetal. Glycoside formation is presumed to take place by initial attack of an 1,2-acyloxonium ion inter-mediate by an alcoholate species, leading to an orthoester derivative,[1] which rapidly rearranges in the presence of the Lewis acid. This glycosidation

---

[1] An orthoester derivative was isolated when the reaction was conducted at $-30°$.

procedure is also applicable to the synthesis of certain disaccharides containing a β-D-ribofuranosyl moiety (15).

Treatment of β-D-glucose pentaacetate (Vol. II [53]) with N,N-dimethylformamide dimethylacetal, in the presence of stannic chloride, gives methyl β-D-glucopyranoside tetraacetate (14). The benzyl glycoside derivative can be similarly prepared. With acetals derived from higher alcohols however, the reaction is much slower and leads to the formation of 2,3,4,6-tetra-O-acetyl-α-D-glucopyranosyl chloride as the preponderant product. Glycoside formation can, however, be effected rapidly and efficiently by treatment of per-O-acylated sugar derivatives with alcohols in the presence of stannic chloride (Vol. VIII [33]).

Procedures

1,2-Orthoesters

3,5-Di-O-benzoyl-α-D-ribofuranose 1,2-(Alkyl orthobenzoates) (11)

To a cooled solution containing 0.48 g (1 mmole) of 2,3,5-tri-O-benzoyl-β-D-ribofuranosyl chloride (12) in 10 ml of dry dichloromethane is added 1 mmole of commercial N,N-dimethylformamide dialkylacetal, and the solution is stirred at 20°–25° for 1–2 h. After evaporation under reduced pressure, the residual syrup is held at 0.1 torr for 1 h to remove traces of residual acetal. The residues containing traces of 2,3,5-tri-O-benzoyl-D-ribofuranose are individually purified by chromatography on 60–200-mesh silica gel using an irrigant of 19:1 v/v benzene–ethyl acetate to give the corresponding orthoesters as colorless syrups; alkyl group: methyl, 96%; [α]D +104° (CHCl$_3$) (19); 2-propyl, 83%; [α]D +75.8° (CHCl$_3$); cyclohexyl, 80%; [α]D +98.3° (CHCl$_3$).

The syrupy orthoesters show evidence of the presence of endo/exo mixtures by nmr. Methanolysis with sodium methoxide in methanol (Vol. II [54]) leads to the corresponding debenzoylated orthoesters as chromatographically homogeneous syrups.

3,4,6-Tri-O-acetyl-α-D-glucopyranose 1,2-(Alkyl orthoacetates) (11)

To a cooled solution of 0.411 g (1 mmole) of 2,3,4,6-tetra-O-acetyl-α-D-glucopyranosyl bromide (Vol. II [55] in 10 ml of dry dichloromethane are successively added 1 mmole of the N,N-dimethylformamide dialkylacetal and 0.257 g (1 mmole) of commercial silver triflate, and the mixture is stirred in the dark for 15 min at 0°. The suspension is filtered through a bed of Celite; the filtrate is washed with aqueous sodium hydrogencarbonate, and the organic phase is processed as usual to give a syrupy residue which is freed from traces of reagent by keeping at 0.1 torr for 1 h. The residues, containing the orthoesters and traces of 2,3,4,6-tetra-O-acetyl-D-glucopyranose, are individually chromatographed on silica gel using an irrigant of 19:1 v/v benzene–ethyl acetate to give the respective orthoesters;[2] alkyl group: methyl (syrup) 87%, [α]D +27° (CHCl$_3$) (6); 2-propyl, mp 118°–119° (ether–pentane), 79% [α]D +30° (CHCl$_3$) (7); 1-propyl (syrup) 76%, [α]D +38.1° (CHCl$_3$) (7); 1-butyl (syrup), 74%, [α]D +33.8° (CHCl$_3$); 2,2-dimethylpropyl, mp 97°–98° (ether pentane), 75%, [α]D +25° (CHCl$_3$).

The syrupy orthoesters show evidence of the presence of endo/exo mixtures by nmr, and can be individually transformed into deacetylated syrupy orthoesters by methanolysis with sodium methoxide in methanol (Vol. II [54]).

Rearrangement of 1,2-Orthoesters into 1,2-*trans* Glycosides (11)

A solution containing 1 mmole of the orthoester derivative in 10 ml of dry dichloromethane is treated at 0° with 0.1 mmole of stannic chloride. After stirring at 0° for 2 h, the solution is treated with aqueous sodium hydrogencarbonate, and the organic phase is processed in the usual manner to give a syrupy residue. Chromatographic purification on silica gel using 19:1 v/v benzene–ethyl acetate as the irrigant gives the glycoside derivative; methyl 2,3,5-tri-O-benzoyl-β-D-ribofuranoside, syrup, 85–90%, [α]D +55° (CHCl$_3$) (17); 2-propyl 2,3,5-tri-O-benzoyl-β-D-ribofuranoside, syrup, 85–90%, [α]D +45° (CHCl$_3$); cyclohexyl 2,3,5-tri-O-benzoyl-β-D-ribofuranoside, syrup, 85–90%, [α]D +23.7°; methyl 2,3,4,6-tetra-O-acetyl-β-D-glucopyranoside, mp 104°–105° (ethanol), ∼60%, [α]D −13° (CHCl$_3$), etc.

Glycosides

Alkyl 2,3,5-tri-O-benzoyl-β-D-ribofuranosides (14)

A cooled solution containing 0.5 g (1 mmole) of 1-O-acetyl-2,3,5-tri-O-benzoyl-β-D-ribofuranose (13) in 10 ml of dry dichloromethane is treated with 0.12 ml (1 mmole) of stannic chloride, and the solution is stirred at 0° for 10 min with exclusion of moisture. The appropriate N,N-dimethylformamide dialkylacetal (1 mmole) is added, and the solution is stirred at 0° for 2 h, after which it is treated with aqueous sodium hydrogencarbonate;

---

[2] Yields pertain to chromatographically homogeneous syrups. Rotations of the 2-propyl and 2,2-dimethylpropyl are those recorded for the crystalline diastereoisomers.

the organic phase is processed as usual. The resulting residues are kept at 0.1 torr for 1 h to remove traces of reagent, the syrups are purified by chromatography over silica gel using 19:1 v/v benzene–ethyl acetate as the irrigant to give the corresponding glycosides; methyl, syrup, 95%, $[\alpha]_D$ +55° (CHCl₃) (*17*); 2-propyl, syrup, 96%, $[\alpha]_D$ +45° (CHCl₃); benzyl, mp 65° (ether), 90%, $[\alpha]_D$ +20° (CHCl₃) (*18*); cyclohexyl, syrup, 90%, $[\alpha]_D$ +23.7° (CHCl₃); 2,2-dimethylpropyl, syrup, 93%, $[\alpha]_D$ +35.5° (CHCl₃).

Debenzoylation of the glycosides give the parent alkyl β-D-ribofuranosides; for example, methyl, mp 76°, $[\alpha]_D$ −55° (H₂O); cyclohexyl, mp 139°, $[\alpha]_D$ −85.7° (MeOH); 2,2-dimethylpropyl, mp 110°, $[\alpha]_D$ −60° (MeOH).

Methyl 2,3,4,6-tetra-O-acetyl-β-D-glucopyranoside (*14*)

A cooled solution of 0.39 g (1 mmole) of β-D-glucose pentaacetate (Vol. II [53]) in 10 ml of dry dichloromethane is treated with 0.12 ml (1 mmole) of stannic chloride. After stirring at 0° for 10 min, 0.13 ml (1 mmole) of *N,N*-dimethylformamide dimethylacetal is added and stirring is continued for 2–3 h. Processing as described above gives the title compound; yield 83% in three crops, mp 104°, $[\alpha]_D$ −16° (CHCl₃).

References

(*1*) For pertinent reviews, see N. K. Kochetkov and A. F. Bochkov, *in* "Recent Developments in the Chemistry of Natural Carbon Compounds," Vol. 4, Akademia Kiado, Budapest, 1972, pp. 77–191; G. Wulff and G. Röhle, *Angew. Chem., Int. Ed. Engl.*, **13**, 157 (1974); S. Hanessian and J. Banoub, *Advan. Chem. Series*, **39**, 36 (1976).

(*2*) N. K. Kochetkov, O. S. Chizhov, and A. F. Bochkov, *in* "International Review of Science," Vol. 7, "Carbohydrates," G. O. Aspinall, ed., University Park Press, Baltimore, Maryland, 1973, pp. 147–190.

(*3*) A. F. Bochkov, V. I. Betanely, and N. K. Kochetkov, *Carbohydr. Res.*, **30**, 418 (1973); *Isvest. Akad. Nauk SSSR*, 1379 (1974); *Chem. Abstr.*, **80**, 451 (1974).

(*4*) R. U. Lemieux, *Chem. Can.*, 14 (1964).

(*5*) P. J. Garegg and I. Kvarnström, *Acta Chem. Scand., Ser. B*, **30**, 655 (1976).

(*6*) N. K. Kochetkov, A. Ya. Khorlin, and A. F. Bochkov, *Tetrahedron*, **23**, 693 (1967).

(*7*) R. U. Lemieux and J. D. T. Cipera, *Can. J. Chem.*, **34**, 906 (1956); R. U. Lemieux and A. R. Morgan, *Can. J. Chem.*, **43**, 2199 (1965).

(*8*) M. Mazurek and A. S. Perlin, *Can. J. Chem.*, **43**, 1918 (1965).

(*9*) B. Helferich and K. Weiss, *Chem. Ber.*, **89**, 314 (1956).

(*10*) See, for example, G. Wulff and W. Kruger, *Carbohydr. Res.*, **19**, 139 (1971); S. E. Zurabyan, M. M. Tikhomirov, V. A. Nesmeyamov, and A. Ya. Khorlin, *Carbohydr. Res.*, **26**, 117 (1973); N. K. Kochetkov, A. F. Bochkov, T. A. Sokolovskaya, and V. I. Snyatkov, *Carbohydr. Res.*, **16**, 17 (1971).

(*11*) S. Hanessian and J. Banoub, *Carbohydr. Res.*, **44**, C14 (1975).

(*12*) S. Hanessian and A. G. Pernet, *Can. J. Chem.*, **52**, 1280 (1974).

(*13*) E. F. Recondo and H. Rinderknecht, *Helv. Chim. Acta*, **42,** 1171 (1959).

(*14*) S. Hanessian and J. Banoub, *Tetrahedron Lett.*, 657 (1976).

(*15*) S. Hanessian and J. Banoub, *Tetrahedron Lett.*, 661 (1976).

(*16*) P. A. J. Gorin, *Can. J. Chem.*, **40,** 275 (1962).

(*17*) J. H. Vanbroeckhoven, J. J. Dierckens, and F. C. Alderweireldt, *Bull. Soc. Chim. Belg.*, **83,** 155 (1974).

(*18*) R. K. Ness, H. W. Diehl, and H. G. Fletcher, Jr., *J. Amer. Chem. Soc.*, **76,** 763 (1954).

[34] Preparation of 1,2-*trans*-Glycosides in the Presence of Stannic Chloride

By S. Hanessian and J. Banoub

*Department of Chemistry, University of Montreal,
Montreal, Quebec, Canada*

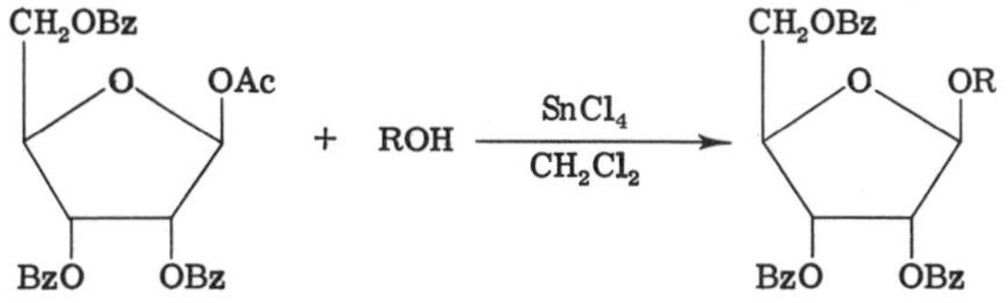

CH$_2$OBz — (sugar ring) OAc + ROH $\xrightarrow[\text{CH}_2\text{Cl}_2]{\text{SnCl}_4}$ CH$_2$OBz — (sugar ring) OR

BzO OBz → BzO OBz

Where R = methyl, 2-propyl, benzyl, cyclohexyl,

2,2-dimethylpropyl, (bicyclic sugar structure with $-$CH$_2$, O, Me, Me, O$-$C$-$Me, Me) etc.

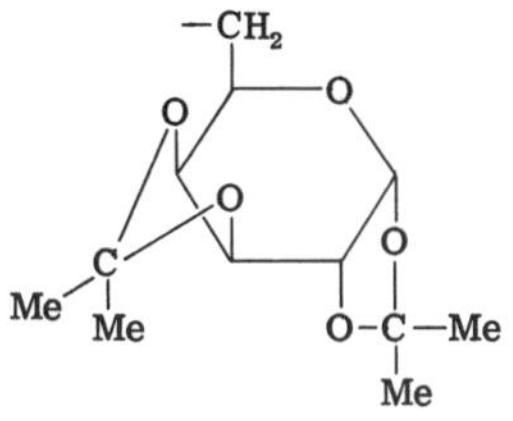

CH$_2$OAc — (sugar ring) OAc + ROH $\xrightarrow[\text{CH}_2\text{Cl}_2]{\text{SnCl}_4}$ CH$_2$OAc — (sugar ring) OR

AcO OAc OAc → AcO OAc OAc

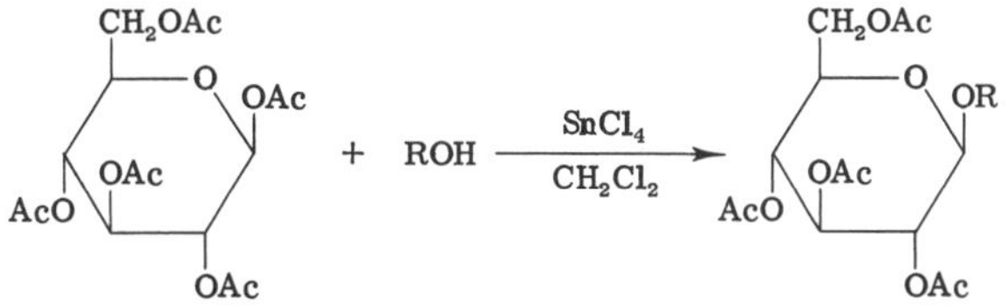

Where R = methyl, 2-propyl, cyclohexyl

Introduction

Simple alkyl furanosides (*1–3*) can be prepared from the parent sugars by the classical Fischer alcoholysis, but for more complex glycosides in this series, the Koenigs–Knorr method and its numerous modifications (*4*, Vol. I [104], [112]; II [87], [88]; VI [87], [89]) have been used. In the presence of a

243

participating group at C-2, this method leads, in general, to 1,2-*trans*-glycosides. Current procedure is to use an appropriately protected glyco-furanosyl halide, which is allowed to react with an alcohol in the presence of a catalyst (usually a silver or mercury salt) and an acid acceptor. Lewis acids such as stannic chloride are efficient reagents for the activation (*5*) of the anomeric site in the readily available per-*O*-acyl sugar derivatives. In the presence of alcohols and stannic chloride, these derivatives give the corresponding 1,2-*trans*-glycosides, presumably via the formation of a 1,2-orthoester intermediate, and subsequent rearrangement (*6*, Vol. III [32]). Thus, a variety of alkyl β-D-ribofuranoside tribenzoates can be obtained conveniently from the treatment of 1-*O*-acetyl-2,3,5-tri-*O*-benzoyl-β-D-ribofuranose with the respective alcohols in the presence of stannic chloride. With an appropriate sugar derivative, the corresponding disaccharide is obtained. (The glycosylation with simple alcohols by this procedure is also applicable in the case of β-D-glucopyranose pentaacetate. In the case of hindered alcohols, 2,3,4,6-tetra-*O*-acetyl-α-D-glucopyranosyl chloride is formed. Previously, methyl (*7, 8*), phenyl (*7, 8*), and steroidal (*9*) β-D-glucopyranoside tetraacetates had been prepared in the presence of stannic chloride.

Procedures

Alkyl β-D-Ribofuranoside Tribenzoates (*6*)

To a solution containing 0.5 g (1 mmole) of 1-*O*-acetyl-2,3,5-tri-*O*-benzoyl-β-ribofuranose (*10*) in 10 ml of dry dichloromethane is added 0.12 ml (1 mmole) of stannic chloride, and the solution is stirred at 0° for 10 min with exclusion of moisture. The alcohol (1 mmole) is then introduced, and the colorless solution is stirred at 25° for 1–4 h. The solution is added dropwise and rapidly into aqueous sodium hydrogencarbonate, and the organic phase is processed as usual to give a syrup. Chromatography on silica gel using 50:1 v/v benzene–ethyl acetate as the irrigant gives the pure glycoside: methyl, syrup (93%), [α]D +58° (chloroform) (*11*); ethyl, syrup (90%), [α]D +50° (chloroform) (*11*); 2-propyl, syrup, (87%, [α]D +46° (chloroform); benzyl, mp 65°, (85%), [α]D +20° (chloroform) (*12*); 2,2-dimethylpropyl, syrup (91%), [α]D +36° (chloroform); cyclohexyl, syrup (82%) [α]D +26° (chloroform).

Alkyl β-D-Glucopyranoside Tetraacetates

A solution of 0.39 g (1 mmole) of β-D-glucopyranose pentaacetate (Vol. II [53]) in 10 ml of dry dichloromethane is treated with 0.12 ml (1 mmole) of stannic chloride, and the solution is stirred at 25° for 10 min. The alcohol (1 mmole) is then added, and the solution is stirred for 2–3 h, after which, the solution is processed as described above. The resulting syrup contains the desired glycoside and traces of 2,3,4,6-tetra-*O*-acetyl-D-glucopyranose. Purification on silica gel using 9:1 v/v benzene–ethyl acetate as the irrigant gives the pure glycoside: methyl, mp 104° (80%), [α]D −16.1° (chloroform) (*7, 13*); 2-propyl, mp 134°–135°, [α]D −21° (chloroform) (*7, 13*); cyclohexyl, syrup, (72%),[1] [α]D −21° (chloroform) (*13*).

6-*O*-(2,3,5-Tri-*O*-benzoyl-*β*-D-ribofuranosyl)-1,2:3,4-
di-*O*-isopropylidene-*α*-D-galactopyranose (*6*)

The above preparation is repeated, except that the alcohol is replaced by 0.26 g (1 mmole) of 1,2:3,4-di-*O*-isopropylidene-*α*-D-galactopyranose (Vol. II [66], [83]) dissolved in the minimum volume of dichloromethane. After stirring the solution at 0° for 4 h, the mixture is processed as described above to give a syrup which is purified by chromatography on silica gel using 19:1 v/v benzene–ethyl acetate as the irrigant to give a syrup (0.52 g, 72%), that crystallizes from ethanol; mp 201°–203°; [α]D +4.3° (chloroform).

References

(*1*)   G. Wulff and G. Röhle, *Angew. Chem., Int. Ed. Engl.*, **13**, 157 (1974).
(*2*)   J. W. Green, *Advan. Carbohydr. Chem.*, **21**, 95 (1966).
(*3*)   R. J. Ferrier, *Topics Curr. Chem.*, **14**, 390 (1970).
(*4*)   K. Igarashi, *Advan. Carbohydr. Chrm. Biochem.*, **34**, 243 (1977).
(*5*)   H. Paulsen, *Advan. Carbohydr. Chem. Biochem.*, **26**, 127 (1971).
(*6*)   S. Hanessian and J. Banoub, *Carbohydr. Res.*, **59**, 261 (1977).
(*7*)   R. U. Lemieux and N. P. Shyluk, *Can. J. Chem.*, **31**, 528 (1953).
(*8*)   T. R. Ingle and J. L. Bose, *Carbohydr. Res.*, **12**, 459 (1970).
(*9*)   K. Honma, K. Nakazima, T. Uematsu, and A. Hamada, *Chem. Pharm. Bull.* (*Tokyo*), **24**, 394 (1976).
(*10*) E. F. Recondo and H. Rinderknecht, *Helv. Chim. Acta*, **42**, 1171 (1959).
(*11*) J. H. Vanbroeckhoven, J. J. Dierckens, and F. C. Alderweireldt, *Bull. Soc. Chim. Belg.*, **82**, 155 (1974).
(*12*) R. K. Ness, H. W. Diehl, and H. G. Fletcher, Jr., *J. Amer. Chem. Soc.*, **76**, 763 (1954).
(*13*) B. Lindberg, *Acta Chem. Scand.*, **3**, 151 (1949).

---

[1] A small amount of 2,3,4,6-tetra-*O*-acetyl-α-D-glucopyranosyl chloride is also formed in this reaction.

[35] Preparation of 1,2-*trans*-Glycosides in the Presence of Silver Trifluoromethanesulfonate

By S. Hanessian and J. Banoub

Department of Chemistry, University of Montreal,
Montreal, Quebec, Canada

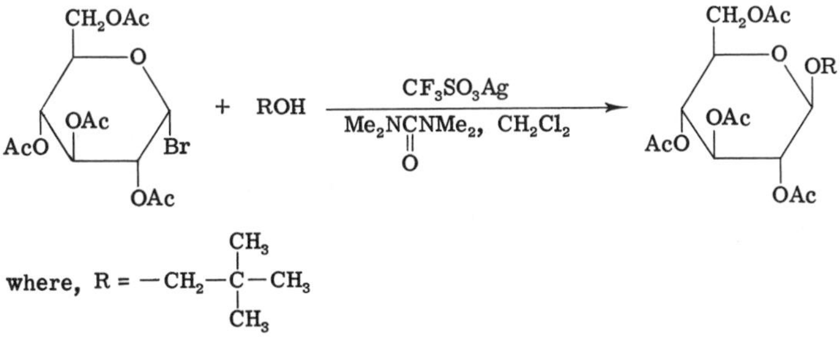

where, R = $-CH_2-\overset{\overset{\displaystyle CH_3}{|}}{\underset{\underset{\displaystyle CH_3}{|}}{C}}-CH_3$

(I)

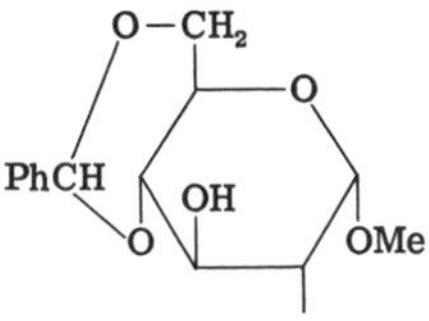

(II)

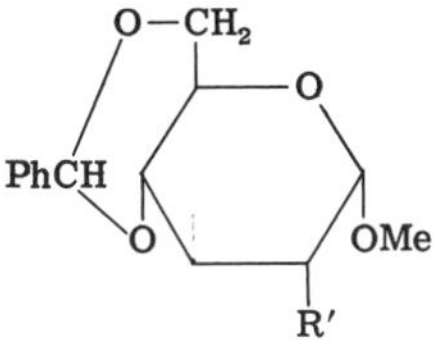

(III, R' = OTs)
(IV, R' = NHCbz)

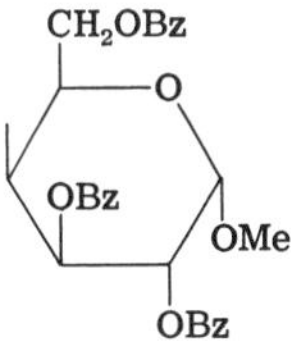

(V)

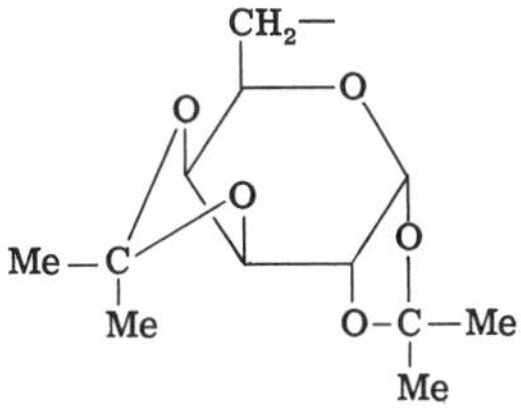

(VI)

METHODS IN CARBOHYDRATE
CHEMISTRY, VOL. VIII

Introduction

A large number of procedures are available for the preparation of simple and complex 1,2-*trans*-glycosides, including di- and oligosaccharides, and related substances (*1*). In general, such glycosides are obtained by the ortho-ester (*2, 3*, Vol. VI [88]), and Koenigs–Knorr (*4*, Vol. I [104], [112]; II [87], [88]; VI [87], [89]) methods and their respective modifications (*1*). Silver trifluoromethanesulfonate (triflate) has been used as an effective catalyst in the methanolysis of derivatives of 2,3,4-tri-*O*-benzyl-α-D-glucopyranosyl bromide (*5*) and in model glycosidations (*6*) leading to glycoside derivatives of the antitumor substances adriamycin and daunorubicin (*7*), and in the synthesis of 1,2-*trans*-disaccharides (*8–10*). Thus, treatment of acylglycosyl halides, such as 2,3,4,6-tetra-*O*-acetyl-α-D-glucopyranosyl bromide (Vol. II [55]) with alcohols, including sugar derivatives, in the presence of silver triflate and 1,1,3,3-tetramethylurea affords the corresponding 1,2-*trans*-glycosides (*8–10*). Using this combination of catalyst and acid acceptor, disaccharide derivatives having 1→2, 1→3, 1→4, and 1→6 linkages have been prepared in preparatively significant yields.

Procedures

2,2-Dimethylpropyl 2,3,4,6-Tetra-*O*-acetyl-β-D-glucopyranoside (I) (*9*)

A solution containing 0.411 g (1 mmole) of 2,3,4,6-tetra-*O*-acetyl-α-D-glucopyranosyl bromide (Vol. II [55]) in 10 ml of dry dichloromethane is treated successively with 88 mg (1 mmole) of 2,2-dimethylpropanol and 0.257 g (1 mmole) of silver triflate, and the mixture is stirred at 0° for 2 h in the dark. The suspension is filtered over Celite; the residue is washed with 20 ml of dichloromethane, and the filtrate is treated with aqueous sodium hydrogencarbonate. The organic phase is processed as usual to give a syrup. Chromatographic purification on silica gel using 5:1 v/v benzene–ethyl acetate as the irrigant gives the title compound as a colorless syrup; yield 0.27 g (64%). Crystallization from ethanol or 2-propyl ether gives the pure product; mp 134°–135°, [α]D −12° (CHCl₃).

Methyl 2-*O*-(2,3,4,6-Tetra-*O*-acetyl-β-D-glucopyranosyl)-4,6-*O*-benzylidene-α-D-glucopyranoside (II) (*10*)

To a solution of 0.411 g (1 mmole) of 2,3,4,6-tetra-*O*-acetyl-α-D-gluco-pyranosyl bromide (Vol. II [55]) in 10 ml of dry dichloromethane is added, in succession, 0.564 g (2 mmoles) of methyl 4,6-*O*-benzylidene-α-D-gluco-pyranoside (Vol. I [30]), 0.33 ml (3 mmoles) of 1,1,3,3-tetramethylurea, and

0.567 g (2.2 mmoles) of silver triflate; and the mixture is stirred at 0° for 4 h in the dark. Processing the mixture as described above gives a syrup which contains a major product, in addition to unreacted starting alcohol. The title compound can be obtained by direct crystallization of such a mixture from methanol. For more consistent recoveries however, the syrup is chromatographed on silica gel using 8:2 v/v benzene–ethyl acetate as the irrigant to give a syrup (0.29 g, 47%), that crystallizes upon trituration with methanol. The title compound is obtained in two crops: total yield 0.255 g (41%), mp 226°–227°; reported mp 227°–228° (*12*), mp 221°–224° (*13*), [α]D +40.1° (CHCl₃).

Methyl 2-*O*-(2,3,4,6-tetra-*O*-acetyl-β-D-galactopyranosyl)-4,6-*O*-benzylidene-α-D-glucopyranoside can be similarly prepared (25°, 4 h) in 15% yield by direct crystallization (the mother liquors account for a further 25–30% of a chromatographically identical syrupy product); mp 197°–198°; reported (*14*) mp 197°–198°; [α]D +54.3° (CHCl₃)

The following disaccharide derivatives can also be prepared by the same general procedure (*10*), except for minor variations in the ratios of substrates and the temperature: methyl 3-*O*-(2,3,4,6-tetra-*O*-acetyl-β-D-glucopyrano-syl)-4,6-*O*-benzylidene-2-*O*-(tolyl-*p*-sulfonyl)-α-D-glucopyranoside (III) (25°, 6 h), colorless foam (86% based on reacted alcohol), [α]D +9.2° (CHCl₃);[1] methyl 3-*O*-(2,3,4,6-tetra-*O*-acetyl-β-D-glucopyranosyl)-4,6-*O*-benzylidene-2-benzyloxycarbonylamino-2-deoxy-α-D-glucopyranoside (IV) (25°, 4 h), colorless foam (82% based on reacted alcohol), [α]D +18.5° (CHCl₃); methyl 4-*O*-(2,3,4,6-tetra-*O*-acetyl-β-D-glucopyranosyl)-2,3,6-tri-*O*-benzoyl-α-D-galactopyranoside (V) (25°, 4 h), colorless foam (72%), [α]D +56.6° (CHCl₃);[2]   6-*O*-(2,3,4,6-tetra-*O*-acetyl-β-D-glucopyranosyl)-1,2:3,4-di-*O*-isopropylidene-α-D-galactopyranose (VI) (25°, 4 h), 60%, mp 140°–141°, [α]D −50° (CHCl₃),[3] reported (*12*) mp 141°.

References

(*1*) For reviews, see for example, G. Wulff and G. Röhle, *Angew. Chem., Int. Ed. Engl.*, **13**, 157 (1974); R. J. Ferrier, *in* "Topics in Current Chemistry," Vol. 14, Springer-Verlag, New York, 1970, pp. 390–429, and references cited therein.

(*2*) N. K. Kochetkov and A. F. Bochkov, *in* "Recent Developments in the Chemistry of Natural Carbon Compounds," Vol. 4, Akademia Kiadó, Budapest, 1972, pp. 77–191.

(*3*) N. K. Kochetkov, O. S. Chisov, and A. F. Bochkov, *in* "MTP International Review of

---

[1] Treatment with sodium amalgam, followed by acetylation gives the known crystalline 2-acetate derivative (*14*).

[2] An acid acceptor is not necessary in this case, provided that the conditions are anhydrous.

[3] The yield of chromatographically pure syrupy product is ∼72%.

Science," Vol. 7, "Carbohydrates," G. O. Aspinall, ed., University Park Press, Baltimore, Maryland, 1973, pp. 147–190.

(4) For a review, see, K. Igarashi, *Advan. Carbohydr. Chem. Biochem.*, **34**, 243 (1977).

(5) F. J. Kronzer and C. Schuerch, *Carbohydr. Res.*, **27**, 379 (1973).

(6) S. Hanessian and S. Penco, unpublished results.

(7) F. Arcamone, S. Penco, S. Redaelli, and S. Hanessian, *J. Med. Chem.*, **19**, 1424 (1976).

(8) S. Hanessian and J. Banoub, *Advan. Chem. Series*, **39**, 36 (1976) and references cited therein.

(9) S. Hanessian and J. Banoub, *Carbohydr. Res.*, **44**, C14 (1975).

(10) S. Hanessian and J. Banoub, *Carbohydr. Res.*, **53**, C13 (1977).

(11) B. Coxon and H. G. Fletcher, Jr., *J. Org. Chem.*, **26**, 2892 (1961).

(12) N. K. Kochetkov, A. J. Khorlin, and A. F. Bochkov, *Tetrahedron*, **23**, 693 (1967).

(13) D. Beck and K. Wallenfells, *Ann.*, **655**, 173 (1962).

(14) A. Ya. Khorlin, A. F. Bochkov, and N. K. Kochetkov, *Izv. Akad. Nauk SSSR, Ser. Khim.*, 168 (1966); *Chem. Abstr.*, **66**, 5272 (1967).

[36] 1,2-*trans*-1-Thioglycosides

By R. J. Ferrier and R. H. Furneaux

Department of Chemistry, Victoria University of Wellington, Wellington, New Zealand

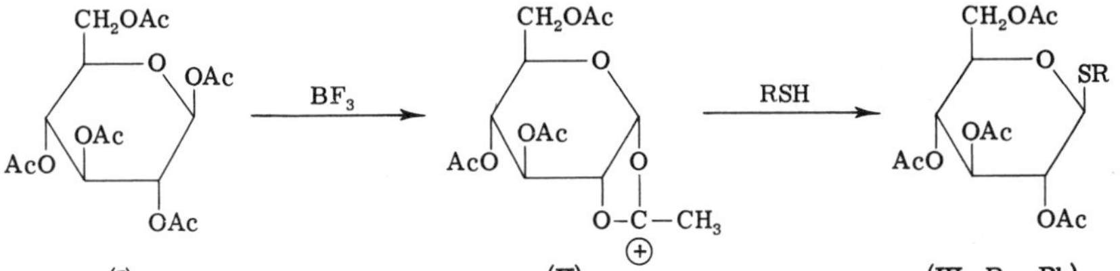

(I) (II) (III, R = Ph) (IV, R = CH_2Ph)

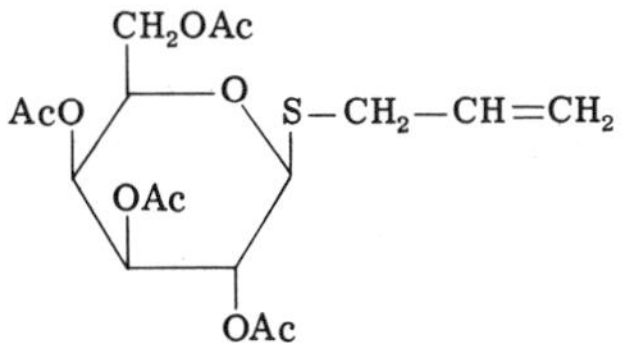

(V)

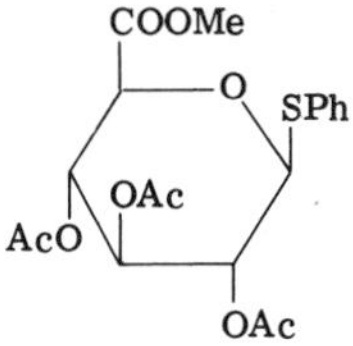

(VI)

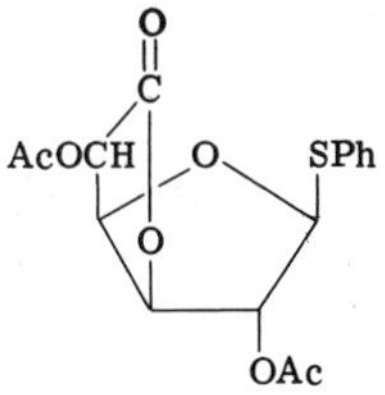

(VII)

Introduction

The need for simple and efficient methods for synthesizing 1-thioglycosides has been highlighted by their recent use as enzyme inhibitors (*1*). This has led to their use in affinity chromatography (*1–3*) and in other biochemical work (*4*). They are also applied in newer approaches to the synthesis of *O*-glycosides (*5–6*). In Vol. II [94], the standard procedures for preparing thioglycosides were surveyed; most work described since then has employed these methods, namely: (a) condensation between glycosyl halide esters and thiolates; (b) *S*-substitution of 1-thiosugars; (c) partial hydrolysis of aldose dialkyl dithioacetals; (d) isomerization of other 1-thioglycosides; and (e) direct, acid-catalyzed thiolysis of free sugars or glycosyl esters. Complications which can arise with the esters were emphasized in the survey rather than their utility which at that time had not been fully exploited.

In Vol. II [90], the Helferich procedure for preparing aryl glycosides by direct condensation between phenols and sugar peresters in the presence of

ISBN 0-12 746208-2

zinc chloride is described. Several reports have now appeared on the application of this approach to the synthesis of 1-thioglycosides using zinc chloride (7–11), tin (IV) chloride (12, 13), or toluene-p-sulfonic acid as acid reagents, but considerable excesses of thiols were usually employed and yields of the required products were not consistently good. Boron trifluoride (14) is recommended here because it permits homogeneous reaction in chloroform with effectively equimolar proportions of thiols.

The boron trifluoride catalyzed reaction is faster with 1,2-trans-related peresters, which indicates that anchimeric assistance is provided by the ester group at C-2 for the establishment of acyloxonium intermediates (for example, II) which are then attacked by nucleophilic sulfur, but 1,2-cis-isomers also undergo thiolysis in the presence of boron trifluoride. Thin-layer chromatography (tlc) (Vol. VI [5]) can be used to follow the progress of the reactions which are stopped as soon as the initial esters have undergone reaction. The reactions are not totally stereospecific and yield small proportions of 1,2-cis-related 1-thioglycosides which are observed as more mobile components by tlc. The procedure is effective with alkyl, alkenyl, and aryl thiols,[1] and with acetates and benzoates of pentoses or hexoses. Reactions of acetylated starting materials proceed smoothly at 20°–25°, but the solutions of benzoates require heating. Compounds IV and V, for example, are prepared directly in 70 and 58% yield from the corresponding β-glycosyl acetates; and furthermore, the procedure is applicable with uronic acid derivatives, compounds VI and VII being similarly obtained directly in 76 and 62% yield, respectively.

Other procedures which have been reported for synthesizing 1-thioglycosides involve: (a) the thiolysis of 3,4,6-tri-O-acetyl-1,2-anhydro-D-glucose (15, 16); (b) pyrolysis of glycosyl xanthate esters (17); (c) photochemical additions of thiols to unsaturated compounds (18) and (d) sulfur extrusions from glycosyl disulfides (19).

Procedure

Phenyl 2,3,4,6-Tetra-O-acetyl-1-thio-β-D-glucopyranoside (III)

Penta-O-acetyl-β-D-glucopyranose (I) (Vol. II [53]) (10 g) is dissolved in 50 ml of chloroform, and to this solution is added 3.4 g (1.2 mole equiv.) of benzenethiol and then 18.3 g (5.0 mole equiv.) of boron trifluoride etherate.

---

[1] Use of p-nitrothiophenol leads to high proportions of cis-related products presumably because of the reduced nucleophilicity of the thiol group and the relative significance of anomerization processes.

The reaction is followed by tlc on silica gel using 1:2 v/v ethyl acetate–hexane as the irrigant, and after 6 h, the solution is washed with saturated aqueous sodium hydrogencarbonate and then with water before being dried with anhydrous sodium sulfate. Removal of the solvent gives a light yellow syrup which crystallizes on treatment with ethanol to afford the product as colorless needles; yield 8.1 g (71%), $[\alpha]_D$ $-14°$ (chloroform). After recrystallization from ethanol, it melts at 117°–118°, $[\alpha]_D$ $-16°$ (chloroform).

From a portion (14%) of the mother liquors, a further quantity of the main product (0.06 g) was obtained by preparative tlc (Vol. VI [6]) together with the α-anomer (0.05 g) and penta-O-acetyl-α-D-glucopyranose (0.11 g).

Phenyl 2,3,4,6-Tetra-O-benzoyl-1-thio-β-D-glucopyranoside

Reaction of 5.0 g of penta-O-benzoyl-β-D-glucopyranose (*20*) in 80 ml of chloroform with 0.94 g (1.2 mole equiv.) of benzenethiol (thiophenol) and 3.04 g (3.0 mole equiv.) of boron trifluoride etherate is complete when the solution is heated at the reflux temperature under nitrogen for 1.5 h. Processing as above and crystallization from acetic acid yields the β-glycoside directly in 62% yield; mp 167°–168°, $[\alpha]_D$ $+34°$ (chloroform).

References

(*1*) E. Steers, P. Cuatrecasas, and H. B. Pollard, *J. Biol. Chem.*, **246**, 196 (1971).
(*2*) P. Cuatrecasas, *Advan. Enzymol.*, **36**, 29 (1972).
(*3*) C. R. Lowe and P. D. G. Dean, "Affinity Chromatography," Wiley-Interscience, London, 1974.
(*4*) S. Chipowsky and Y. C. Lee, *Carbohydr. Res.*, **31**, 339 (1973).
(*5*) R. J. Ferrier, R. W. Hay, and N. Vethaviyasar, *Carbohydr. Res.*, **27**, 55 (1973)
(*6*) P. J. Pfaffli, S. H. Hixson, and L. Anderson, *Carbohydr. Res.*, **23**, 195 (1972).
(*7*) R. U. Lemieux, *Can. J. Chem.*, **29**, 1079 (1951).
(*8*) R. U. Lemieux and C. Brice, *Can. J. Chem.*, **33**, 109 (1955).
(*9*) M. L. Chawla and O. P. Bahl, *Carbohydr. Res.*, **32**, 25 (1974).
(*10*) C. S. Jones, R. H. Shah, D. J. Kosman, and O. P. Bahl, *Carbohydr. Res.*, **36**, 241 (1974).
(*11*) J. Schneider, H. H. Liu, and Y. C. Lee, *Carbohydr. Res.*, **39**, 156 (1975).
(*12*) A. B. Landge, T. R. Ingle, and J. L. Bose, *Indian J. Chem.*, **7**, 1200 (1969).
(*13*) T. R. Ingle and J. L. Bose, *Carbohydr. Res.*, **12**, 459 (1970).
(*14*) R. J. Ferrier and R. H. Furneaux, *Carbohydr. Res.*, **52**, 63 (1976).
(*15*) F. Weygand and H. Ziemann, *Ann.*, **657**, 179 (1962).
(*16*) H. Frenzel, P. Nuhn, and G. Wagner, *Arch. Pharm.*, **302**, 62 (1969).
(*17*) M. Sakata, M. Haga, and S. Tejima, *Carbohydr. Res.*, **13**, 379 (1970).
(*18*) Y. Araki, K. Matsuura, Y. Ishido, and K. Kushida, *Chem. Letters*, 383 (1973).
(*19*) D. N. Harpp and J. G. Gleason, *J. Amer. Chem. Soc.*, **93**, 2437 (1971).
(*20*) R. K. Ness, H. G. Fletcher, Jr., and C. S. Hudson, *J. Amer. Chem. Soc.*, **72**, 2200 (1950).

[37] Primary Glycopyranosylamines

BY HORACE S. ISBELL AND HARRIET L. FRUSH

Department of Chemistry, The American University, Washington, D.C.

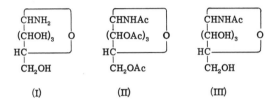

Introduction

Reaction of ammonia or suitable amines with sugars produces glyco-pyranosylamines by replacement of one amino hydrogen with a glycosyl group. A variety of glycosylamines and related compounds has been prepared from sugars by reaction with numerous primary and secondary aromatic and aliphatic amines (1, Vol. II [27]). Methods are described here for the preparation of the primary glycopyranosylamines (I), their acetates (II) and N-acetyl derivatives (III).

The glycosylamines, as a class, are not particularly stable, although they show considerable variation in their stability. Their melting points (decomposition points) are influenced by the rate of heating. On storage over an acidic desiccant, or even in solution, the primary glycosylamines tend to form diglycosylamines by elimination of ammonia. In aqueous solution, they show a striking sensitivity to hydrolysis (Fig. 1), being stable in highly acidic or basic solutions, but readily hydrolyzed within a limited range, with a maximum rate near pH 5 (2–4). In contrast to the mutarotation of the sugars, the mutarotation of the primary glycosylamines is strongly catalyzed by acids, but only slightly, if at all, by bases.

Unlike the glycopyranosylamines, their acyl derivatives are stable compounds, with sharp melting points. The primary glycopyranosylamines may be completely acetylated, and the resulting amide–esters selectively hydrolyzed, to yield stable N-acetylglycopyranosylamines. Some of these latter compounds have also been prepared by other methods (5). Ammonolysis of certain acetylated or benzoylated monosaccharides has yielded N-acetylglycofuranosylamines (6, 7). Crystalline isopropylideneglycofuran-

METHODS IN CARBOHYDRATE
CHEMISTRY, VOL. VIII

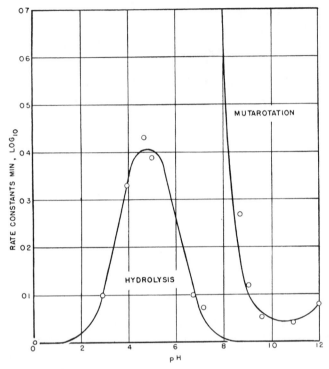

FIG. 1.—Rate constants for the mutarotation and hydrolysis of α-D-galacto-pyranosylamine.

osylamine toluene-*p*-sulfonates have been prepared by reaction of the corresponding glycopyranosylamines, acetone, 2,2-dimethoxypropane and toluene-*p*-sulfonic acid; and the compounds have been used in further syntheses (*8*).

The uronic acids, as lactones or esters, react with ammonia, forming 1-amino-amides which, on selective hydrolysis of the amino group, yield uronic amides (*9, 10*).

Procedure

β-D-Mannopyranosylamine (I) Monohydrate (*11*)

D-Mannose (20 g) and 0.5 g of ammonium chloride are added to 50 ml of methanol; and the mixture, at 0°, is treated with ammonia gas until all the sugar has dissolved. The solution is stored at 0° until satisfactory crystallization has occurred. The first crystallization may be slow and require storage for as long as a month. However, with the aid of seed crystals, the period of crystallization is shortened to a few days. The crystals are separated and washed with methanol. The crude product ($\sim$17 g) is dissolved in an equal weight of water, and the solution is diluted with an approximately ten-fold volume of methanol, followed by sufficient ethanol to produce incipient turbidity. After about 2 days, the crystals are separated, washed with methanol, and stored over sodium hydroxide in an atmosphere of ammonia. (See Table I for constants.)

TABLE I

Primary Glycopyranosylamines of the Pentoses and Hexoses, with Their Acetylated Derivatives

	Mp	$[\alpha]_D$	Solvent	Ref.[a]
α-L-Arabinopyranosylamine	124°–125°	+86.3°	H_2O	4, 3, 12
N-Acetyl-tri-O-acetyl-	177°–178°	+89.6°	$CHCl_3$	3
N-Acetyl-	222°–224°	+69.1°	H_2O	3
β-D-Xylopyranosylamine	128°–129°	−19.6°	H_2O	11, 12
N-Acetyl-tri-O-acetyl-	172°–173°	+28.5°	$CHCl_3$	11
N-Acetyl-	213°–214°	−0.7°	H_2O	11
D-Ribopyranosylamine	128°–129°	−35.3°	H_2O	13
N-Acetyl-tri-O-acetyl-	128°–130°	+35.3°	$CHCl_3$	13
N-Acetyl-α-	189°–200°	+17.8°	H_2O	13
N-Acetyl-β-	195°–197°	−23.4°	H_2O	13
D-Lyxosylamine	142°–143°	−44.5°	H_2O	14, 15
N-Acetyl-β-	166°–167°	−47.0°	H_2O	16
β-D-Glucopyranosylamine	125°–127°	+20.8°	H_2O	17, 18
N-Acetyl-tetra-O-acetyl-	163°–164°	+17.4°	$CHCl_3$	19
N-Acetyl-	260°[b]	−22.8°	H_2O	19
β-D-Mannopyranosylamine (monohydrate)[c]	93°–94°	−11.6°	H_2O	11
N-Acetyl-tetra-O-acetyl-	188°–189°	−16.5°	$CHCl_3$	11
N-Acetyl- (monohydrate)	203°–204°	−47.4°	H_2O	11
α-D-Galactopyranosylamine (NH$_3$ complex)[c]	107°–109°	+138°	H_2O	4, 12
N-Acetyl-tetra-O-acetyl-	172°–173°	+117.4°	$CHCl_3$	4
N-Acetyl-	179°–180°	+194.9°	H_2O	4
β-D-Galactopyranosylamine	134°–136°	+62.2°	H_2O	4, 12
N-Acetyl-tetra-O-acetyl	173°–174°	+34.7°	$CHCl_3$	4
N-Acetyl-	233°	+9.8°	H_2O	4
(L-Rhamnosylamine)$_2$ · CH_3OH[d]	116°	+38°	H_2O	12
(L-Rhamnosylamine)$_2$ · C_2H_5OH[d]	80°	+28°	H_2O	12
N-Acetyl-tri-O-acetyl-β-	135°–137°	+6.2°	$CHCl_3$	16
N-Acetyl-β-	210°–211°	+65.7°	H_2O	16

[a] Reference is to the original preparation and/or an early one. In some cases, the specific rotations and melting points differ somewhat from those reported in the literature.

[b] Gradual decomposition from 230°.

[c] Hydrate or complex applies only to the amine.

[d] Solvent of crystallization applies only to the amine.

N-Acetyl-tetra-O-acetyl-β-D-mannopyranosylamine (II) (11)

Finely powdered, crystalline β-D-mannopyranosylamine (10 g) is added to a previously cooled solution of 100 ml of pyridine and 50 ml of acetic anhydride in a flask equipped with a magnetic stirrer and immersed in a mixture of ice and salt. Stirring is continued in the ice–salt bath until the crystals dissolve. After standing for a few hours at 20°–25°, the solution is poured into 1 liter of ice and water; the mixture is stirred for 0.5 h and then extracted with chloroform. Evaporation of the chloroform yields a crude product (~16 g) which is recrystallized twice from chloroform, with the addition of petroleum ether, to give pure N-acetyl-tetra-O-acetyl-β-D-mannopyranosylamine.

N-Acetyl-β-D-mannopyranosylamine (III) Monohydrate (11)

N-Acetyl-tetra-O-acetyl-β-D-mannopyranosylamine (10 g) is dissolved in 100 ml of anhydrous methanol (Vol. VII [3]) containing 10 ml of 1.2 M barium methoxide. After the solution has stood for 1 h at 20°–25°, dilute sulfuric acid, exactly equivalent to the barium methoxide, is added. Barium sulfate is removed by filtration of the mixture on a Buchner funnel heavily coated with diatomaceous earth and decolorizing carbon. The filtrate is concentrated under reduced pressure to a syrup, from which crystalline N-acetyl-β-D-mannopyranosylamine monohydrate is separated; yield ~6 g. The crude product is recrystallized by dissolving it in two parts of water and adding about 12 parts of methanol, followed by ethanol to incipient turbidity.

References

(1)   G. P. Ellis and J. Honeyman, *Advan. Carbohydr. Chem.*, **10**, 95 (1955).
(2)   H. S. Isbell and H. L. Frush, *J. Amer. Chem. Soc.*, **72**, 1043 (1950).
(3)   H. S. Isbell and H. L. Frush, *J. Res. Nat. Bur. Stds.*, **46**, 132 (1951).
(4)   H. L. Frush and H. S. Isbell, *J. Res. Nat. Bur. Stds.*, **47**, 239 (1951).
(5)   K. Onadera and S. Kitaoka, *J. Org. Chem.*, **25**, 1322 (1960) and references cited therein.
(6)   A. S. Cerezo and V. Deulofeu, *Carbohydr. Res.*, **2**, 35 (1966).
(7)   R. C. Hockett and L. B. Chandler, *J. Amer. Chem. Soc.*, **66**, 957 (1944).
(8)   N. J. Cusack, D. H. Robinson, P. W. Rugg, G. Shaw, and (in part) R. Lofthouse, *J. Chem. Soc., Perkin Trans.*, **1**, 73 (1974).
(9)   H. L. Frush and H. S. Isbell, *J. Res. Nat. Bur. Stds.*, **41**, 11 (1948).
(10) H. L. Frush and H. S. Isbell, *J. Res. Nat. Bur. Stds.*, **41**, 609 (1948).
(11) H. S. Isbell and H. L. Frush, *J. Org. Chem.*, **23**, 1309 (1958).
(12) C. A. Lobry de Bruyn and F. H. Van Leent, *Rec. trav. chim.*, **14**, 134 (1895).
(13) R. S. Tipson, *J. Org. Chem.*, **26**, 2462 (1961).
(14) P. A. Levene and F. B. LaForge, *J. Biol. Chem.*, **22**, 333 (1915).

(*15*) P. A. Levene, *J. Biol. Chem.*, **24,** 62 (1916).
(*16*) S. Delpy, *Ph.D. Thesis*, Buenos Aires, Argentina, 1962.
(*17*) C. A. Lobry de Bruyn, *Rec. trav. chim.*, **14,** 98 (1895).
(*18*) A. R. Ling and D. R. Nanji, *J. Chem. Soc.*, **121,** 1682 (1922).
(*19*) P. Brigl and H. Kepler, *Z. physiol. Chem.*, **180,** 38 (1929).

[38] Production of Sugar Nucleotides by Fermentation

By Kichitaro Kawaguchi

Mergenthaler Laboratory for Biology, The Johns Hopkins University, Baltimore, Maryland

Hiroyasu Kawai

Department of Food Science and Nutrition, Nara Women's University, Nara, Japan

AND

Tatsurokuro Tochikura

Department of Food Science and Technology, Kyoto University, Kyoto, Japan

Introduction

Sugar nucleotides play important roles in carbohydrate metabolism, and more than 70 of these compounds have been isolated from nature. Sugar nucleotides have also been prepared by chemical and enzymic syntheses (*1*). However, the preparative isolation of individual sugar nucleotides on a large scale from natural sources is at times difficult unless starting material having a high content of sugar nucleotides is available, and the chemical synthesis is generally laborious.

We have recently developed a new method for the fermentative production of sugar nucleotides (*2*). The principle is simply illustrated in Scheme 1. It is

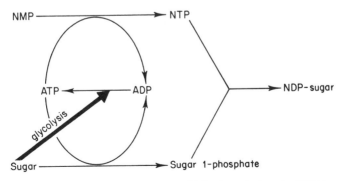

possible to produce GDP-D-mannose (GDPMan) (*3–5*), UDP-D-glucose (UDPGlc) (*6*), UDP-D-galactose (UDPGal) (*7–8*), and UDP-2-acetamido-2-

METHODS IN CARBOHYDRATE
CHEMISTRY, VOL. VIII

deoxy-D-glucose (UDPGlcNAc) (9) in high yields and on a large scale by incubating the corresponding nucleoside 5′-monophosphate and D-hexose in the presence of high concentrations of potassium phosphate buffer and magnesium ion with dried cells of yeast.

Procedures

Preparation of Dried Cells of Yeasts

Commercial, pressed baker's yeast is crushed on a large Petri dish and dried at room temperature for ~13 h with an electric fan. The air-dried cells are further dried over P_2O_5 in vacuo to bring the water content to 4–5%. This drying step is indispensable to the achievement of high yields of sugar nucleotides by the fermentative procedure (4, 5, 8, 9).

Acetone-dried cells of brewer's yeast are prepared as follows: 100 g of pressed cells are washed twice with water, suspended in 50 ml of cold water, and poured into 2 liters of cold acetone ($-20°$) with vigorous stirring for 5 min. The cells are collected by filtration, suspended in 2 liters of cold acetone, and stirred for 5 min. The cells are again collected by filtration and washed with ether. Then the cells are spread on a large Petri dish and left to stand at room temperature for 3 h to eliminate the organic solvents.

Torulopsis candida IFO 0768 is grown on a medium which contains 50 g of D-galactose or lactose, 5 g of peptone, 2 g of yeast extract, 2 g of KH_2PO_4, 2 g of $(NH_4)_2HPO_4$, and 1 g of $MgSO_4 \cdot 7H_2O$ in 1 liter of tap water, pH 6.0. The yeast is cultured on a reciprocal shaker at 28° for 24 h in a 2-L shaking flask containing 500 ml of the medium. The cells are harvested by centrifugation, washed twice with water, and air-dried at room temperature, followed by drying over P_2O_5 in vacuo. The dried cells of yeasts prepared should be kept at $-20°$ until use and can be kept without loss of activities for at least 6 months.

Analyses of Nucleotides

In order to produce sugar nucleotides in maximum yields, the time course of the reaction should first be investigated. Several test tubes containing 2.5 ml of the reaction mixtures shown below are incubated at 28° with shaking for the period indicated in the figures. The reaction is terminated by immersing the test tube in boiling water for 3 min. The cells are removed by centrifugation, and the supernatant solution is employed for the determination of nucleotides. An aliquot of the supernatant is paper-chromatographed with a solvent system of 15:6 v/v 95% ethanol–1 M ammonium acetate (pH 7.5) (*10*). The amounts of sugar nucleotide, nucleoside 5′-mono-, di- and triphosphates, nucleoside, and base are determined by measuring absorbances at 260 nm after extracting these compounds from the spots with 0.1 *M* HCl.

UDPGlc is also determined enzymically with UDPGlc dehydrogenase by the method of Strominger *et al.* (*11*). Because the paper-chromatographic separation of UDPGal from UDPGlc is unsuccessful with the solvent system described above, the amount of UDPGal is calculated as the difference between the amount of total UDP-sugars (UDPGal + UDPGlc) determined by paper chromatography and the amount of UDPGlc determined enzymically. UDPGlcNAc is determined by a color reaction with the Ehrlich reagent according to the method of Reissig *et al.* (*12*).

Reaction products are separated by column chromatography with Dowex 1 X2 (Cl⁻, 200–400 mesh) by the method of Cohn and Carter (*13*). Nucleotides are concentrated by adsorption on charcoal, followed by elution with 10:1:9 v/v ethanol–ammonium hydroxide–water. Column fractions are monitored by measuring the absorbance at 260 nm and by paper chromatography with the solvent system described above. The use of a ^{14}C-labeled tracer of each sugar nucleotide is recommended during the isolation process (Vol. VI [80]).

Production of GDPMan by Dried Cells of Baker's Yeast (*3, 4*)

The production of GDPMan is accomplished through the cooperative action of the many enzymes in dried cells of baker's yeast as indicated below:

$$\text{D-Glucose} \xrightarrow{\text{glycolysis}} \longrightarrow \text{D-fructose 6-phosphate} \xrightarrow[\text{isomerase}]{\text{phosphomannose}}$$

$$\text{D-mannose 6-phosphate} \xrightarrow[\text{mutase}]{\text{phosphogluco}} \text{D-mannose 1-phosphate}$$

$$\text{GMP} \xrightarrow[\text{kinase}]{\text{nucleoside monophosphate}} \text{GDP} \xrightarrow[\text{kinase}]{\text{nucleoside diphosphate}} \text{GTP}$$

$$\text{GTP} + \text{D-mannose 1-phosphate} \xrightarrow{\text{GDPMan pyrophosphorylase}} \text{GDPMan} + \text{PP}_i$$

$$\text{PP}_i + \text{H}_2\text{O} \xrightarrow{\text{inorganic pyrophosphatase}} 2\,\text{P}_i$$

A typical time course of the production of GDPMan by dried cells of baker's yeast is shown in Figure 1. GDPMan is produced in 45% yield (based on GMP added) by an 8-h incubation.

The reaction mixture contains 19.9 g of disodium GMP, 288 g of D-glucose, 9.86 g of MgSO$_4$ · 7H$_2$O, 400 ml of 1.8 *M* potassium phosphate buffer, pH 7.0, and 200 g of dried cells of baker's yeast in a total volume of 2 liters.

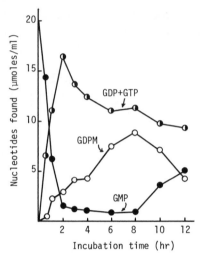

Fig. 1.—Time course of GDPMan (GDPM) production by dried cells of baker's yeast.

The reaction is done at 28° for 7 h in four 2-L shaking flasks containing 500 ml of the reaction mixture. The reaction is terminated by immersing the flasks in boiling water for 15 min with occasional shaking, and the flasks are cooled in ice. After the cells are removed by centrifugation, the resulting supernatant solution is acidified with 1 M HCl to pH 3.8; and the precipitates are removed by centrifugation. Approximately 5 g of charcoal is added to the supernatant solution for each mmole of nucleotide as calculated from the absorbance at 260 nm. The charcoal is removed by filtration, and then nucleotides are eluted with ammonia in ethanol until the recovery of nucleotides reaches at least 70%. The eluate is then concentrated under diminished pressure to half its volume at 30°. The concentrated solution is adjusted to pH 8.0 with 2 M ammonium hydroxide and applied to a Dowex 1 × 2 (Cl⁻, 2.5 × 50 cm). Nucleotides are eluted with 0.01 M HCl containing stepwise increments of NaCl. GMP is eluted at 0.08 M NaCl, GDPMan and GDP separately at 0.12 M NaCl, and GTP at 0.4 M NaCl. Charcoal is again added to these fractions, and the nucleotides are again eluted with ammonia in ethanol. Each eluate is concentrated to a small volume under reduced pressure at 30°, then lyophilized. About 7.2 g of GDPMan 1.5 g of GDP, and 3.1 g of GTP are obtained as their ammonium salts. An improved method for the production of GDPMan using dried cells of *Hansenula jadinii* IFO 0987 is available (5).

Production of UDPGlc by Acetone-dried Cells of Brewer's Yeast (6)

The production of UDPGlc is accomplished by the following enzymes present in acetone-dried cells of brewer's yeast.

$$\text{D-Glucose} \xrightarrow{\text{hexokinase}} \text{D-glucose 6-phosphate} \xrightarrow{\text{phosphoglucomutase}} \text{D-glucose 1-phosphate}$$

$$\text{UMP} \xrightarrow[\text{kinase}]{\text{nucleoside monophosphate}} \text{UDP} \xrightarrow[\text{kinase}]{\text{nucleoside diphosphate}} \text{GTP}$$

$$\text{UTP} + \text{D-glucose 1-phosphate} \xrightarrow{\text{UDPGlc pyrophosphorylase}} \text{UDPGlc} + \text{PP}_i$$

$$\text{PP}_i + \text{H}_2\text{O} \xrightarrow{\text{inorganic pyrophosphatase}} 2\,\text{P}_i$$

Figure 2 shows a typical time course of the production of UDPGlc by acetone-dried cells of brewer's yeast. UDPGlc is produced in 60% yield (based on UMP added) by a 3-h incubation.

The reaction mixture contains 37 g of disodium UMP, 360 g of D-glucose, 500 ml of 1.8 M potassium phosphate buffer, pH 7.0, 15 g of MgSO$_4$ · 7H$_2$O, and 450 g of acetone-dried cells of brewer's yeast in a total volume of 5 L. The reaction is performed with shaking at 28° for 4.5 h in ten 2-L shaking flasks containing 500 ml of the reaction mixture. The incubated mixture is heated at 100° for 5 min and cooled in ice. The cells are removed by centrifu-

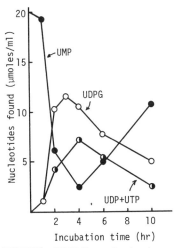

Fig. 2.—Time course of UDPGlc (UDPG) production by acetone-dried cells of brewer's yeast.

gation, and the supernatant is acidified to pH 3.8 with 1 M HCl. About 500 g of charcoal is added to the supernatant, and the mixture is stirred at 4° overnight. The charcoal is collected by filtration, and washed with a small volume of cold water. Nucleotides are eluted several times with ammonia in ethanol until the recovery reaches at least 70%, and the eluate is concentrated by evaporation under reduced pressure at 30°. The concentrate is adjusted to pH 8.0 with 2 M ammonium hydroxide and applied to a column of Dowex 1 X2 (Cl⁻, 10 × 43 cm). UDPGlc is eluted with 0.01 M HCl containing stepwise increments of NaCl. UDPGlc is eluted at 0.15 M NaCl. The fractions containing UDPGlc are pooled and concentrated by adsorption of the nucleotide onto and elution from charcoal. The concentrate is lyophilized; yield 23.5 g of UDPGlc is obtained as its ammonium salt.

Production of UDPGal by Dried Cells of *Torulopsis candida* (*7, 8*)

The production of UDPGal is performed by the cooperation of many enzymes in dried cells of *T. candida* as indicated below:

$$\text{D-galactose} \xrightarrow{\text{galactokinase}} \text{D-galactose 1-phosphate}$$

$$\text{Intracellular glycogen} + \text{H}_2\text{O} \xrightarrow{\text{phosphorylase}} \text{D-glucose 1-phosphate}$$

$$\text{UMP} \xrightarrow[\text{kinase}]{\text{nucleoside monophosphate}} \text{UDP} \xrightarrow[\text{kinase}]{\text{nucleoside diphosphate}} \text{UTP}$$

$$\text{UTP} + \text{D-glucose 1-phosphate} \xrightarrow{\text{UDPGlc pyrophosphorylase}} \text{UDPGlc} + \text{PP}_i$$

$$\text{UDPGlc} + \text{D-galactose 1-phosphate} \xrightarrow[\text{uridylyl transferase}]{\text{D-galactose 1-phosphate}} \text{UDPGal} + \text{D-glucose 1-phosphate}$$

$$\text{PP}_i + \text{H}_2\text{O} \xrightarrow{\text{inorganic pyrophosphatase}} 2\,\text{P}_i$$

Figure 3 shows a time course of the fermentative production of UDPGal by dried cells of *T. candida*. The final yield of UDPGal is about 70% (based on added UMP). In order to reduce the amount of UDPGlc formed without loss of UDPGal, an incubation longer than 12 h is recommended.

The reaction mixture contains 13.2 g of disodium UMP, 54 g of D-galactose, 300 ml of 1 M potassium phosphate buffer, pH 7.0, 4.44 g of MgSO$_4$ · 7H$_2$O, and 150 g of dried cells of *T. candida* in a total volume of 1.5 L. The reaction is done with shaking at 28° for 20 h using three 2-L shaking flasks containing 500 ml of the reaction mixture. The reaction is terminated by immersing the flasks in boiling water for 15 min. The cells are removed by

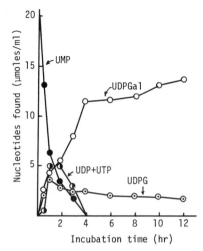

FIG. 3.—Time course of UDPGal production by dried cells of *Torulopsis candida*.

centrifugation, and the supernatant solution is adjusted to pH 3.8 with 2 M HCl. To the supernatant is added 150 g of charcoal, and the mixture is stirred for a few hours. The charcoal is collected by filtration and washed with a small volume of 0.001 M HCl. Nucleotides are eluted several times with ammonia in ethanol until recovery of nucleotides reaches at least 70%, and the eluate is concentrated to half its volume under reduced pressure at 30°. The concentrated eluate is adjusted to pH 8.0 with 2 M NH₄OH and applied on a column of Dowex 1 X2 (Cl⁻, 5 × 52 cm). Nucleotides are eluted with 0.01 M HCl containing stepwise increments of NaCl. UDPGal is eluted at 0.2 M NaCl. The fractions containing UDPGal are pooled and concentrated by treatment with charcoal. The concentrate is lyophilized, and ∼6 g of UDPGal is obtained as its ammonium salt. It is possible to raise the UMP concentration in the reaction mixture to 140 μmoles/ml and to convert 80% of UMP to UDPGal (51.5 mg/ml).

Production of UDPGlcNAc by Dried Cells of Baker's Yeast (9)

UDPGlcNAc is produced by the cooperative action of the many enzymes present in dried cells of baker's yeast.

$$\text{D-Glucosamine} \xrightarrow{\text{hexokinase}} \text{D-glucosamine 6-phosphate} \xrightarrow[\text{N-acetylase}]{\text{D-glucosamine 6-phosphate}} N\text{-acetyl-D-}$$

$$\text{glucosamine 6-phosphate} \xrightarrow[\text{phosphomutase}]{\text{N-acetylglucosamine}} N\text{-acetyl-D-glucosamine 1-phosphate}$$

$$\text{UMP} \xrightarrow[\text{kinase}]{\text{nucleoside monophosphate}} \text{UDP} \xrightarrow[\text{kinase}]{\text{nucleoside diphosphate}} \text{UTP}$$

$$\text{UTP} + N\text{-acetyl-D-glucosamine 1-phosphate} \xrightarrow{\text{UDPGlcNAc pyrophosphorylase}} \text{UDPGlcNAc} + \text{PP}_i$$

$$\text{PP}_i + \text{H}_2\text{O} \xrightarrow{\text{inorganic pyrophosphatase}} 2\,\text{P}_i$$

A time course of the fermentative production of UDPGlcNAc by dried cells of baker's yeast is shown in Figure 4. The maximum yield of UDPGlc-NAc is about 40% (based on added UMP) by an 8 h incubation.

The reaction mixture contains 0.944 g of disodium UMP, 3.6 g of D-glucose, 4.3 g of D-glucosamine hydrochloride, 20 ml of 1 M potassium phosphate buffer, pH 7.4, 43 mg of MgCl$_2$, and 10 g of dried cells of baker's yeast in a total volume of 100 ml. The reaction is done with shaking at 28° in a 500-ml shaking flask. The reaction is terminated by heating the incubated mixture at 100° for 5 min. Charcoal (9 g) is added to the supernatant solution which is acidified to pH 3.8 with 2 M HCl and stirred for a few hours. The charcoal is collected by filtration, and the nucleotides are eluted with ammonia in ethanol until the recovery of nucleotides reaches at least 70%. The eluate is concentrated until its pH gets to 8.0, and then subjected to column chromatography using Dowex 1 X2 (Cl$^-$, 2.2 × 22 cm). Nucleotides are eluted with 0.01 M HCl containing stepwise increments of NaCl. UDPGlc-

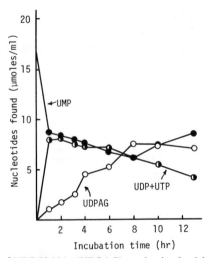

FIG. 4.—Time course of UDPGlcNAc (UDPAG) production by dried cells of baker's yeast.

NAc which is eluted at 0.1 M NaCl is concentrated by treatment with charcoal and elution with ammonia in ethanol. The eluate is concentrated and lyophilized. Finally, 253 μmoles of UDPGlcNAc is obtained.

References

(1) N. K. Kochtkov and V. N. Shibaev, *Advan. Carbohydr. Chem. Biochem.*, **28**, 307 (1973).
(2) T. Tochikura, H. Kawai, K. Kawaguchi, Y. Mugibayashi, and K. Ogata, *Proc. Intern. Ferment. Symp.*, *4th, Kyoto, Japan, 1972, Ferment. Technol. Today*, (1972) 463.
(3) T. Tochikura, K. Kawaguchi, T. Kano, and K. Ogata, *J. Ferment. Technol.*, **47**, 564 (1968).
(4) K. Kawaguchi, K. Ogata, and T. Tochikura, *Agr. Biol. Chem.*, **34**, 908 (1970).
(5) K. Kawaguchi, S. Tanida, Y. Mugibayashi, Y. Tani, and K. Ogata, *J. Ferment. Technol.*, **49**, 195 (1971).
(6) T. Tochikura, H. Kawai, S. Tobe, K. Kawaguchi, M. Osugi, and K. Ogata, *J. Ferment. Technol.*, **46**, 957 (1968).
(7) T. Tochikura, K. Kawaguchi, H. Kawai, Y. Mugibayashi, and K. Ogata, *J. Ferment. Technol.*, **46**, 970 (1968).
(8) T. Tochikura, Y. Mugibayashi, H. Kawai, K. Kawaguchi, and K. Ogata, *Hakko To Taisha*, **22**, 144 (1970).
(9) T. Tochikura, H. Kawai, and T. Gotan, *Agr. Biol. Chem.*, **35**, 163 (1971).
(10) A. C. Paladini and L. F. Leloir, *Biochem. J.*, **51**, 426 (1952).
(11) J. L. Strominger, E. S. Maxwell, J. Axelrod, and H. M. Kalckar, *J. Biol. Chem.*, **224**, 79 (1957).
(12) J. L. Reissig, J. L. Strominger, and L. F. Leloir, *J. Biol. Chem.*, **217**, 959 (1955).
(13) W. E. Cohn and C. E. Carter, *J. Amer. Chem. Soc.*, **72**, 4273 (1950).

ESTERS AND AMIDES

[39] O-Chloroacetate Derivatives of Sugars as Synthetic Intermediates

BY CORNELIS P. J. GLAUDEMANS

National Institutes of Health, Bethesda, Maryland

AND

MAURICE J. BERTOLINI

Cutter Laboratories, Berkeley, California

Introduction

Chloroacetyl groups can be used for the reversible protection of hydroxyl groups. Both the acylation reaction and the dechloroacetylation with thiourea are readily achieved.

In the field of carbohydrate synthesis, there is great potential utility for a hydroxyl blocking-group which can be removed under neutral and mild conditions. In the synthesis of 1-O-acyl-aldoses, the blocking groups employed on the carbohydrate moiety have to be of such a nature that their removal does not involve treatment either with acid or alkali. Thus, the benzyl group has been frequently used as a blocking group in this type of synthesis. For example, 2,3,4,6-tetra-O-benzyl-D-glucose was used in the condensation reactions leading to the preparation of the 1-O-galloyl and 1-O-(p-hydroxybenzoyl) derivatives of D-glucose (1–4). Similarly in his extensive work on benzyl derivatives of sugars, Fletcher, together with Tejima, used tri-O-benzyl-L-arabinopyranose and furanose to prepare the four possible 1-O-benzoyl-L-arabinoses (5). The use of benzyl ethers as O-blocking groups in syntheses of this type may, however, not be applied when the 1-O-acyl-aglycon is sensitive to hydrogenation. An example of that would be the synthesis of 1-O-acylglycoses in which the acyl group bears an alkenic bond. The use of O-chloroacetates as blocking groups in syntheses of that type would circumvent the use of acid, alkali, or reductive procedures to deblock the 1-O-acyl-aldose. Advantages of the O-chloroacetate as a blocking group are its ease of formation and its smooth removal by treatment with thiourea.

271

ISBN 0-12 746208-2

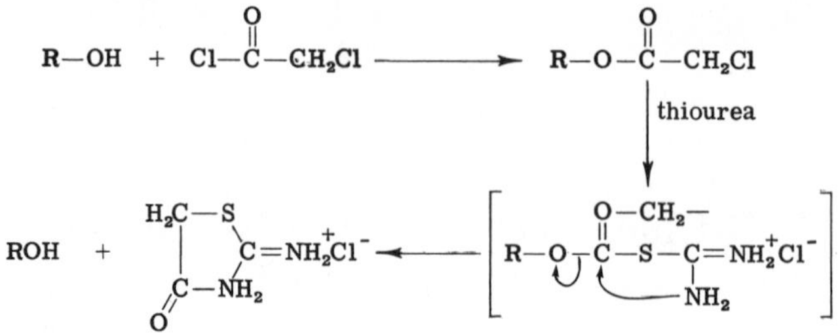

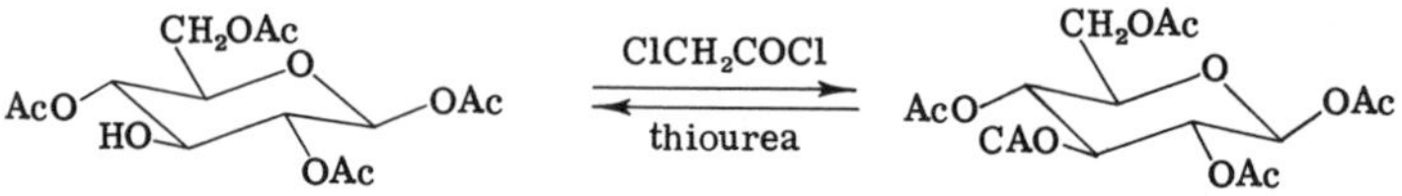

(I)

(II)
(CA = COCH₂Cl)

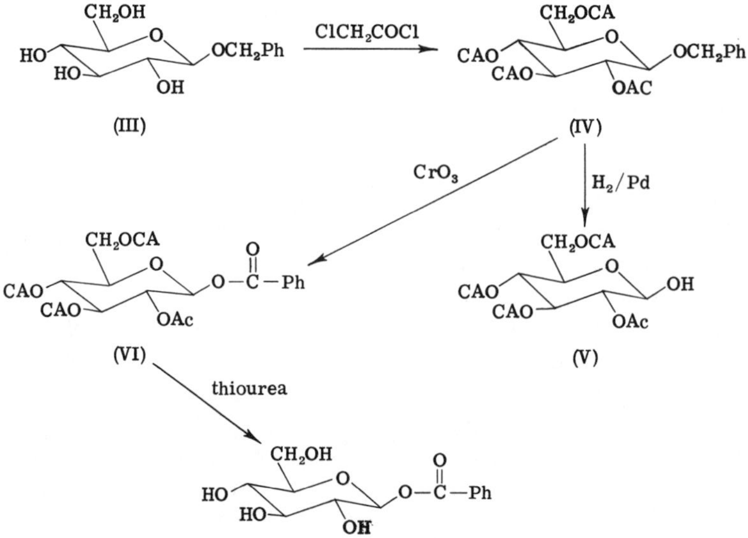

(III)

(IV)

(V)

(VI)

(VII)

Mulder (6) and Maly (7) discovered the reaction of thiourea with N-chloro-acyl and O-chloroacyl derivatives, supposedly yielding thiohydantoins, but later showed to yield pseudothiohydantoins (8). Masaki *et al.* (9) and Fontana and Scoffone (10) have studied some reactions of N- and O-chloro-acetates of amino acids with thiourea, and Glaudemans and co-workers (11, 12) have studied O-chloroacetates of carbohydrate derivatives.

Below are a few examples for the preparation of chloroacetate-derivatives of glucose and also a few examples illustrating the selective removal of these groups under mild conditions, and their stability under reducing or oxidizing conditions.

Procedures

1,2,4,6-Tetra-*O*-acetyl-3-*O*-chloroacetyl-*β*-D-glucopyranose (II)

1,2,4,6-Tetra-*O*-acetyl-*β*-D-glucopyranose (I) (300 mg) (*13*) is dissolved in 150 ml of ether. Pyridine[1] (300 μl) is added, and the solution is cooled in ice. A solution of 300 μl of chloroacetyl chloride in 5 ml of ether is added dropwise; and after the addition is complete, the mixture is kept at 20°–25° for 4 h. After dilution with ether, the reaction mixture is washed with cold water, cold 1 *M* HCl, and cold aqueous sodium hydrogencarbonate. The solution is dried with sodium sulfate and evaporated to a syrup. The product is crystallized from ethanol; yield 215 mg (58.8%), mp 117°–119°, $[\alpha]^{20}$D +3.1° (*c* 1.0, chloroform).

Dechloroacetylation of I to yield 1,2,4,6-Tetra-O-acetyl-α-D-glucose (I)

II (200 mg) is dissolved in 20 ml of warm methanol and 40 mg of thiourea is added. The reaction is stirred at 20°–25° for 24 h, at which time thin-layer chromatography on silica gel using 1:1 v/v benzene–ether as the irrigant (Vol. VI [6]) shows that nearly all of I has been converted. The solvent is evaporated, and the residue is extracted with ether. The extract is evaporated, and the concentrate is put on a column of silica gel. The products are eluted with 4:1 v/v benzene–acetone. The major fraction is evaporated and swirled with benzene–hexane to give 1,2,4,6-tetra-*O*-acetyl-*β*-D-glucose (I); yield 70 mg (43%), mp 126°–127°.

Methyl 2,3,4,6-tetra-*O*-chloroacetyl-α-D-glucopyranoside

Methyl α-D-glucopyranoside (10 g) is suspended in 100 ml of acetone containing 25 ml of pyridine. The mixture is cooled in a Dry Ice–acetone bath. A 1:1 v/v solution of chloroacetyl chloride in 50 ml of ether is added dropwise while 300 ml of cold ether is added batchwise during the addition of the acid chloride. After the additions are complete, the mixture is allowed to

---

[1] Although no difficulty has been found in using pyridine in these reactions as described here, we have evidence that *prolonged* contact between pyridine and chloroacetyl chloride may lead to complications in that quarternized pyridinium compounds may be formed. If prolonged reaction times are required, this difficulty can be obviated by the use of 2,6-lutidine in the place of pyridine.

reach 20°–25° and is filtered. The yellow filtrate is washed with 0° water and 0° sodium hydrogencarbonate solution. It is dried with sodium sulfate, filtered, and evaporated to a yellow syrup, which crystallizes. The material is recrystallized from chloroform–petroleum ether to give pure title compound; yield 5 g (19%), mp 95°–97° $[\alpha]^{20}$D −106° (c 2.0, chloroform).

Benzyl 2,3,4,6-Tetra-O-chloroacetyl-α-D-glucopyranoside (IV)

Benzyl α-D-glucopyranoside (III) (2.0 g) is suspended in 50 ml of acetone containing 6 ml of pyridine, and cooled in a Dry Ice–acetone bath. Chloroacetyl chloride (6 ml) in 20 ml of cold ether is added dropwise, after which 750 ml of cold ether is added; and the mixture is allowed to warm to 20°–25°. It is worked up as in the preparation of methyl 2,3,4,6-tetra-O-chloroacetyl-α-D-glucopyranoside, and the thick syrup obtained is dissolved in chloroform. Addition of ligroin to slight turbidity causes crystallization. Recrystallization from the same solvent gives pure IV; yield 2.0 g (45%), mp 81°–83°, $[\alpha]^{20}$D −33° (c 2.0, chloroform).

2,3,4,6-Tetra-O-chloroacetyl-D-glucose (V)

To a solution of 1 g of IV in 50 ml of 2:1 v/v benzene–ether is added 0.5 g of palladium black. The suspension is stirred vigorously for 12 h under a hydrogen atmosphere. The suspension is filtered, and the filtrate is evaporated to a syrup. Upon the addition of ether, it crystallizes to yield V; yield 500 mg (60%), mp 104°–106°, $[\alpha]^{20}$D +24.5° (initial) → +72.5° (constant, c 1, chloroform containing 1 drop of pyridine per ml).

1-O-Benzoyl-2,3,4,6-tetra-O-chloroacetyl-β-D-glucopyranose (VI)

Compound IV (1 g) dissolved 20 ml of in acetic acid is treated with 2.3 g of dried chromium trioxide at 20°–25° with stirring overnight. Cold chloroform is added, and the resultant solution is washed, first with 0° water and then with aqueous sodium hydrogen carbonate solution. The organic layer is dried with sodium sulfate, decolorized with charcoal, and filtered. The filtrate is concentrated to a syrup which is crystallized from ether–hexane. Recrystallization from warm ethanol gives pure VI; yield 360 mg (35%), mp 113°–114°, $[\alpha]^{20}$D −17.8° (c 1, chloroform).

1-O-Benzoyl-β-D-glucopyranose (VII)

Compound VI (300 mg) is dissolved in 5 ml of chloroform and 35 ml of ethanol was added. Thiourea (200 mg) is added, and the suspension is stirred at 35° for 24 h, concentrated, and passed through a column of Amberlite IR-120 (H⁺) cation-exchange resin. After washing the column with methanol, the washings and eluate are combined and evaporated to a syrup. This product is fractionated on a silica gel column by elution with 6:1 v/v ether–methanol. The appropriate fractions are pooled and concentrated to a syrup which crystallized from ether–chloroform; yield 20 mg (14%), mp and mixed mp 188°–189°, $[\alpha]^{20}$D $-28°$ (c 0.5, water).

References

(*1*) O. T. Schmidt and H. Reuss, *Ann.*, **649**, 137 (1961).
(*2*) O. T. Schmidt and H. Schmadel, *Ann.*, **649**, 149 (1961).
(*3*) O. T. Schmidt and H. Schmadel, *Ann.*, **649**, 157 (1961).
(*4*) O. T. Schmidt, T. Auer, and H. Schmadel, *Chem. Ber.*, **93**, 556 (1960).
(*5*) S. Tejima and H. G. Fletcher, Jr., *J. Org. Chem.*, **28**, 2999 (1963).
(*6*) E. Mulder, *Ber.*, **8**, 1261 (1875).
(*7*) R. Maly, *Ann.*, **189**, 380 (1877).
(*8*) C. Lieberman and A. Lange, *Ann.*, **207**, 123 (1881).
(*9*) M. Masaki, T. Kitahara, H. Kurita, and M. Ohta, *J. Amer. Chem. Soc.*, **90**, 4508 (1968).
(*10*) A. Fontana and E. Scoffone, *Gazz. Chim. Ital.*, **98**, 1261 (1968).
(*11*) M. Bertolini and C. P. J. Glaudemans, *Carbohydr. Res.*, **15**, 263 (1970).
(*12*) N. Roy and C. P. J. Glaudemans, *Carbohydr. Res.*, **45**, 299 (1975).
(*13*) K. Freudenberg and E. Plankenhorn, *Ann.*, **536**, 257 (1938).

[40] Cyclic Thionocarbonates

By D. Trimnell, B. S. Shasha, and W. M. Doane

Northern Regional Research Center, U.S. Department of Agriculture,
Peoria, Illinois

Introduction

Cyclic thionocarbonate esters are useful in the preparation of unsaturated sugars (Vol. VI [52]), for formation of acyclic thiocarbonyl derivatives (1), and in structural studies (2). Methods have been reported for preparing thionocarbonates via dithiobis(thioformates) (Vol. VI [77]) and using bis-(imidazol-1-yl)thione (Vol. VI [52]). A general procedure, which gives cyclic thionocarbonates from cis- and trans-vicinal diols of acyclic and pyranoid sugars, is based on the use of thiophosgene in pyridine. This one-step procedure has been used with sugars containing either acid- or base-labile protecting groups to afford yields in the range of 50–90%. To prepare some thionocarbonates, 2,4,6-trimethylpyridine is the preferred solvent.

277

METHODS IN CARBOHYDRATE
CHEMISTRY, VOL. VIII

Procedure

Methyl 4,6-*O*-Benzylidene-α-D-glucopyranoside 2,3-Thionocarbonate (II) (*3*, compare Vol. VI [77])

A solution of 1.0 g of methyl 4,6-*O*-benzylidene-α-D-glucopyranoside (I) (Vol. I [30]) in 20 ml of *p*-dioxane is warmed to 65° in a water bath. After the solution is stirred, 1 ml of thiophosgene is added, followed by dropwise addition of 3 ml of dry pyridine (Vol. II [43], [53], [63], [73]; Vol. IV [73]; Vol. VII [2]). When pyridine is added, a dark-colored solution and precipitate form. After being kept 10 min, the mixture is cooled to ~25°, and 100 ml of chloroform is added. The filtrate from the mixture is extracted with three 20-ml portions of water and dried with sodium sulfate. Evaporation of the chloroform under reduced pressure gives a syrup, which is dissolved in a small volume of chloroform and adsorbed onto an *n*-hexane-saturated mixture of 140 g of silicic acid (Mallinckrodt,[1] 100 mesh) and 20 g of activated carbon (Darco) contained in a 100 × 80-mm, coarse, fritted-glass filter. Ether–hexane (1:1 v/v) is percolated through the adsorbent mixture. The first 500 ml of percolate is discarded. The next 2.5 liters is collected and evaporated to a syrup. Crystallization from carbon disulfide gives crystals of methyl 4,6-*O*-benzylidene-α-D-glucopyranoside 2,3-thionocarbonate (II); yield 1.01 g (88%), mp 137°–138°, $[\alpha]^{25}$D +0° (*c* 3.0, chloroform), $\lambda^{\text{MeOH}}_{\text{max}}$ 238 nm (ε 14,500).

Methyl 2,3-Di-*O*-methyl-α-D-glucopyranoside 4,6-Thionocarbonate (IV) (*4*)

A solution of 5.0 g of methyl 2,3-di-*O*-methyl-α-D-glycopyranoside (*5*) in 2,4,6-trimethylpyridine is cooled to 5° and stirred rapidly while adding 5 ml of thiophosgene dropwise. The mixture is then warmed to ~25° and kept 30 min before adding 80 ml of chloroform and decanting into 100 ml of ether. The solid filtrate is extracted first with 20 ml of 1.4 *M* hydrochloric acid and then with 20 ml of 0.6 *M* sodium hydrogen carbonate. The aqueous extracts are combined and extracted with 100 ml of ether. The combined organic phases are dried with sodium sulfate, filtered, and evaporated under reduced pressure to a syrup. The syrup is dissolved in 3 ml of chloroform and adsorbed onto a mixture of 140 g of silicic acid and 20 g of carbon. Elutions are made with 1 liter of 1:1 v/v ether–hexane, 1 liter of ether, and

---

[1] The mention of firm names or trade products does not imply that they are endorsed or recommended by the Department of Agriculture over other firms or similar products not mentioned.

3 liters of 1:1 v/v chloroform–ether. The ether and chloroform–ether eluates contain **IV**, obtained as a syrup on evaporation of the solvents and crystallized from chloroform–hexane; yield 2.9 g (49%), mp 120°–122° $[\alpha]^{22}D$ +21.6° (c 0.97, ether), λ_{max}^{ether} 246–247 nm (ε 14,200).

In this procedure, when 2,4,6-trimethylpyridine is replaced by pyridine, the yield of thionocarbonate is inferior and side reactions occur. Some of these side reactions may be due to a reaction between pyridine and thiophosgene (*6*).

References

(*1*) W. M. Doane, B. S. Shasha, E. I. Stout, C. R. Russell, and C. E. Rist, *Carbohydr. Res.*, **11**, 321 (1969).

(*2*) A. H. Haines and C. S. P. Jenkins, *Chem. Commun.*, 350 (1969).

(*3*) E. I. Stout, W. M. Doane, B. S. Shasha, C. R. Russell, and C. E. Rist, *Carbohydr. Res.*, **3**, 354 (1967).

(*4*) D. Trimnell, W. M. Doane, C. R. Russell, and C. E. Rist, *Carbohydr. Res.*, **17**, 319 (1971).

(*5*) D. Trimnell, W. M. Doane, C. R. Russell, and C. E. Rist, *Carbohydr. Res.*, **11**, 497 (1969).

(*6*) R. Hull, *J. Chem. Soc. C*, 1777 (1968).

[41] Desulfation of Polysaccharides

By Elizabeth Percival

Department of Chemistry, Royal Holloway College,
(University of London), Egham, Surrey, England

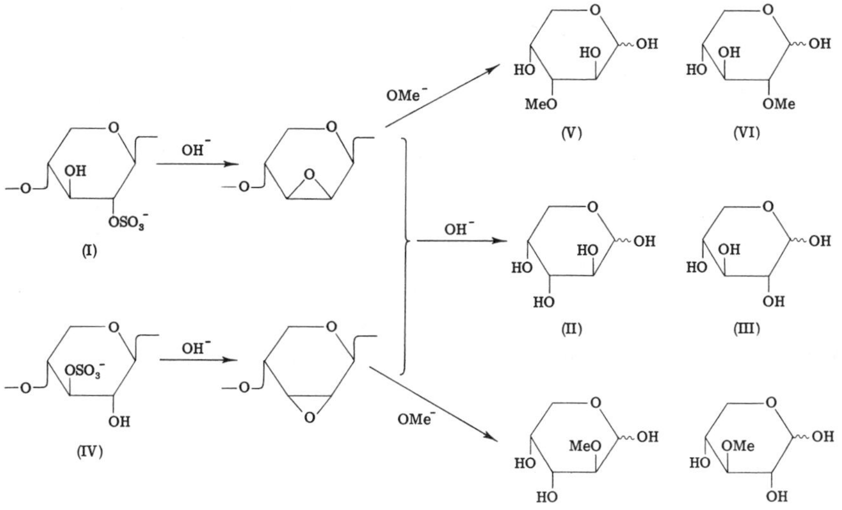

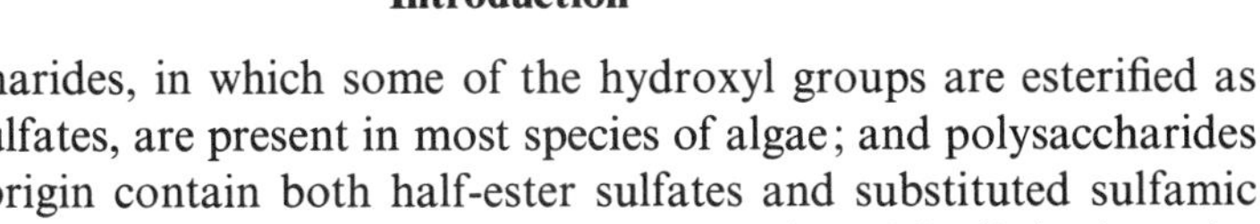

Introduction

Polysaccharides, in which some of the hydroxyl groups are esterified as half-ester sulfates, are present in most species of algae; and polysaccharides of animal origin contain both half-ester sulfates and substituted sulfamic acids. Conditions suitable for hydrolysis of the glycosidic links in polysaccharides generally remove the sulfate groups at the same time, but in order to remove sulfate groups with little or no degradation of the polysaccharides, different methods have to be used depending on the material being studied.

281

Many polysaccharides have sulfate groups on different sugars and in different positions on them, and by the choice of a suitable method of desulfation, it may be possible to remove one type of sulfate selectively.

The greatest difference in the behavior of various types of sulfate groups is found in their susceptibility to hydrolysis by alkali. It has been established (1, 2) from experiments on model monosaccharide sulfates that the sulfate ester groups are extremely stable to hot alkali solutions unless (a) the sulfate group is adjacent and trans to a free hydroxyl group, or (b) the sulfate is linked to C-6 or C-3 and the hydroxyl group on C-3 or C-6 is free.

Desulfation is not a hydrolysis but a nucleophilic attack on a carbon atom by a suitably placed oxygen atom. In the case of adjacent groups, an epoxide ring is first formed. This ring can be broken in two ways giving rise to two sugars. Thus the polysaccharide from *Ulva lactuca* which has $(1 \rightarrow 4)$-linked β-D-xylose groups sulfated in the two-position (I) yields D-arabinose (II) and D-xylose (III) on treatment with sodium hydroxide followed by hydrolysis of the glycosidic linkages (3). The isolation of these two sugars does not, however, prove that the sulfate group is in the C-2 position, for II and III could equally be derived from a D-xylose 3-sulfate residue (IV). However, ring opening with sodium methoxide yields mono-O-methyl sugars, so that the methoxyl group is found in the same position as the sulfate in the original sugar (VI) and in the neighboring position in the new sugar (V).

If there is a free hydroxyl group adjacent and trans to the epoxide ring, ring migration can take place. Thus the formation of trace quantities of D-lyxose in the hydrolysis products of alkali-treated *U. lactuca* polysaccharide was attributed to epxoide ring migration from D-xylose 2-sulfate end groups (3). In a hexose with a free hydroxyl group in the C-6 position, there is also the possibility of 3,6-anhydro ring formation.

D-Galactose residues with sulfate groups in the C-6 position (VII) are found in extracts from red algae; they readily form 3,6-anhydrosugar residues (VIII) on treatment with alkali. In some carrageenan-type products $(1 \rightarrow 4)$-linked D-galactose residues are sulfated at C-6 and C-2; in this case, only the C-6 sulfate group is removed by the treatment with alkali. When polysaccharides are subjected to alkaline conditions, the inclusion of sodium borohydride in the reaction mixture prevents degradation (4).

Very dilute hydrogen chloride in dry methanol removes sulfate ester groups, and if the polysaccharide is stable to acid conditions as are, for example, those containing uronic acid groups, little depolymerization takes place. This procedure was originally used on chondroitin sulfate (5, Vol. V [33]) where repeated treatment with 0.06 M HCl in methanol at 20°–25° completely removed the sulfate groups. With other polysaccharides, trials are necessary to determine conditions which will remove sulfate but are mild enough not to cause depolymerization. Suitable conditions have not been found for the sulfated fucan from brown seaweeds. A modification using Dowex 50(H$^+$) cation-exchange resin and sodium chloride in methanol removed only the equatorial sulfate groups from a carrageenan fraction from *Gigartina skottsbergii* (6).

The acid desulfation is to some extent selective as C-6 sulfate groups are less readily removed (7) than those in other positions.

Heating the pyridinium salts of polysaccharide sulfates in such solvents as pyridine, *p*-dioxane, *N,N*-dimethylformamide, and methyl sulfoxide has, in some cases, given better yields of desulfated polymers than has acid desulfation (8). The solvolytic reaction can be considered to be a reversal of the commonly used method of sulfation of carbohydrates with the sulfur trioxide–pyridine complex in one of these solvents (see, for example, Vol. IV [68]). A comparative study (9) of the desulfation of the acidic polysaccharide from *Acetabularia mediterranea* with methanolic hydrogen chloride and as the pyridinium salt with 5% pyridine in dimethyl sulfoxide gave 73% recovery of carbohydrate and reduction of sulfate from 21.4% to 7.7% under the former conditions and a yield of 90% and a reduction of sulfate to 0.6% with the latter. Inclusion of 5% water or methanol in the dimethyl sulfoxide used in the desulfation of heparin has been found to remove the *N*-sulfate groups while leaving the *O*-sulfate groups combined in the polymer (10, Vol. VIII [42]).

Reductive desulfation of monosaccharide sulfates takes place with lithium aluminum hydride in *p*-dioxane, but reduction has not yet been found effective with polysaccharide sulfates (11).

A number of enzymes which can remove sulfate ester groups from monosaccharides have been found in molluscs, in bacteria, in fungi, and in the livers of fishes and other animals (2), but in general, they have little effect on polysaccharide sulfates. Enzymic desulfation of chondroitin sulfates has been reported (12), but there is considerable concomitant breakdown of the polymer.

Procedure

Removal of Alkali-labile Sulfate Groups

Treatment with Sodium Hydroxide

The following method has been used with the sulfated galactans extracted from red algae [removal of C-6 sulfate (4) to give 3,6-anhydrogalactose residues] and with the polysaccharide from *Ulva lactuca* [removal of C-2 sulfate (*3*) and formation of new sugar residues].

The polysaccharide (3 g) is dissolved in 500 ml of water and 0.4 g of sodium borohydride is added. After storage for 48 h at 20°–25°, 20 g of sodium hydroxide and 3 g of sodium borohydride are added, and the loosely stoppered flask is maintained at ∼80°. After 4 h, an additional 3 g of sodium borohydride is added, and after 10 h, the solution is cooled and made slightly acid with 60 ml of conc. hydrochloric acid. The mixture is dialyzed (Vol. V [15]) for 3 days until free from chloride, concentrated, and freeze-dried (Vol. V [17]); yield from galactans ∼85% and from *Ulva* ∼72% by weight.

The 3,6-anhydrogalactose content is determined and compared with that of the starting material, or the product is hydrolyzed and the sugars present are compared with those from the original polysaccharide.

Treatment with Sodium Methoxide (3)

The sulfated polysaccharide from *Ulva* (2 g) is dried at 60° *in vacuo* over phosphorus pentaoxide for 1 week and soaked in dry methanol (Vol. VII [3]) for a further 2 days. The dried material is rapidly filtered and added to a solution of 0.2 g of lithium borohydride and 6 g of sodium in 250 ml of dry methanol, and the mixture is heated to reflux for 24 h. The insoluble polysaccharide (after filtration and washing with methanol) is hydrolyzed with 0.5 M sulfuric acid at 100° for 4 h. The isolation of 2-*O*-methyl-D-xylose shows the presence of 2-*O*-sulfate in the sulfated polysaccharide.

Desulfation with Dilute Methanolic Hydrogen Chloride (*5*)

Dry chondroitin sulfate (2.4 g) is shaken for 1 day with 400 ml of 0.06 M hydrogen chloride in dry methanol (Vol. VII [3]). The process is repeated with fresh reagent for two additional days. The polysaccharide is collected by filtration, dissolved in water, purified by dialysis, and precipitated by ethanol. An average yield of 80% is obtained. Desulfation is complete.

Solvolytic Desulfation of the Pyridinium Salt (*8*), see also Vol. VIII [41])

The sulfated galactan extracted from *Liangia pacifica* (300 mg) is converted into the acid form by treatment with Dowex 50(H⁺) cation-exchange resin and neutralized with pyridine. The pyridinium salt is freeze-dried and dried overnight at 60° *in vacuo* over phosphorus pentaoxide and then held at 100° in 170 ml of 5:12 v/v dimethyl sulfoxide–pyridine for 9 h. The product is dialyzed and freeze-dried; yield 90%.

References

(*1*) E. G. V. Percival, *Quart. Rev.* (London), **3**, 369 (1949).

(*2*) J. R. Turvey, *Advan. Carbohydr. Chem.*, **20**, 183 (1965).

(*3*) E. Percival and J. K. Wold, *J. Chem. Soc.*, 5459 (1963).

(*4*) D. A. Rees, *J. Chem. Soc.*, 5168 (1961).

(*5*) T. G. Kantor and M. Schubert, *J. Amer. Chem. Soc.*, **79**, 152 (1957).

(*6*) A. S. Cerezo, *Carbohydr. Res.*, **36**, 201 (1974).

(*7*) D. A. Rees, *Biochem. J.*, **88**, 343 (1963).

(*8*) A. I. Usov, K. S. Adamyants, L. I. Miroshnikova, A. A. Shaposhnikova, and N. K. Kochetkov, *Carbohydr. Res.*, **18**, 336 (1971).

(*9*) B. Smestad-Paulsen, unpublished results.

(*10*) Y. Inoue and K. Nagasawa, *Carbohydr. Res.*, **46**, 87 (1976).

(*11*) D. Grant and A. Holt, *Chem. Ind.* (London), 1492 (1959).

(*12*) K. Atsumi, Y. Kawai, N. Seno, and K. Anno, *Biochem. J.*, **128**, 982 (1972).

[42] Desulfation of Glycosaminoglycuronan Sulfates

By KINZO NAGASAWA AND YUKO INOUE

School of Pharmaceutical Sciences, Kitasato University, Tokyo, Japan

Introduction

The conventional methods of desulfation for glycosaminoglycuronan sulfates are mainly based on catalytic desulfation in anhydrous methanolic hydrogen chloride according to the procedure of Kantor and Schubert (1). However, they usually require quite an amount of time for completing the reaction as well as tedious procedures involving repeated desulfation in heterogeneous medium and subsequent saponification (1–3).

Recently, a new method for desulfation of glycosaminoglycuronan sulfates based on the solvolysis in dimethyl sulfoxide containing a small amount of water or methanol has been developed (4, 5). This new method is superior to the conventional method because desulfation takes place with a minimal degree of depolymerization. Furthermore, it is not only simple, but the reaction is completed rapidly in homogeneous medium.

Desulfation of chondroitin sulfate and heparin according to the procedure of Kantor and Schubert has been described in Vol. V [33] and [85], respectively. In this chapter, we shall describe the preparation of chondroitin from chondroitin 4- or 6-sulfates and desulfation of other glycosaminoglycuronan sulfates[1] by solvolysis in dimethyl sulfoxide.

Procedure

Pyridinium Salts of Chondroitin 4-Sulfate and of Other Glycosaminoglycuronan Sulfates

The sodium salt of chondroitin 4-sulfate (1.40 g) is dissolved in 40 ml of water, and the solution is passed through a column of Dowex 50WX8(H^+, 20–50 mesh) at 4°. The effluent and washings are combined, neutralized with

---

[1] Preparation of de-N-sulfated heparin and completely desulfated N-acetylated heparin according to the solvolytic method is described in Vol. VIII [43] of this series.

METHODS IN CARBOHYDRATE
CHEMISTRY, VOL. VIII

TABLE I

Analytical Data and Yields of Desulfated Products of Chondroitin Sulfates

			Desulfated product	
Starting material	$S (\%)^a$	Condition of desulfation	$S (\%)^a$	Yield (%)
Chondroitin 4-sulfate	5.97	10% H_2O in DMSO	1.04	86.0
Chondroitin 4-sulfate	5.97	10% MeOH in DMSO	0.07	82.8
Chondroitin 6-sulfate	6.36	10% H_2O in DMSO	1.05	87.4
Chondroitin 6-sulfate	6.36	10% MeOH in DMSO	0.02	86.6

[a] Sulfur contents were obtained on the sodium salts of samples which had been dried over phosphorus pentaoxide for 2 h at 100° under reduced pressure.

TABLE II

Analytical Data and Yields of Desulfated Products of Glycosaminoglycuronan Sulfates[a]

		Desulfated product	
Starting material	$S (\%)^b$	$S (\%)^b$	Yield
Dermatan sulfate	6.67	1.75	83.9%
Heparin[c]	12.11	6.53	93.0%
N-Acetylated product of N-desulfated heparin[d]	8.07	1.76	97.0%
Whale heparin[e]	8.41	4.23	50 mg → 33 mg
Heparan sulfate[f]	5.33	2.78	50 mg → 36 mg
Keratan polysulfate[g]	7.46	1.16	50 mg → 25 mg

[a] Desulfated with 10% H_2O in DMSO for 5 h at 80°.
[b] See footnote to Table I.
[c] Hog-mucosal heparin, 165 U.S.P. units/mg.
[d] Prepared by N-acetylation of hog-mucosal heparin N-desulfated with 5% H_2O in DMSO for 90 min at 50° (see Vol. VIII [43]).
[e] Prepared from whale intestine, 147 U.S.P. units/mg.
[f] A fraction eluted with 1.25 M NaCl through a column of Dowex-1(Cl⁻).
[g] Prepared from shark cartilage.

pyridine, and lyophilized to give the pyridinium salt of chondroitin 4-sulfate as a white power; yield 1.47 g, 94.5%.[2]

The pyridinium salts of chondroitin 6-sulfate and of other glycosaminoglycuronan sulfates are all obtainable in a comparable yield with that of chondroitin 4-sulfate by the procedure described above.

---

[2] Yield of this product and those of products of subsequent experiments are given as the weight obtained on samples dried with phosphorus pentaoxide overnight at 20°–25° under reduced pressure.

Desulfation of Chondroitin Sulfates and Other Glycosaminoglycuronan Sulfates

The pyridinium salt of chondroitin 4-sulfate (0.53 g) is dissolved in 50 ml of dimethyl sulfoxide containing 10% of water or methanol, and the solution is kept for 5 h at 80°. The reaction mixture is diluted with an equal volume of water and adjusted to pH 9.0–9.5 by the addition of 1 M sodium hydroxide. The solution is dialyzed against running tap water overnight, and then against distilled water for 20 h. Filtration and lyophilization of the dialyzate gives the sodium salt of chondroitin as a white powder.

The pyridinium salts of chondroitin 6-sulfate and other glycosaminoglycuronan sulfates are similarly desulfated with dimethyl sulfoxide containing 10% water or methanol using the procedure described above.

Analytical data and yields of the products are shown in Tables I and II.

References

(1)   T. G. Kantor and M. Schubert, *J. Amer. Chem. Soc.*, **79,** 152 (1957).
(2)   I. Danishefsky, H. B. Eiber, and J. J. Carr, *Arch. Biochem. Biophys.*, **90,** 114 (1960).
(3)   M. L. Wolfrom, J. R. Vercellotti, and G. H. S. Thomas, *J. Org. Chem.*, **29,** 536 (1964).
(4)   K. Nagasawa, Y. Inoue, and T. Kamata, *Carbohydr. Res.*, **58,** 47 (1977).
(5)   K. Nagasawa, Y. Inoue, and T. Tokuyasu, *J. Biochem.* (Tokyo), **86,** 1323 (1979).

[43] De-*N*-sulfation

By Kinzo Nagasawa and Yuko Inoue

School of Pharmaceutical Sciences, Kitasato University, Tokyo, Japan

Introduction

Heparin and heparan sulfate are glycosaminoglycans which contain *N*-sulfate groups (*1*). Previous methods of de-*N*-sulfation have been based on controlled hydrolysis in dilute acid (*2–4*, Vol. V [85]), and it has been suggested that hydrolysis of *N*-sulfate groups in this way is accompanied by some cleavage of the glycosidic linkages and *O*-sulfate groups (*5–7*).

Recently, it has been reported that the pyridinium salt of 2-deoxy-2-sulfamino-D-glucose is desulfated much more rapidly than that of D-glucose 6-sulfate in dimethyl sulfoxide containing a small amount of water or methanol (*8*), and a new method for de-*N*-sulfation of heparin based on this solvolysis has been developed (*8–10*).

Procedure

Pyridinium Salt of Heparin

The sodium salt of purified heparin[1] (3 g) is dissolved in 25 ml of water, and the solution is passed through a column of Dowex 50WX8(H$^+$, 20–50 mesh) at 4°. The effluent and washings are combined, neutralized with pyridine, and lyophilized to give the pyridinium salt of heparin as a white powder; yield 2.9 g.

Completely De-*N*-sulfated Heparin

The pyridinium salt of heparin (2.0 g) is dissolved in 100 ml of dimethyl sulfoxide containing 5% of water, and the solution is kept for 90 min at 50°, then diluted with an equal volume of water. The pH of the solution was adjusted to 9.0–9.5 by the addition of 0.1 M sodium hydroxide. The solution is dialyzed (Vol. V [15]) against running tap water overnight, and then

---

[1] The starting heparin used in these experiments was prepared from porcine intestinal mucosa, and its chemical and biological data are shown in Table I.

METHODS IN CARBOHYDRATE
CHEMISTRY, VOL. VIII

TABLE I

Chemical and Biological Data for Completely or Partially De-N-sulfated Heparins and of Completely Desulfated N-Acetylated Heparins[a]

Compound	Total S[b] (%)	N-Sulfate[c] (%)	Mol			N-Acetyl[d] (mol)	Hexosamine[e] (%)	Anticoagulant activity (I.U.)[f]
			Total S	N-S	O-S			
Heparin	12.13	4.47	2.25	0.83	1.42	0.12	24.6	165
Heparin completely de-N-sulfated at 50° for 90 min	8.87	0.60	1.38	0.07	1.31			<1
Heparin partially de-N-sulfated								
At 20° for 5 min	11.48	4.10	2.06	0.73	1.33			134
At 20° for 10 min	11.35	4.04	2.02	0.72	1.30			127
At 20° for 20 min	11.10	3.75	1.95	0.66	1.29			134
At 20° for 60 min	10.84	3.28	1.88	0.57	1.31			125
At 20° for 120 min	10.55	2.79	1.80	0.48	1.32			102
Completely desulfated, N-acetylated heparin	0.18		0.023			0.99[g]	40.76	

[a] All analytical data were obtained on the samples which had been dried under reduced pressure for 2 h at 100°.

[b] Total sulfate content as sulfur was determined by turbidimetry (12).

[c] N-Sulfate content as sulfur was determined by turbidimetry of the inorganic sulfate liberated after treatment of the samples with nitrous acid (13).

[d] Determined by the nmr method (9).

[e] Determined by a modification of the Elson–Morgan (Vol. I [74], [140]; Vol. VIII [11]) procedure (14), after hydrolysis in 4 M hydrochloric acid for 14 h at 100°.

[f] Determined by the USP method.

[g] Determined by the glc method (10).

against distilled water for 20 h. Filtration and lyophilization of the dialyzate gives the sodium salt of completely de-N-sulfated heparin as a white power; yield 1.62 g.[2]

Analytical data for the product are shown in Table I.

Partially De-N-sulfated Heparin

A solution of the pyridinium salt of heparin (1 g) in 2.5 ml of water is diluted with 47.5 ml of dimethyl sulfoxide below 20°, and the solution is kept for 60 min at 20°. The reaction mixture is treated by the method just described to give 982 mg of the sodium salt of partially de-N-sulfated heparin.

Analytical and biological data of the product are shown in Table I together with those of other partially de-N-sulfated heparins obtained in a similar yields by varying reaction periods.

Completely Desulfated N-Acetylated Heparin

N-Acetylation of completely de-N-sulfated heparin can be done by the method of Danishefsky *et al.* (Vol. V [85], *11*). To 30 ml of an aqueous solution of 300 mg of the sodium salt of completely de-N-sulfated heparin (total S: 8.87%), 5 ml of methanol and 0.9 ml of acetic anhydride are added; and the mixture is kept for 2 h at pH 6.5 by adding 10% sodium hydrogencarbonate under stirring at 3°–4°. The reaction mixture is dialyzed against running tap water overnight, and then against distilled water for 20 h. Filtration and lyophilization of the dialyzate gives the sodium salt of completely de-N-sulfated, N-acetylated heparin as a white powder; yield 270 mg, absorbance (360 nm) of its 2,4,6-trinitrophenylated amino groups 0.0076/mg.

The sodium salt prepared above is converted to the pyridinium salt by the procedure described above. The pyridinium salt (100 mg) is dissolved in 15 ml of dimethyl sulfoxide containing 10% of methanol, and the solution is heated for 2 h at 100° with stirring. The reaction mixture is treated by the procedure described above to give the sodium salt of completely desulfated, N-acetylated heparin as a white powder; yield 63.3 mg.

References

(*1*) R. W. Jeanloz, *Advan. Exp. Med. Biol.*, **52**, 3 (1975).
(*2*) M. L. Wolfrom and W. H. McNeely, *J. Amer. Chem. Soc.*, **67**, 748 (1945).
(*3*) A. B. Foster, E. F. Martlew, and M. Stacey, *Chem. Ind.* (London), 899 (1953).

---

[2] Yield of this product, and those of products of subsequent experiments, is given as the weight including moisture.

(4)   A. G. Lloyd, G. Embery, and L. J. Fowler, *Biochem. Pharmacol.*, **20**, 637 (1971).

(5)   J. R. Helbert and M. A. Marini, *Biochemistry*, **2**, 1101 (1963).

(6)   J. A. Cifonelli, *Carbohydr. Res.*, **2**, 150 (1966).

(7)   A. H. Johnson and J. R. Baker, *Biochim. Biophys. Acta*, **320**, 341 (1973).

(8)   K. Nagasawa and Y. Inoue, *Carbohydr. Res.*, **36**, 265 (1974).

(9)   Y. Inoue and K. Nagasawa, *Carbohydr. Res.*, **46**, 87 (1976).

(10)   K. Nagasawa, Y. Inoue, and T. Kamata, *Carbohydr. Res.*, **58**, 47 (1977).

(11)   I. Danishefsky, H. B. Eiber, and J. J. Carr, *Arch. Biochem. Biophys.*, **90**, 114 (1960).

(12)   K. S. Dodgson, *Biochem. J.*, **78**, 312 (1961); **84**, 106 (1962).

(13)   Y. Inoue and K. Nagasawa, *Anal. Biochem.*, **71**, 46 (1976).

(14)   L. A. Elson and W. T. J. Morgan, *Biochem. J.*, **27**, 1824 (1933).

[44] N-Deacetylation of Polysaccharides

By Lennart Kenne and Bengt Lindberg

Department of Organic Chemistry, Stockholm University,
Stockholm, Sweden

Introduction

Many polysaccharides and glycoconjugates contain 2-acetamido-2-deoxyhexose residues. The corresponding 2-amino-2-deoxyhexose residues, obtained by N-deacetylation, are suitable points of attack in specific degradations, performed as part of the structural work. The 2-amino-2-deoxyhexosidic linkage is resistant to acid-catalyzed hydrolysis; and disaccharides, with the 2-amino-2-deoxy sugar glycosidically linked to the subsequent sugar, are readily isolated. When the amino group occupies an equatorial position in the most stable chain form, as in 2-amino-2-deoxy-D-glucose and -D-galactose residues, deamination by treatment with nitrous acid results in cleavage of the glycosidic linkage with the formation of a 2,5-anhydrohexose residue. The oligosaccharides resulting from such degradations may be further specifically degraded by β-elimination under alkaline conditions. After reduction with borodeuteride, the oligosaccharide and the alkali degradation product can be subjected to methylation analysis for determination of the mode of substitution in the oligosaccharide and the 2,5-anhydrohexitol residue. Examples of the β-elimination and the methylation analysis are given in Reference *1*. Even if the deamination to 2,5-anhydrohexose residues is the main reaction, 2-deoxy-2-formylhexofuranoside residues are also formed (*2*).

N-Deacetylation may be performed with a strong base in aqueous solution or with hydrazine (Vol. VI [34]). Anhydrous hydrazine, and small amounts of hydrazine sulfate, have been reported to be especially efficient (*3–5*). Even with the latter reagent, however, it is sometimes difficult to obtain complete N-deacetylation. Polysaccharides containing 2-acetamido-2-deoxy-D-glucose residues, linked through O-3, have proved to be especially resistant.

With the procedure described below, using sodium hydroxide–sodium thiophenoxide in aqueous dimethyl sulfoxide (*6*), it has been possible to achieve complete N-deacetylation when the other methods have failed. The sodium thiophenoxide acts as an oxygen scavenger and also exerts some catalytic effect. The method has been applied to lipopolysaccharides from

METHODS IN CARBOHYDRATE
CHEMISTRY, VOL. VIII

Shigella, Klebsiella, and *E. coli* and to pneumococcal polysaccharides (*6, 7*). The depolymerization is generally not serious and the material can be recovered by dialysis. The reaction conditions, in order to get complete *N*-deacetylation, may vary considerably for different polysaccharides. For some, 2 h at 80° is sufficient while others may require 15 h at 100°. The lipopolysaccharide from *E. coli* 0:111 was especially resistant and required 15 h at 125°. During these conditions, degradation was extensive, and only 40% of the material was recovered. The reason for these differences in reactivity is not well understood, but steric effects may play an important role. In the [^{1}H] nmr spectrum, *N*-acetyl groups give a signal at approximately δ 2.1, which should consequently be absent in the *N*-deacetylated product.

Procedure

The polysaccharide (10–50 mg), 400 mg of sodium hydroxide, and 100 mg of benzenethiol are dissolved in 1 ml of water. Dimethyl sulfoxide (5 ml) is added, and the stirred mixture is heated under nitrogen in a sealed tube or serum vial. A reaction temperature of 80°–100°, and a reaction time between 2 and 15 h is generally sufficient, and the mildest possible conditions should be aimed at. After complete reaction, the solution is cooled, centrifuged, dialyzed against tap water for 24 h, centrifuged, and lyophilized. The recovery is high, generally almost quantitative.

Nmr of original and modified polysaccharide in deuterium oxide, preferably using PFT, gives information on the completeness of the reaction.

References

(*1*) L. Kenne, B. Lindberg, K. Petersson, and E. Romanowska, *Carbohydr. Res.*, **56**, 363 (1977).
(*2*) C. Erbing, B. Lindberg, and S. Svensson, *Acta Chem. Scand.*, **27**, 3699 (1973).
(*3*) Z. Yosizawa, T. Sato, and K. Schmid, *Biochim. Biophys. Acta*, **121**, 417 (1966).
(*4*) B. A. Dmitriev, Yu. A. Knisel, and N. K. Kochetkov, *Carbohydr. Res.*, **29**, 451 (1973).
(*5*) B. A. Dmitriev, Yu. A. Knirel, and N. K. Kochetkov, *Carbohydr. Res.*, **30**, 45 (1973).
(*6*) C. Erbing, K. Granath, L. Kenne, and B. Lindberg, *Carbohydr. Res.*, **47**, C5 (1976).
(*7*) Unpublished results from the authors' laboratory.

[45] N-Trifluoroacetylation

Methyl 6-Deoxy-6-trifluoroacetamido-α-D-glucopyranoside and Its N-Detrifluoroacetylation

BY ROY L. WHISTLER AND ABUL KASHEM M. ANISUZZAMAN

Department of Biochemistry, Purdue University, Lafayette, Indiana

Introduction

Aminodeoxy sugars are constituents of many biologically important and medicinal compounds. In the synthetic field, it is often necessary to protect the amino group selectively by acylation under conditions that do not affect the hydroxyl groups. Conversion of the amino group to the N-acetamido or N-benzamido group has the disadvantage that these groups are difficult to remove, usually requiring drastic acid or alkaline hydrolysis. Moreover, these groups have a tendency to participate in displacement reactions (1). However, N-trifluoroacetyl derivatives are easy to prepare, and this group is conveniently removed without neighboring group participation. The N-trifluoroacetyl group may be introduced through reaction with trifluoroacetic anhydride. Unfortunately, racemization of an optically active center

297

may take place in the presence of excess of the reagent (*2*), and the generated trifluoroacetic acid may bring about secondary changes such as cleavage of the glycosidic linkage or *O*-isopropylidene groups (*3*). A mild, but efficient, method for *N*-trifluoroacetylation makes use of *S*-ethyl trifluorothioacetate (*2*). The reagent was first used to block amino groups in amino acids (*2*) and later adopted to carbohydrates (*4*, *5*). Acylation can occur in a variety of solvents (*2*, *4–7*) including water (*5*, *6*).

The *N*-trifluoroacetyl group is easily removed at ambient temperatures by hydrolysis with aqueous sodium hydroxide (*2*), methanolic barium methoxide (*6*), methanolic ammonia (*4*), and Dowex 1-X-8(OH⁻) resin (*5*). A method for the preparation of methyl 6-deoxy-6-trifluoroacetamido-α-D-glucopyranoside using *S*-ethyl trifluorothioacetate and the hydrolysis of the acetamido derivative to methyl 6-amino-6-deoxy-α-D-glucopyranoside illustrates the application of the reagent.

The starting amino compound is prepared from the corresponding 6-halogen derivative by displacement with azido anion followed by catalytic hydrogenation.

Procedures

Methyl 6-Deoxy-6-trifluoroacetamido-α-D-glucopyranoside

Methyl 6-Azido-6-deoxy-α-D-glucopyranoside (II)

Methyl 6-bromo-6-deoxy-α-D-glucopyranoside (Vol. VIII [30]) (2.57 g) is dissolved in 50 ml of *N,N*-dimethylformamide, and 1.8 g of sodium azide is added. After heating the mixture at 105° for 18 h, solvent is removed as the azeotrope with xylene (150 ml) by evaporation under reduced pressure. The residue is then extracted with dry acetone; evaporation of the filtered extract gives syrupy methyl 6-azido-6-deoxy-α-D-glucopyranoside (II); yield 2.1 g (98%). This product is homogeneous by tlc[1] with solvent A (6:1 v/v chloroform–methanol; R_f 0.45). Acetylation of 220 mg of II with acetic anhydride in pyridine gives methyl 2,3,4-tri-*O*-acetyl-6-azido-6-deoxy-α-D-glucopyranoside; yield 326 mg (95%), mp 103°.

Methyl 6-Amino-6-deoxy-α-D-glucopyranoside (III)

Compound II (2.2 g) in 100 ml of ethanol containing 2.5 g of Raney nickel[2] is hydrogenated at 25° and 3 atm pressure for 12 h. The mixture is

---

[1] Tlc is performed on silica gel G (E. Merck, Darmstadt, Germany) coated glass plates.

[2] Nickel catalyst grade 986 is obtained from Davison Chemical, Div. W. R. Grace and Company, Baltimore Maryland.

filtered and the filtrate is evaporated to dryness to give III as a syrup; yield 1.8 g (95%). This material is homogeneous by tlc[1] with solvent B (1:1 v/v chloroform–methanol; R_f 0.15). On reaction with hydrochloric acid, III gives crystalline hydrochloride; mp 195°–200° (decomp).

Methyl 6-Deoxy-6-trifluoroacetamido-α-D-glucopyranoside (IV)

Compound III (1.93 g) is dissolved in 25 ml of methanol to which 4 ml of *S*-ethyl trifluorothioacetate (Aldrich Chemical Co., Inc., Milwaukee, Wisconsin) is added. The resulting solution is kept at 25° for 18 h, when tlc[1] with solvent B indicates complete conversion of III to IV. The solution is diluted with 30 ml of water and extracted with two 30-ml portions of hexane. The aqueous layer is evaporated under reduced pressure to a crystalline residue which is crystallized from ether; yield of IV 2.6 g (90%), mp 192°, $[\alpha]^{25}$D +121.5° (*c* 1.1, methanol), R_f 0.90 (solvent B).

Hydrolysis of IV

To a solution of IV (290 mg) in 10 ml of methanol is added 10 ml of 0.1 *M* methanolic barium methoxide. After 18 h at 25°, the solution is evaporated to a residue which is mixed with 10 ml of water and brought to neutrality with carbon dioxide. The mixture is then filtered through a pad of Celite, and the filtrate is passed through a 15-cm × 10-mm column of Amberlite 1R-45(OH⁻) resin to remove trifluroacetic acid. The eluent fractions containing III are evaporated to a syrup; yield 189 mg (98%). This material is homogeneous by tlc[1] with solvent B and identical with III prepared from II.

References

(1) S. Hanessian, *J. Org. Chem.*, **32**, 163 (1967).
(2) E. E. Schallenberg and M. Calvin, *J. Amer. Chem. Soc.*, **77**, 2779 (1955).
(3) M. Chmielewski and R. L. Whistler, *J. Org. Chem.*, **40**, 639 (1975).
(4) M. L. Wolfrom and P. J. Conigliaro, *Carbohydr. Res.*, **11**, 63 (1969).
(5) C. Chipowski and Y. C. Lee, *Carbohydr. Res.*, **31**, 339 (1973).
(6) R. Barker, K. W. Olsen, J. H. Sharper, and R. L. Hill, *J. Biol. Chem.*, **247**, 7135 (1972).
(7) D. Levy and R. A. Paselk, *Biochim. Biophys. Acta*, **310**, 398 (1973).

[46] Nucleophilic Reaction of the Tosyl Group in Enolic Sugar Derivatives

By J. Lehmann and M. Brockhaus

Chemisches Laboratorium der Universität Freiburg, Freiburg, Germany

Introduction

The possibility of activating a leaving group such as a methylsulfonyl (mesyl) group or a tolyl-*p*-sulfonyl (tosyl) group by deliberately placing it in an allylic position of an unsaturated substrate is rearely used in preparative carbohydrate chemistry (*1, 2*). Cases are known where the ease of a displacement is explained by the formation of an intermediate allylic tosylate (*3*). The method of Helferich and Himmen (*4*) to introduce enolic double bonds under very mild conditions by shaking 6-iodo or 6-bromo hexoside derivatives with silver fluoride in pyridine enables one to obtain an allyl-activated 4-*O*-tosyl derivative of any hexopyranoside from the corresponding 2,3-di-*O*-acetyl-6-deoxy-6-halo-4-*O*-(tolyl-*p*-sulfonyl)-hexopyranoside (*5*). Methyl 2,3-di-*O*-acetyl-6-deoxy-4-*O*-(tolyl-*p*-sulfonyl)-β-D-*xylo*-hex-5-enopyranoside (II) obtained from methyl 2,3-di-*O*-acetyl-6-deoxy-6-iodo-4-*O*-(tolyl-*p*-

METHODS IN CARBOHYDRATE
CHEMISTRY, VOL. VIII

sulfonyl)-β-D-glucopyranoside (I) reacts with ease (6). Solvolysis in aqueous pyridine yields the 4-epimerized intermediates IIIa and IIIb which can be acetylated to give methyl 2,3,4-tri-O-acetyl-6-deoxy-α-L-*arabino*-hex-5-eno-pyranoside (IV) (7). II is also readily converted into methyl 3,4-anhydro-6-deoxy-α-L-*arabino*-hex-5-enopyranoside (V) which is an excellent substrate for regio- and stereo-selective ring opening by many nucleophiles under very mild reaction conditions (8). The place of attack is always the allylic carbon atom. The primary product may either be stable (VIII) or undergo subsequent rearrangement until a stable product is formed (IX, X). Treatment with aqueous sodium cyanide or sodium azide solution gives immediately unstable products, VI and VII respectively. VI rearranges into stable methyl 4-cyano-4,6-dideoxy-α-L-*threo*-hex-4-enopyranoside (IX), and VII rearranges into the equally stable methyl 6-azido-4,6-dideoxy-α-L-*threo*-hex-4-enopyranoside (X). The reaction of V with aqueous ammonia yields methyl 4-amino-4,6-dideoxy-β-D-*xylo*-hex-5-enopyranoside (VIII).

Procedure

Methyl 2,3-Di-O-acetyl-6-deoxy-6-iodo-4-O-(tolyl-p-sulfonyl)-β-D-glucopyranoside (I) (5)

Methyl 2,3-di-O-acetyl-4,6-di-O-(tolyl-p-sulfonyl)-β-D-glucopyranoside (10.0 g) (5) and 3 g of sodium iodide are stirred in 50 ml of acetic anhydride at 120°. After 20 min, the mixture is cooled to 20°–25°, diluted with 200 ml of acetone, and filtered with suction. The filtrate is evaporated under diminished pressure, and the residue is dissolved in 250 ml of ether. The ether solution is washed with two 100-ml portions of aqueous sodium hydrogencarbonate solution to which ~ 100 mg of sodium bisulfite has been added. The almost colorless ether solution is washed with three 100-ml portions of water, dried with anhydrous magnesium sulfate, and evaporated to about 15 ml. The product crystallizes when petroleum ether (60°–70°) is added dropwise. Recrystallization from ether–petroleum ether (60°–70°) gives I; yield 83%, mp 157°, $[\alpha]_{578}^{25}$ $-21°$ (c 1.0, chloroform).

Methyl 2,3-Di-O-acetyl-6-deoxy-4-O-(tolyl-p-sulfonyl)-β-D-*xylo*-hex-5-enopyranoside (II) (6)

Methyl 2,3-di-O-acetyl-6-deoxy-6-iodo-4-O-(tolyl-p-sulfonyl)-β-D-gluco-pyranoside (I) (5 g) is dissolved in 30 ml of anhydrous pyridine (Vol. II [43], [53], [63], [73]; IV [73]; VII [2]) and shaken vigorously at 20°–25° for 3 h with 5 g of crude commercial silver fluoride (Vol. II [105]). The dark slurry is then added dropwise under vigorous mechanical stirring to 1 liter of ether.

The ether solution is decanted from the tarlike residue, washed with 200 ml of a 2% aqueous sodium thiosulfate solution and then with three 150-ml portions of water. The organic layer is dried with anhydrous magnesium sulfate to which a little activated charcoal is added. On evaporation of the filtrate, the product crystallizes. Recrystallization from ether–petroleum ether (60°–70°) gives II; yield 66%, mp 90°, $[\alpha]_{578}^{25}$ $-62°$ (c 1.0, chloroform).

Methyl 2,3,4-Tri-O-acetyl-6-deoxy-α-L-arabino-hex-5-enopyranoside (IV) (7)

Methyl 2,3-di-O-acetyl-6-deoxy-4-O-(tolyl-p-sulfonyl)-β-D-xylo-hex-5-enopyranoside (II) (2.5 g) is dissolved in a solution of 10 ml of pyridine and 10 ml of water and stirred for 8 h at 35°. The reaction mixture is then evaporated under reduced pressure, and the residual mixture is acetylated with 10 ml of acetic anhydride in 15 ml of pyridine for 10 h (Vol. II [53]). After evaporation under diminished pressure, the oily residue is dissolved in 100 ml of ether and washed with 50 ml of saturated aqueous sodium hydrogen carbonate solution, then with two 50-ml portions of water. The organic layer is dried with anhydrous magnesium sulfate and then reduced to a volume of about 10 ml. On dropwise addition of n-hexane, crystalline IV starts separating. Recrystallization from ether–petroleum ether (60°–70°) gives pure IV; yield 84%, mp 83°, $[\alpha]_{578}^{25}$ $-46°$ (c 1.0, chloroform).

Methyl 3,4-Anhydro-6-deoxy-α-L-arabino-hex-5-enopyranoside (V) (6)

To a solution of 2.0 g of methyl 2,3-di-O-acetyl-6-deoxy-4-O-(tolyl-p-sulfonyl)-β-D-xylo-hex-5-enopyranoside (II) in 25 ml of dry acetone (Vol. II [83]), ~5 ml 1 M sodium methoxide in methanol is added dropwise until the solution remains slightly alkaline (an excess has to be avoided). Precipitated sodium toluene-p-sulfonate is removed by filtration, and the filtrate is evaporated under diminished pressure. The residue is extracted with two 400-ml portions of petroleum ether (60°–70°). On cooling, pure V crystallizes; yield 81%, mp 90°–91°, $[\alpha]_{578}^{25}$ $-180°$ (c 1.0, chloroform).

Methyl 4-Cyano-4,6-dideoxy-α-L-threo-hex-4-enopyranoside (IX) (8)

Methyl 3,4-anhydro-6-deoxy-α-L-arabino-hex-5-enopyranoside (V) (508 mg) and 2.0 g of sodium cyanide are dissolved in 10 ml of water, and the solution is stored for 3 h at 20°–25°. The aqueous solution is then extracted continuously for 48 h with 100 ml of dichloromethane. The organic layer is dried with anhydrous magnesium sulfate and evaporated under diminished pressure. The crystalline residue is recrystallized from ether–petroleum ether (60°–70°) to give pure IX; yield 78%, mp 112° $[\alpha]_{578}^{25}$ $-57°$ (c 1.0, chloroform).

Methyl 6-Azido-4,6-dideoxy-α-L-*threo*-hex-4-enopyranoside (X) (*8*)

Methyl 3,4-anhydro-6-deoxy-α-L-*arabino*-hex-5-enopyranoside (V) (1.1 g) and 10 g of sodium azide are dissolved in 40 ml of water and stored for 1 h at 20°–25°. The reaction solution is then extracted continuously for 48 h with 150 ml of dichloromethane. The organic solution is dried with magnesium sulfate and evaporated under diminished pressure. The remaining syrup crystallizes on scratching after adding a few drops of ether. Recrystallization from ether–petroleum ether (60°–70°) affords pure X; yield 75%, mp 71°–72°, $[\alpha]_{578}^{25}$ $-178.5°$ (c 1.0, chloroform).

Methyl 4-Amino-4,6-dideoxy-β-D-*xylo*-hex-5-enopyranoside (VIII) (*8*)

Methyl 3,4-anhydro-6-deoxy-α-L-*arabino*-hex-5-enopyranoside (V) (500 mg) is dissolved in 5 ml of conc. aqueous ammonia and kept at 20°–25° for 8 h. On evaporation under diminished pressure, a syrup is obtained which crystallizes on scratching with a little methanol. Recrystallization from methanol–benzene gives pure VIII; yield 64%, mp 124°, $[\alpha]_{578}^{25}$ $-104.5°$ (c 1.0, ethanol).

References

(*1*) J. Lehmann, *Angew. Chem., Int. Ed. Engl.*, **4**, 874 (1965).
(*2*) D. M. Ciment, R. J. Ferrier, and W. G. Overend, *J. Chem. Soc. C.* 446 (1966).
(*3*) A. K. Al-Radhi, J. S. Brimacombe, L. C. N. Tucker, and O. A. Chiny, *J. Chem. Soc C*, 2305 (1971).
(*4*) B. Helferich and E. Himmen, *Ber. Deut. Chem. Ges.*, **61**, 1825 (1925).
(*5*) J. W. Oldham and J. K. Rutherford, *J. Amer. Chem. Soc.*, **54**, 366 (1932).
(*6*) M. Brockhaus and J. Lehmann, *Justus Liebigs Ann. Chem.*, 1675 (1974).
(*7*) J. Lehmann and W. Weckerle, *Carbohydr. Res.*, **22**, 23 (1972).
(*8*) M. Brockhaus, W. Gorath, and J. Lehmann, *Justus Liebigs Ann. Chem.*, 89 (1976).

ETHERS

[47] Allyl Ethers as Protecting Groups

Synthesis of 3-*O*-[6-*O*-(α-D-Galactopyranosyl)-β-D-galacto-pyranosyl]-1,2-di-*O*-stearoyl-L-glycerol, a "Digalactosyl Diglyceride"

By Patricia A. Manthorpe (née Gent) and Roy Gigg

Laboratory of Lipid and General Chemistry, National Institute for Medical Research, London, England

Introduction

The rearrangement of allyl (prop-2-enyl) ethers to *cis*-prop-1-enyl ethers by strong bases has been investigated by several workers (*1*) and was shown in 1961 to be particularly rapid and quantitative with potassium *t*-butoxide in dimethyl sulfoxide (*2*). The application of allyl ethers as protecting groups, using this facile rearrangement and subsequent cleavage of the labile prop-1-enyl group by dilute acid hydrolysis, ozonolysis, or oxidation with alkaline permanganate, was introduced by our laboratory in 1964 (*3–5*). It was later shown that the prop-1-enyl group could be removed under neutral conditions using mercury(II) chloride in the presence of mercury(II) oxide (*6*). Other methods for the removal of allyl groups are as follows: (a) by the action of selenium dioxide (*7*), (b) by isomerization of the allyl group with tris(triphenylphosphine)rhodium(I) chloride and subsequent hydrolysis of the prop-1-enyl group as described above (*8*), (c) by the addition of diethylazodicarboxylate to the allyl ether to give a vinyl ether which can be hydrolyzed as described above (*9, 10*), (d) by sodium in liquid ammonia (*11, 12*), (e) by the action of organometallic reagents (*13, 14*), and (f) by isomerization of the allyl group with palladium on carbon and subsequent hydrolysis of the prop-1-enyl group (*15*).

We showed (*16*) that 3-methylallyl (but-2-enyl) ethers were cleaved by the action of potassium *t*-butoxide in dimethyl sulfoxide (to regenerate the alcohol) at a higher rate than the allyl ethers were isomerized and that the 1-methylallyl (*16*) and 2-methylallyl (*17*) ethers were isomerized more slowly than the allyl ethers under these conditions. We have found these protecting groups particularly useful in the preparation (*18, 19*) of specifically benzylated carbohydrate derivatives, which are valuable intermediates for the

METHODS IN CARBOHYDRATE
CHEMISTRY, VOL. VIII

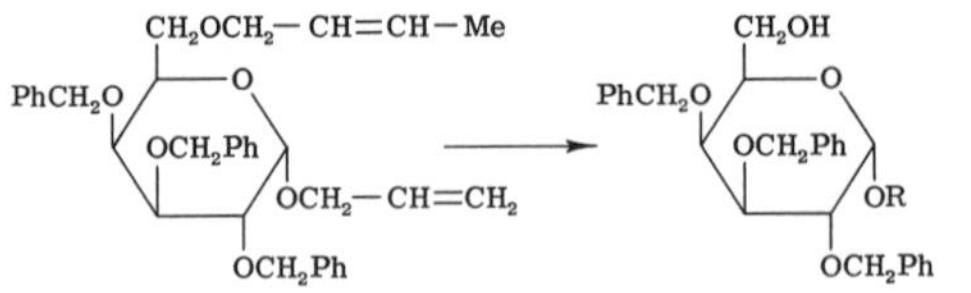

CH_2OCH_2—CH=CH—Me

$PhCH_2O$

OCH_2Ph

OCH_2—CH=CH$_2$

OCH_2Ph

(I)

CH_2OH

$PhCH_2O$

OCH_2Ph

OR

OCH_2Ph

(II, R = CH_2—CH=CH$_2$)
(III, R = CH=CH—Me)

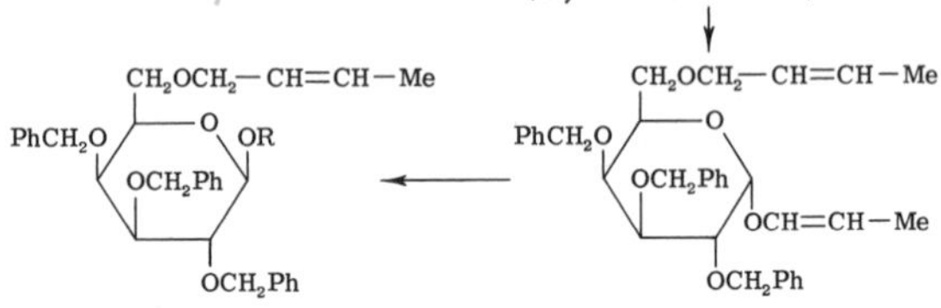

CH_2OCH_2—CH=CH—Me

$PhCH_2O$

OR

OCH_2Ph

OCH_2Ph

(V, R = H)
(VI, R = COPh[p-NO$_2$])

CH_2OCH_2—CH=CH—Me

$PhCH_2O$

OCH_2Ph

OCH=CH—Me

OCH_2Ph

(IV)

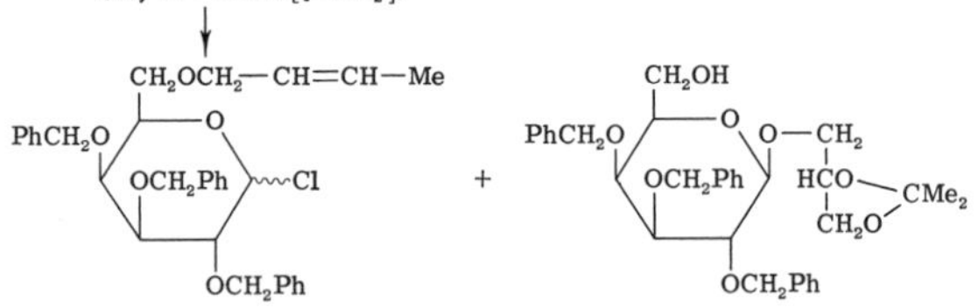

CH_2OCH_2—CH=CH—Me

$PhCH_2O$

OCH_2Ph

Cl

OCH_2Ph

(VII)

+

CH_2OH

$PhCH_2O$

OCH_2Ph

O—CH_2
HCO
CH_2O
CMe$_2$

OCH_2Ph

(VIII)

CH_2OR

$PhCH_2O$

OCH_2Ph

OCH_2Ph

O—CH_2

$PhCH_2O$

OCH_2Ph

O—CH_2
HCO
CH_2O
CMe$_2$

OCH_2Ph

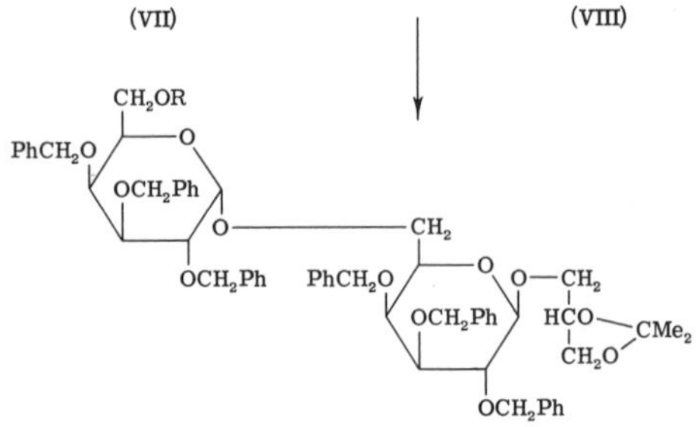

(IX, R = CH_2—CH=CH—Me)
(X, R = H)
(XI, R = CH_2Ph)

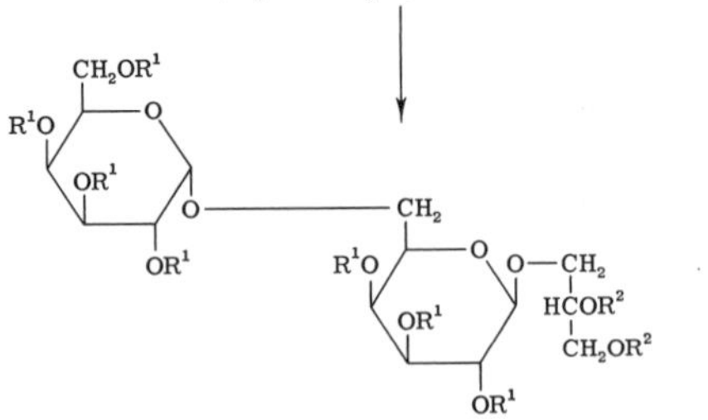

CH_2OR^1

R^1O

OR^1

OR^1

CH_2

R^1O

OR^1

O—CH_2
HCOR^2
CH_2OR^2

OR^1

(XII, R^1 = CH_2Ph; R^2 = H)
(XIII, R^1 = CH_2Ph; R^2 = CO[CH_2]$_{16}$Me)
(XIV, R^1 = H; R^2 = CO[CH_2]$_{16}$Me)

synthesis of glycosides, and have developed methods for the syntheses of di- (*19–23*) and oligosaccharides (*24, 25*) using partially allylated and benzylated carbohydrate derivatives. The 2-*O*-allyl group (like the 2-*O*-benzyl group) serves as a suitable nonparticipant for 1,2-*cis*-glycoside synthesis (*21, 22*).

Our particular interest in the use of these derivatives for the preparation of oligosaccharides is in the syntheses of glycolipids derived from microorganisms and mammalian tissues (*26*). As an example of the use of the allyl and but-2-enyl protecting groups, we describe the synthesis of a digalactosyl diglyceride (XIV) by a method which should be generally useful for the preparation of related compounds.

Allyl 2,3,4-tri-*O*-benzyl-6-*O*-(but-2-enyl)-α-D-galactopyranoside (I) (*20*) was treated with potassium *t*-butoxide in dimethyl sulfoxide which removed the but-2-enyl group and isomerized the allyl group to give prop-1-enyl 2,3,4-tri-*O*-benzyl-α-D-galactopyranoside (III). This compound was treated with "crotyl bromide" and sodium hydride to give the but-2-enyl ether (IV), and the prop-1-enyl group was subsequently hydrolyzed with dilute acid to give the free sugar (V). Compound V was converted *via* the crystalline *p*-nitrobenzoate (VI) into the glycosyl chloride (VII) which was condensed with a derivative (VIII) (*27*) of β-D-galactopyranosylglycerol to give the crystalline digalactosylglycerol derivative IX in which the terminal galactose residue is joined in an α-D linkage. Removal of the but-2-enyl group from compound IX gave the alcohol X which was converted into the benzyl ether XI. Hydrolysis of the isopropylidene group from compound XI, followed by acylation and hydrogenolysis of the benzyl groups, gave the digalactosyl diglyceride XIV.

Digalactosyl diglycerides were first isolated from plants but are now known to be present in various microorganisms and in mammalian tissues (*26*). Compound X also served as an intermediate for the synthesis of a trigalactosyl diglyceride (*25*) by the condensation of another molecule of the chloride VII.

Procedures

2,3,4-Tri-*O*-benzyl-6-*O*-(but-2-enyl)-1-*O*-(*p*-nitrobenzoyl)-β-D-galactopyranose (VI) (*20, 27*)

Allyl 2,3,4-tri-*O*-benzyl-6-*O*-(but-2′-enyl)-α-D-galactopyranoside (I) (*18*) (10 g) is treated with 10 g of potassium *t*-butoxide in 200 ml of dry dimethyl sulfoxide[1] at 50°. The reaction followed by tlc on silica gel using an irrigant

---

[1] Vol. VI [64]; Vol. VII [26]; Vol. VIII [6].

of 1:1 v/v ether–light petroleum[2] which indicates rapid conversion of compound I (R_f 0.8) into the alcohol II (R_f 0.3) and the subsequent conversion of compound II into the prop-1-enyl glycoside (III) (R_f 0.4). When the isomerization is complete (2 h), the solution is cooled and diluted with 200 ml of water, and compound III is extracted with ether; yield 8 g. Compound III is then treated with 5 ml of a mixture of 1-bromobut-2-ene and 3-bromobut-1-ene ('crotyl bromide') (14) and 5 g of sodium hydride in benzene under reflux until tlc (as above) shows complete conversion of compound III into the but-2-enyl ether IV (R_f 0.9). Methanol is then added to destroy the excess of sodium hydride, and the solution is washed with water, dried with potassium carbonate, and evaporated to give the glycoside IV as a syrup. A solution of the product in 100 ml of 9:1 v/v acetone–1 M HCl is heated under reflux for 15 min to remove the prop-1-enyl group. Sodium hydrogencarbonate is added to neutralize the acid; the solvents are removed by evaporation, and the 2,3,4-tri-O-benzyl-6-O-(but-2-enyl)-D-galactopyranose (V) is extracted with chloroform; yield 7 g. A solution of 1.8 g of p-nitrobenzoyl chloride in 4.6 ml of dry dichloromethane is added to a solution of 3 g of compound V in 23 ml of dry dichloromethane containing 1 ml of dry pyridine.[3] The mixture is stirred at 20° for 6 h when tlc (as above) shows complete conversion of compound V (R_f 0.4) into the product (R_f 0.8). Water (1 ml) is added, and the solution is stirred at 20° for 2 h to convert the excess of chloride into p-nitrobenzoic acid, and the solution is then washed with water, 1 M HCl, and sodium hydrogencarbonate solution and dried with magnesium sulfate. The product is crystallized from ethanol to give 2,3,4-tri-O-benzyl-6-O-(but-2'-enyl)-1-O-p-nitrobenzoyl-β-D-galactopyranose (VI); yield 2.3 g, mp 94°–97°, [α]D −40° (c 0.9, CHCl$_3$).

2,3,4-Tri-O-benzyl-6-O-(but-2-enyl)-D-galactopyranosyl Chloride (VII) (20)

Dry hydrogen chloride is passed through a solution of 2.25 g of the p-nitrobenzoate (VI) in 20 ml of dry ether and 30 ml of dry dichloromethane for 4 h. Tlc on silica gel using an irrigant of 1:2 v/v ether–light petroleum[2] then indicates complete conversion of compound VI (R_f 0.3) into a product (R_f 0.7). The solvents are removed by evaporation; 25 ml of dichloromethane is added, and the solution is washed with saturated sodium hydrogencarbonate solution to remove the p-nitrobenzoic acid and traces of hydrogen chloride. The solution is dried with magnesium sulfate and evaporated to give the chloride VII as a syrup.

[2] Vol. VI [6].

[3] Vol. II [43], [53], [63], [73]; Vol. IV [73]; Vol. VII [2].

3-O-[2,3,4-Tri-O-benzyl-6-O-(2,3,4-tri-O-benzyl-6-O-(but-2'-enyl)-α-D-galactopyranosyl)-β-D-galactopyranosyl]-1,2-O-isopropylidene-L-glycerol (IX) (*24*)

3-O-(2,3,4-Tri-O-benzyl-β-D-galactopyranosyl)-1,2-O-isopropylidene-L-glycerol (VIII) (*27*) (1.1 g), 2.1 g of 2,3,4-tri-O-benzyl-6-O-(but-2-enyl)-D-galactopyranosyl chloride (VII), 0.72 g of dry tetraethylammonium chloride, and 1.2 ml of dry triethylamine in 20 ml of dry dichloroethane are heated at the reflux temperature under dry nitrogen for 17.5 h. Tlc on silica gel using an irrigant of 2:1 v/v ether–light petroleum[2] then indicates the presence of a small amount of the alcohol VIII (R_f 0.1) and the chloride VII (R_f 0.85), a major product (R_f 0.6) and minor products (R_f 0.75 and 0.5) together with some free sugar derived from hydrolysis of the chloride VII (R_f 0.4). Water (0.5 ml) is added, and the mixture is heated under reflux for 45 min, cooled, washed with water, and dried with magnesium sulfate; yield 3.6 g.

The crude product is chromatographed on alumina. Elution with 1:1 v/v ether–light petroleum removes the by-product (R_f 0.75), and elution with 2:1 v/v ether–light petroleum gives the product (IX) (1.3 g) (R_f 0.6) which is recrystallized from light petroleum (bp 60°–80°); mp 94°–95°, [α]D +33.3° (*c* 0.5, CHCl₃).

3-O-[2,3,4-Tri-O-benzyl-6-O-(2,3,4-tri-O-benzyl-α-D-galactopyranosyl)-β-D-galactopyranosyl]-1,2-O-isopropylidene-L-glycerol (X) (*24*)

A solution of 1 g of compound IX and 1 g of potassium *t*-butoxide in 20 ml of dry dimethyl sulfoxide[1] is kept at 20° for 4 h when tlc on silica gel using an irrigant of 2:1 v/v ether–light petroleum[2] shows complete conversion of compound IX (R_f 0.6) into a single product (R_f 0.1). Water (100 ml) is added, and the product is isolated by extraction with ether. The ether solution is dried with potassium carbonate. The product is recrystallized from ethyl acetate–light petroleum (bp 60°–80°) to give compound X; yield 0.54 g, mp 132°–134°, [α]D +24.3° (*c* 0.5, CHCl₃).

3-O-[6-O-(α-D-Galactopyranosyl)-β-D-galactopyranosyl]-1,2-di-O-octadecanoyl-L-glycerol (XIV) (*24*)

Compound X (0.33 g) is treated with 0.5 ml of benzyl chloride and 0.5 g of sodium hydride in 20 ml of dry benzene under reflux for 2.5 h; tlc on silica gel using an irrigant of 2:1 v/v ether–light petroleum[2] indicates complete conversion of compound X (R_f 0.2) into the perbenzyl ether XI (R_f 0.65). Methanol is added to destroy the excess of sodium hydride, and the solution is washed with water, dried with potassium carbonate, and evaporated. The product (0.4 g) is heated under reflux in 10 ml of 1:9 v/v 1 M HCl–dioxane for 10 min when tlc on silica gel using an irrigant of 4:1 v/v toluene–acetone[2] indicates complete conversion of compound XI (R_f 0.8) into the diol XII (R_f 0.3). An excess of sodium hydrogencarbonate is added; the solvents are removed by evaporation, and the product (XII) is extracted from the residue with chloroform. A solution of 0.35 g of compound XII in 10 ml of dry pyridine[3] and 0.5 ml of octadecanoyl chloride is kept at 20° for 4.5 h; tlc (as above) then indicates complete conversion of compound XII into a single product (R_f 0.9). Water (0.5 ml) is added, and the mixture is stirred at 20° for 1 h. The solution is poured into 200 ml of 1 M hydrochloric acid, and the products are removed by extraction with ether. The extract is washed with water and dried with magnesium sulfate. The crude product (XIII, contaminated with stearic acid) is chromatographed on neutral alumina. Elution with 1:1 v/v toluene–ether gives the pure ester XIII; yield 0.33 g. Compound XIII is dissolved in 10 ml of 1:1 v/v chloroform–methanol and treated with hydrogen over 10% palladium on charcoal (previously washed with acetic acid to neutralize basic components which can hydrolyze the ester bonds) at atmospheric pressure until tlc on silica gel using an irrigant of 2:1 v/v chloroform–methanol[2] shows the presence of a major product (R_f 0.8) and only traces of other products. The crude product (0.2 g) is chromatographed on silica gel. Elution with 20:1 v/v chloroform–methanol removes some less polar contaminants and elution with 10:1 v/v chloroform–methanol gives the major product (110 mg) which is recrystallized from methanol to give the digalactosyl diglyceride XIV; softening point 77°, forms a meniscus at 225°–230°; [α]D +36.3° (c 1, pyridine).

References

(1)   T. J. Prosser, *J. Amer. Chem. Soc.*, **83**, 1701 (1961) and references cited therein.

(2)   C. C. Price and W. H. Snyder, *J. Amer. Chem. Soc.*, **83**, 1773 (1961).

(3)   J. Cunningham, R. Gigg, and C. D. Warren, *Tetrahedron Lett.*, 1191 (1964).

(4)   R. Gigg and C. D. Warren, *J. Chem. Soc.*, 2205 (1965).

(5)   J. Gigg and R. Gigg, *J. Chem. Soc. C*, 82 (1966).

(6)   R. Gigg and C. D. Warren, *J. Chem. Soc. C*, 1903 (1968).

(7)   K. Kariyone and H. Yazawa, *Tetrahedron Lett.*, 2885 (1970).

(8)   E. J. Corey and J. W. Suggs, *J. Org. Chem.*, **38**, 3224 (1973).

(9)   T.-L. Ho and C. M. Wong, *Synth. Commun.*, **4**, 109 (1974).

(10) E. J. Corey and J. W. Suggs, *Tetrahedron Lett.*, 3775 (1975).

(11) C. M. Stevens and R. Watanabe, *J. Amer. Chem. Soc.*, **72**, 725 (1950).

(12) S. W. Baldwin and J. C. Tomesch, *J. Org. Chem.*, **39**, 2382 (1974).

(13) C. M. Hill, D. E. Simmons, and M. E. Hill, *J. Amer. Chem. Soc.*, **77**, 3889 (1955) and references cited therein.

(14) C. D. Broaddus, *J. Org. Chem.*, **30**, 4131 (1965).

(15) R. Boss and R. Scheffold, *Angew. Chem. Int. Ed. Engl.*, **15,** 558 (1976).

(16) P. A. Gent, R. Gigg, and R. Conant, *J. Chem. Soc., Perkin Trans. 1*, 1535 (1972).

(17) P. A. Gent, R. Gigg, and R. Conant, *J. Chem. Soc., Perkin Trans. 1*, 1858 (1973).

(18) P. A. Gent and R. Gigg, *Carbohydr. Res.*, **49,** 325 (1976).

(19) P. A. Gent and R. Gigg, *J. Chem. Soc., Perkin Trans. 1*, 1446 (1974).

(20) P. A. Gent and R. Gigg, *J. Chem. Soc., Perkin Trans. 1*, 1835 (1974).

(21) P. A. Gent and R. Gigg, *J. Chem. Soc., Perkin Trans. 1*, 361 (1975).

(22) P. A. Gent and R. Gigg, *Chem. Phys. Lipids*, **17,** 111 (1976).

(23) P. A. Gent, R. Gigg, and A. A. E. Penglis, *J. Chem. Soc., Perkin Trans. 1*, 1395 (1976).

(24) P. A. Gent and R. Gigg, *J. Chem. Soc., Perkin Trans. 1*, 1521 (1975).

(25) P. A. Gent and R. Gigg, *J. Chem. Soc., Perkin Trans. 1*, 1779 (1975).

(26) R. H. Gigg, *in* "Rodd's Chemistry of Carbon Compounds," S. Coffey, ed., Elsevier, Amsterdam, Vol. 1E, 1976, p. 349.

(27) P. A. Gent and R. Gigg, *J. Chem. Soc., Perkins Trans. 1*, 364 (1975).

CYCLIC ACETALS

[48] Methyl 4,6-O-Benzylidene-α- and -β-D-glucopyranosides Via Acetal Exchange

By Michael E. Evans

*The Australian Wine Research Institute, Private Mail Bag,
Glen Osmond, South Australia, Australia*

Introduction

The various methods described for synthesis of the title compounds have been reviewed (*1, 2*). The preparations described here involve reaction of equimolar amounts of a glucoside and benzaldehyde dimethyl acetal in acidified *N*,*N*-dimethylformamide, a modification of a previously described procedure (*3*).

Procedure

α,α-Dimethoxytoluene (Benzaldehyde dimethyl acetal)

A solution of 21.2 g of benzaldehyde and 24.0 g of trimethyl orthoformate in 100 ml of methanol containing 1 g of Amberlite IR-120(H$^+$) cation-exchange resin (20–50 mesh, washed with methanol) is boiled under reflux for 3 h. The resin is removed by filtration, and the filtrate is concentrated at <35° on a rotary evaporator. The product is distilled at 104°–108°/40 torr (or 195–199°/760 torr) to give α,α-dimethoxytoluene, which should show no C=O stretching in its i.r. spectrum; yield 26.5 g (88%).

Methyl 4,6-O-Benzylidene-α-D-glucopyranoside

Methyl α-D-glucopyranoside (4.85 g, dried to constant weight at 100°), 3.8 g of α,α-dimethoxytoluene, 17.5 ml of dry[1] *N*,*N*-dimethylformamide (DMF),[1] and 0.015 g of toluene-*p*-sulfonic acid hydrate are placed in a

---

[1] DMF may be dried very effectively by distillation until a few ml have been collected at a stillhead temperature of 152°–154°, then allowing the undistilled bulk of the liquid to cool in a tightly stoppered flask. Drying DMF with phosphorus pentoxide makes the DMF acidic and is not recommended because it can lead to hydrolysis of the product if insufficient base is used in the neutralization step.

METHODS IN CARBOHYDRATE
CHEMISTRY, VOL. VIII

150-ml round-bottomed flask, which is then attached via a greased seal to a rotary evaporator, rotated, evacuated with a water aspirator, and lowered into a water bath at 60° ± 5° so that DMF refluxes in the steam duct. The solid should dissolve in 1–2 min; if the vacuum is such that the DMF distils at 60°, the evacuated evaporator may be isolated from the pump for a short time until the glucoside dissolves, then reconnected. After 40 min, a short path evaporation adaptor (3) is fitted between the flask and the steam duct and the solvent is removed, raising the bath temperature to 90°. The product is cooled; 40 ml of 0.05 M sodium hydrogencarbonate is added, and the mixture is distilled until no more benzaldehyde comes over (2–5 ml of distillate). The resulting solution is filtered hot through a porosity 3 sinterred filter, and the filter is rinsed with two 10-ml portions of 95°–100° water. The filtrate and washings are combined and concentrated at 100° to ~40 ml, then cooled to 3°. The product is collected by filtration, washed twice with water at 0°–5°, and then dried, initially at 25° and finally at 100°; yield of title compound 5.78 g (82%), mp 165°–166°, [α]D +105° (c 1.1, chloroform). Recrystallization from 15 ml of 1-propanol gives material melting at 166°–167°; yield 4.48 g (64%).

Methyl 4,6-O-Benzylidene-β-D-glucopyranoside

Dry[2] methyl β-D-glucopyranoside (1.94 g), 1.52 g of α,α-dimethoxytoluene, 7 ml of dry[1] DMF, and 0.010 g of toluene-p-sulfonic acid hydrate are placed in a 50-ml round-bottomed flask. This is attached to a rotary evaporator, and benzylidenation and removal of the solvent is done as described for methyl α-D-glucopyranoside. The product is broken up and transferred to a 150-ml flask. Sodium hydrogencarbonate solution (60 ml, 0.03 M) is added, and the mixture is distilled until no more benzaldehyde comes over. 1-Propanol (10 ml) is added, and the mixture is boiled until the solid dissolves. The solution is then filtered hot through a porosity 3 sintered filter and the filter is washed with 15 ml of 95°–100° water. The filtrate and washing are combined and cooled to 3°. The product is collected by filtration, washed, and dried as in the preparation of methyl 4,6-O-benzylidene-α-D-glucopyranoside to give the title compound; yield 2.26 g (80%), mp 203°–205°, and [α]D −76° (c 1.0, methanol). Recrystallization from n-butyl acetate (24 ml per gram) gives material melting at 205°–207° with 72% recovery.

---

[2] Methyl β-D-glucopyranoside is somewhat hygroscopic and should be dried to constant weight at 100°, preferably in vacuo, otherwise the yield of product is reduced considerably.

References

(*1*)   A. N. deBelder, *Advan. Carbohydr. Chem.*, **20**, 219 (1965).
(*2*)   J. W. Van Cleve, *Carbohydr. Res.*, **17**, 461 (1971).
(*3*)   M. E. Evans, *Carbohydr. Res.*, **21**, 473 (1972).

[49] Benzylidenation of Diols with α,α-Dihalotoluenes in Pyridine

BY P. J. GAREGG AND C.-G. SWAHN

*Department of Organic Chemistry, Arrhenius Laboratory,
University of Stockholm, Stockholm, Sweden*

Introduction

Benzylidene acetals of diols are normally produced using benzaldehyde and zinc chloride or some other acidic catalyst (*1*). In a newer method, α,α-dimethoxytoluene and a catalytic amount of toluene-*p*-sulfonic acid in *N,N*-dimethylformamide (DMF) is used; the reaction equilibrium is displaced in favor of the product by co-distilling off the methanol formed with DMF under reduced pressure (Vol. VIII [48]).

Both these methods require acidic catalysis. Since the reaction is reversible under these conditions, the configuration at the benzylidene carbon is thermodynamically controlled (*1*). Baggett and co-workers have described

317

the synthesis of 4,6-*O*-benzylidene acetals of hexopyranosides using α,α-dibromotoluene and potassium *tert*-butoxide. Because this reaction is irreversible, 4,6-*O*-benzylidene derivatives with the phenyl group situated both axially and equatorially were obtained (*2, 3*). The preparation was subsequently modified as exemplified by the reaction of benzaldehyde and potassium *tert*-butoxide with methyl 2,3-di-*O*-methyl-6-*O*-(tolyl-*p*-sulfonyl)-α-D-glucopyranoside to give the 4,6-*O*-benzylidene acetals (*4*). The yields obtained under basic conditions are, however, moderate (*2–4*).

An alternative method is benzylidenation with α,α-dihalotoluenes (chloro- or bromo-) in pyridine at the reflux temperature (*5, 6*). Although the reaction most probably is irreversible, the more stable stereoisomers predominate. To facilitate the work-up, remaining free hydroxyl groups are usually acetylated *in situ*. Good yields are frequently obtained. The reaction conditions are reasonably compatible with the presence of acid-labile triphenylmethyl groups and alkali-labile acetyl groups at other positions, but in these cases, the yields are lower and chromatographic purification is necessary (*6*).

Procedure

Methyl 2,3-Di-*O*-acetyl-4,6-*O*-benzylidene-α-D-glucopyranoside (II) (*6*)

A solution of 0.50 g of methyl α-D-glucopyranoside (I) and 0.75 g of α,α-dibromo-toluene in 10 ml of pyridine is heated at the reflux temperature for 1 h. Acetic anhydride (3 ml) is added. After standing at 20°–25° overnight, the solution is poured into water. The mixture is extracted with chloroform. The combined chloroform phases are shaken with water, aqueous sodium hydrogencarbonate, and then water. After drying with magnesium sulfate and filtration, the solution is concentrated to dryness under reduced pressure. Pyridine is removed by repeated co-distillation with toluene under reduced pressure; yield of II 0.78 g (86%), mp 95°–100°.

Methyl 3,4-*O*-Benzylidene-6-deoxy-β-D-galactopyranoside (IV) (*7*)

A solution of 2.5 g of methyl 6-deoxy-β-D-galactopyranoside (III) (*7*) and 3.3 ml of α,α-dibromotoluene in 45 ml of pyridine is heated at the reflux temperature for 2 h. The reaction mixture is diluted with chloroform, and the chloroform solution is shaken with water, dried with sodium sulfate, filtered, and concentrated under reduced pressure to a syrup. The major component is isolated by silica gel column chromatography (column length 50 cm, diameter 5 cm) using 1:1 v/v toluene–ethyl acetate as eluant to give a mixture of stereoisomers; yield 2.45 g (66%), mp 97°–99°, $[\alpha]^{20}$D +2°

(*c* 0.6, chloroform). [^{1}H] nmr indicates that the isomer with the *exo* phenyl group predominates.

Methyl 2,6-Di-*O*-acetyl-3,4-*O*-benzylidene-β-D-galactopyranoside (VI) (6)

A solution of 1.77 g of methyl 6-*O*-acetyl-β-D-galactopyranoside (V) (6) and 1.61 g of α,α-dichlorotoluene in 25 ml of pyridine is heated at the reflux temperature for 5 h. More α,α-dichlorotoluene (1.61 g) is added, and the solution is heated to reflux for an additional 3 h. Acetic anhydride (5 ml) is added to the still warm solution, which is then allowed to stand at 20°–25° overnight. The solution is diluted with toluene, and the toluene solution is shaken with water, aqueous sodium hydrogencarbonate, and then water. After drying with sodium sulfate and filtration, the solution is concentrated to dryness under reduced pressure. Pyridine is removed by repeated co-distillation with toluene under reduced pressure. The crude crystalline product is washed with light petroleum (60°–80°) in order to remove excess α,α-dichlorotoluene. The remaining product (2.74 g) is purified by silica gel column chromatography (column length 50 cm, diameter 5 cm) using 9:1 v/v chloroform–ethyl ether as eluant to give VI; yield 1.60 g (58%), mp 113°–117°. [^{1}H] nmr indicates that the isomer with the *exo* phenyl group predominates.

References

(*1*)   A. N. de Belder, *Advan. Carbohydr. Chem.*, **20**, 219 (1965).

(*2*)   N. Baggett, J. M. Duxbury, A. B. Foster, and J. M. Webber, *Chem. Ind. (London)*, 1832 (1964).

(*3*)   N. Baggett, J. M. Duxbury, A. B. Foster, and J. M. Webber, *Carbohydr. Res.*, **1**, 22 (1965).

(*4*)   N. Baggett, M. D. Mosihuzzaman, and J. M. Webber, *Carbohydr. Res.*, **11**, 263 (1969).

(*5*)   P. J. Garegg, L. Maron, and C.-G. Swahn, *Acta Chem. Scand.*, **26**, 518 (1972).

(*6*)   P. J. Garegg and C.-G. Swahn, *Acta Chem. Scand.*, **26**, 3895 (1972).

(*7*)   K. Eklind, P. J. Garegg, and B. Gotthammar, *Acta Chem. Scand.*, **B29**, 633 (1975).

[^{1} H] nmr indicates that the isomer with the *exo* phenyl group predominates.

OXIDIZED PRODUCTS

[50] Photochemical Oxidation of Carbohydrates

1,2:3,4-Di-O-isopropylidene-α-D-galacto-hexodialdo-1,5-pyranose

BY ROGER W. BINKLEY

Department of Chemistry, Cleveland State University, Cleveland, Ohio

Introduction

Considerable progress has been made during the past decade in development of methods for oxidizing carbohydrates. Several excellent reviews are devoted in part or in total to this subject (1–3). Recently, a photochemical approach to the hydroxyl-to-carbonyl oxidation process has been reported (4, 5). This new alternative (photochemical oxidation), which has been successful in oxidizing a number of carbohydrates (4, 6) and non-carbohydrates (5, 7), has as its chemical basis the reactions shown in equations one and two.

$$HOCH_2R + CH_3-\overset{O}{\underset{\|}{C}}-\overset{O}{\underset{\|}{C}}-Cl + \underset{N}{\bigcirc} \longrightarrow CH_3-\overset{O}{\underset{\|}{C}}-\overset{O}{\underset{\|}{C}}-OCH_2R + \underset{\underset{Cl^-}{\overset{\underset{H}{N^+}}{\bigcirc}}}{} \quad (1)$$

$$CH_3-\overset{O}{\underset{\|}{C}}-\overset{O}{\underset{\|}{C}}-OCH_2R \overset{h\nu}{\longrightarrow} CH_3\overset{O}{\underset{\|}{C}}H + CO + H\overset{O}{\underset{\|}{C}}R \quad (2)$$

The following four aspects of photochemical oxidation are important to its utilization as a synthetic technique: (a) product yields are good, ranging from 50% to 100%; (b) primary and secondary alcohols are oxidized with equal ease; (c) oxidation is not complicated by competing reactions; and (d) oxidation is conducted under mild conditions, that is, at or below room temperature in an inert solvent and in the absence of acids or inorganic ions. The last of these features deserves special emphasis since the conditions for photochemical oxidation are sufficiently mild to permit successful oxidation in situations where the reactant or the product is relatively unstable; for example, 2,3,4,6-tetra-O-acetyl-β-D-glucopyranose is oxidized by this method to 2,3,4,6-tetra-O-acetyl-D-glucono-1,5-lactone without further re-

321

METHODS IN CARBOHYDRATE
CHEMISTRY, VOL. VIII

action even though the lactone easily experiences further change (for example, weak bases such as pyridine, triethylamine, and acetate cause rapid elimination of the elements of acetic acid from the lactone at room temperature).

The technique for conducting photochemical oxidation is illustrated by applying this method to the oxidation of 1,2:3,4-di-O-isopropylidene-α-D-galactopyranose (I). Other oxidizing agents which have been used in this conversion include chromium trioxide/pyridine (8), methyl sulfoxide/dicyclohexylcarbodiimide (9–12; see also Vol. VI [55]) and methyl sulfoxide/acetic anhydride (13, see also Vol. VI [55], [56], [57]).

Five carbohydrates in addition to 2,3,4,6-tetra-O-acetyl-β-D-glucopyranose and 1,2:3,4-di-O-isopropylidene-α-D-galactopyranose have been oxidized via photochemical oxidation (7). These are 1,2:5,6-di-O-isopropylidene-α-D-glucofuranose; 1,2:4,5-di-O-isopropylidene-β-D-fructopyranose; 2,3:4,5-di-O-isopropylidene-β-D-fructopyranose; methyl 2,3-O-isopropylidene-β-D-ribofuranoside; 1,2,3,4-tetra-O-acetyl-β-D-glucopyranose; and 1,3,4,6-tetra-O-acetyl-α-D-glucopyranose.

Procedure

Photochemical Oxidation of 1,2:3,4-Di-O-isopropylidene-α-D-galactopyranose (I)

1,2:3,4-Di-O-isopropylidene-α-D-galactopyranose (14) (1.0 g, 3.8 mmole) and 1.3 g (16 mmole) of pyridine are dissolved in 50 ml of anhydrous benzene. Pyruvoyl chloride (2-ketopropanoyl chloride) (15) (0.42 g, 4.0 mmole) in 50 ml of anhydrous benzene is added to this solution in a dropwise manner with constant stirring. Cooling the reaction mixture with cold water is necessary during addition to keep the temperature below 10°. (Note: It is not necessary to distill the pyruvoyl chloride prior to use. Satisfactory results were obtained by mixing 0.71 g (8.0 mmole) of pyruvic acid and 0.92 g (8.0 mmole) of α,α-dichloromethyl methyl ether (16), maintaining the temperature of the reaction mixture at 50° for 30 min, and using directly the pyruvol chloride thus formed.) After stirring for 15 min, most of the pyridinium hydrochloride is removed by filtration, and the benzene is distilled *in vacuo* to yield the pyruvate ester contaminated with pyridinium hydrochloride. The contaminant was removed by shaking the reaction mixture in 50 ml of carbon tetrachloride, allowing it to stand for a few hours, and removing the insoluble material by filtration. When the carbon tetrachloride is evaporated from the filtrate, a quantitative yield of 1,2:3,4-di-*O*-isopropylidene-6-*O*-pyruvoyl-α-D-galactopyranose remained. This compound in chloroform-d had pmr absorptions (60 MHz) at δ 5.53 (1H, doublet, J = 3 Hz), 4.80–3.97 (6H, multiplet), 2.48 (3H, singlet), 1.55 (3H, singlet), 1.48 (3H, singlet), and 1.37 (6H, singlet). (Upon standing for two days in 5 ml of carbon tetrachloride the product crystallized; mp 82°–86°; however, since the non-crystalline ester was quite pure, its crystallization offered no improvement to the oxidation process.) The pyruvate ester is dissolved in 350 ml of dry benzene and placed in a model 6515 Ace photochemical reaction vessel, and the solution is purged with nitrogen for 1 h. The nitrogen purge is continued during Pyrex-filtered irradiation with a 450-W medium-pressure, Hanovia, mercury lamp. After 1 h, the irradiation is stopped; the benzene is removed by distillation using a water aspirator, and the residual liquid is distilled *in vacuo* using a Buchi/Brinkman micro distillation oven. The fraction boiling at 110°–113° (1 torr) is pure 1,2:3,4-di-*O*-isopropylidene-α-D-*galacto*-hexodialdo-1,5-pyranose (II); yield 0.7 g (70%). Compound II in carbon tetrachloride had pmr absorptions (60 MHz) at δ 9.51 (1H, broad singlet), 5.58 (1H, doublet, J = 6 Hz), 2.79–2.39 (4H, multiplet), 1.51 (3H, singlet), 1.45 (3H, singlet), and 1.33 (6H, singlet). The proton magnetic resonance spectrum of II has been published (9).

References

(1) R. F. Butterworth and S. Hanessian, *Synthesis*, 70 (1971).
(2) J. S. Brimacombe, *Angew. Chem., Int. Ed. Engl.*, **8**, 401 (1969).
(3) H. Grisebach and R. Schmid, *Angew. Chem., Int. Ed. Engl.*, **11**, 159 (1972).
(4) R. W. Binkley, *Carbohydr. Res.*, **48**, C1 (1976).
(5) R. W. Binkley, *Synth. Commun.*, **6**, 281 (1976).
(6) R. W. Binkley, *J. Org. Chem.*, **41**, 3030 (1976).
(7) R. W. Binkley, *J. Org. Chem.*, **42**, 1216 (1977).
(8) R. E. Arrick, D. C. Baker, and D. Horton, *Carbohydr. Res.*, **26**, 441 (1969).
(9) D. Horton, M. Nakadate, and J. M. J. Tronchet, *Carbohydr. Res.*, **7**, 56 (1968).
(10) D. J. Ward, W. A. Szarek, and J. K. N. Jones, *Carbohydr. Res.*, **21**, 305 (1972).
(11) D. Horton, J. B. Hughes, and J. M. J. Tronchet, *Chem. Commun.*, 481 (1965).
(12) G. B. Howarth, D. G. Lance, W. A. Szarek, and J. K. N. Jones, *Can. J. Chem.*, **47**, 75 (1969).
(13) R. F. Butterworth and S. Hanessian, *Can. J. Chem.*, **49**, 2755 (1971).
(14) A. L. Raymond and E. F. Schroeder, *J. Amer. Chem. Soc.*, **70**, 2785 (1948).
(15) H. C. J. Ottenheijm and J. H. M. de Man, *Synthesis*, 163 (1975).
(16) Aldrich Chemical Company, Milwaukee, Wisconsin.

Glossary

Bio-Gel. Trade name of poly(acrylamide) products for gel filtration of Bio-Rad Laboratories, Richmond, California.

Bio-Rad. Trade name of products of Bio-Rad Laboratories, Richmond, California.

Celite. Trade name of a hydrated, amorphous silica (diatomaceous earth) filter aid; a product of Johns-Manville Co., New York, N.Y.

Chromasorb. Trade name of flux-calcined diatomite solid support for gas chromatography; a product of Johns-Manville Co., New York, N.Y.

Dowex. Trade name of ion-exchange resins of the Dow Chemical Co., Midland, Michigan.

Dry Ice. Solid carbon dioxide.

Light petroleum. Low-boiling fraction of petroleum.

Parafilm. A laboratory film product of Marathon Products, American Can Co., Neenah, Wisconsin.

PTFE. Poly(tetrafluoroethylene).

Sephadex. Trade name of cross-linked destran products for gel filtration of Pharmacia, Uppsala, Sweden.

Teflon. Trade name of poly(tetrafluoroethylene); a product of E. I. DuPont de Nemours and Co., Wilmington, Delaware.

Tris. 2-Amino-2-hydroxymethyl-1,3-propanediol-[tris(hydroxymethylamino)methane].

Author Index

Numbers in parentheses are footnote reference numbers and indicate that an author's work is referred to although the name is not cited in the text.

A

Abdel-Akher, M., 74(2), *76*
Achmatowicz, O., 219, *225*
Ackers, G. K., 46(13), *52*
Acton, E. M., 207(3), 208(3), *209*
Adam, H., 21(4), *32*
Adamyants, K. S., 283(8), 284(8), *285*
Adbullah, M., 140(9), *144*
Ahmed, Z. F., 143(20), *144*
Alderweireldt, F. C., 240(17), 241(17), *242*, 244(11), *245*
Allen, W. S., *96*
Almquist, R. G., 228(4), *231*
Al-Radhi, A. K., 301(3), *304*
Altgelt, K. H., 46(7), *52*
Amako, H., 21(3), *32*
Amar, C., 46(47), *53*
Amaral, D., 94(15), *96*, 131(3, 4, 7, 8), 132(7, 10), *132, 133*, 139(5), *144*
Ames, B. N., 24, *32*
Anderson, B., *96*
Anderson, D. M. W., 74(3), *76*
Anderson, J. S., 107(6), 112(6), 113(6), *116*, 211(3), 213(11), 216(11), *217*
Anderson, L., 251(6), *253*
Anderson, S. A., 46(49), *53*
Andre, F., 46(31), *52*
Anet, E. F. L. J., 70(12), *71*
Angyal, S. J., 117, *122*, 176(7), *176*, 233(1, 2), 234(5), *235*
Anisuzzaman, A. K. M., 228(21), *231*
Anno, K., 81(6, 11, 12), 86(27), *87, 88*, 283(12), *285*
Antonopoulos, C. A., 89(8), *96*
Araki, Y., 252(18), *253*
Aramaki, Y., 46(60), *53*
Arcamone, F., 248(7), *250*
Arrick, R. E., 322(8), *323*
Asensio, C., 131(3, 7), 132(7, 10), *132, 133*, 139(5), *144*

Asnis, R. E., 74(4), *76*
Atsumi, K., 283(12), *285*
Auer, T., 272(4), *275*
Autio, S., 120(8), *122*
Avery, O. T., 211, *217*
Avigad, G., 131(3, 7), 132(7, 10), *132, 133*, 139(5), *144*
Axelrod, J. 263(11), *269*

B

Bab'eva, I. P., 123(5), 124(10, 11), 125(11), *125*
Bachhawat, B. K., 89(3, 4, 5, 6), *96*
Baggett, N., 318(2, 3, 4), *319*
Bagdian, G., 118(4), *122*
Bahl, O. P., 252(9, 10), *253*
Baker, D. C., 322(8), *323*
Baker, J. R., 291(7), *294*
Baldwin, S. W., 305(12), *310*
Ballou, C. E., 112(10), *116*, 124(12, 13), *125*
Banaszek, A., 185(11), *194*
Banks, J., 79(9), *80*
Banoub, J., 238(1, 14), 239(11, 14), 240(11), 241(14), *241, 242*, 244(6), 245(6), *254*, 248(8, 9, 10), 249(10), *250*
Banzhof, M., 35(22), *43*
Barfield, M., 157(18), *164*
Barker, R., 101(21), 102(21), *104*, 154(8, 9, 10), 156(8), 157(17), 158(8), 159(29), 161(8), *164, 165*, 298(6), *299*
Barker, S. A., 18(4), *19*, 74(8), *76*
Barondes, S. H., 46(40, 50), *52, 53*
Bartnicki-Garcia, S., 46(27), *52*
Bathgate, G. N., 46(61), 51(61), *53*
Battino, R., 35(21, 22), *43*
Bauer, H. F., 142(13), *144*
Beck, D., 249(13), *250*
Bekker, P. I., 46(62), 51(62), *53*
Belue, G. P., 46(22), *52*
Benmanan, J. D., 91(11), *96*
Bennich, H., 49(80), *53*

327

Subject Index

A

Acetals, amide, glycosidation, 237–242
Acetolysis, of heteroglycans, 112
Acetylation, of carbohydrates, 55
Agar diffusion, of glycans, 215
Alditols
  acetates, gas-liquid chromatography-mass spectrometry of partially methylated, 57
  oxidation of acetylated, with chromium trioxide, 121
    by glucose oxidase and by galactose oxidase, 141
  2-acetamido-2-deoxy-, gas-liquid chromatography-mass spectrometry, 56
  2-amino-2-deoxy-, gas-liquid chromatography-mass spectrometry, 56
Aldonic acids, 1-carbon-13 labeled, coupling patterns, 157
Allopyranoses, carbon-13 nuclear magnetic resonance spectroscopy, 152
Allyl ethers, as protecting groups, 305–311
Altropyranoside
  methyl α-D-, preparation, 169, 170
  methyl 3,4-isopropylidene-α-D-, preparation, 169, 171
  methyl 4,6-O-isopropylidene-α-D-, preparation, 169
Amide acetals, see Acetals
Amino sugars
  carbon-13 nuclear magnetic resonance spectroscopy, 152
  gas-liquid chromatography, 55–65
  mass spectrometry, 55–65
  synthesis, 208
Antigens, isolation and purification of carbohydrate, 211–217

B

Benzylidenation, of diols, 317–319
Boron trifluoride, catalyst in 1-thioglycoside preparation, 252

C

Calcium chloride, complex with methyl β-D-mannofuranoside, 234
Carbohydrates, see also Disaccharides; Monosaccharides; Oligosaccharides; Polysaccharides
  acetates, oxidation with chromium trioxide in acetic acid, 117–122
  acetylation, 55
  automatic analyzer, 4–12
  branched-chain, synthesis, 185–194
  chloroacetates, as synthetic intermediates, 271–275
  complexes with metal cations, 233–235
  fluorimetric assay, 11
  high-performance liquid chromatography, 33–43
  identification by microenzymic procedures, 139–144
  isotopically enriched, carbon-13 nuclear magnetic resonance spectroscopy, 151–165
  ligands, affinity for proteins, 145–149
  methylation, 55, 56
  photochemical oxidation, 321–323
  structure determination by carbon-13 nuclear magnetic resonance spectroscopy, 97–105, 151–165
  synthesis, methoxymercuration, 207–209
  trimethylsilylation, 55
Carbon-13 nuclear magnetic resonance spectroscopy
  of isotopically enriched carbohydrates, 151–165
  polysaccharide structure determination, 97–105
Cellulose, O-carboxymethyl-, determination of carboxymethyl ether groups, 127–129
Chloroacetyl group, blocking group, 271–275
Chondroitinases, preparation, 81, 82
Chondroitin sulfates
  desulfation, 288, 289
  enzymic assay, 81
  paper chromatographic separation of unsaturated disaccharides, 83, 84